adsorptive
bubble
separation
techniques

CONTRIBUTORS

J. AROD

DIBAKAR BHATTACHARYYA

STANLEY E. CHARM

W. DAVIS, Jr.

ALBERT J. de VRIES

DAVID W. ECKHOFF

D. W. FUERSTENAU

ROBERT B. GRIEVES

P. A. HAAS

T. W. HEALY

C. JACOBELLI-TURI

DAVID JENKINS

BARRY L. KARGER

ROBERT LEMLICH

F. MARACCI

A. MARGANI

M. PALMERA

T. A. PINFOLD

V. V. PUSHKAREV

ALAN J. RUBIN

ELIEZER RUBIN

T. SASAKI

JAN SCHERFIG

adsorptive bubble separation techniques

Edited by **Robert Lemlich**

Department of Chemical
and Nuclear Engineering
University of Cincinnati
Cincinnati, Ohio

1972

Academic Press
New York and London

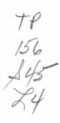

ACADEMIC PRESS, INC.
111 Fifth Avenue, New York, New York 10003

United Kingdom Edition published by
ACADEMIC PRESS, INC. (LONDON) LTD.
24/28 Oval Road, London NW1 7DD

LIBRARY OF CONGRESS CATALOG CARD NUMBER: 75-154398

PRINTED IN THE UNITED STATES OF AMERICA

List of Contributors

Numbers in parentheses indicate the pages on which the authors' contributions begin.

J. AROD (243), Service d'Analyse et de Chimie Appliquée, Centre d'Études Nucléaires de Cadarache, France

DIBAKAR BHATTACHARYYA (183), Department of Chemical Engineering, University of Kentucky, Lexington, Kentucky

STANLEY E. CHARM (157), New England Enzyme Center, Tufts University Medical School, Boston, Massachusetts

W. DAVIS, Jr. (279), Chemical Technology Division, Oak Ridge National Laboratory, Oak Ridge, Tennessee

ALBERT J. de VRIES (7), Pechiney-Saint-Gobain, Centre de Recherches de la Croix-de-Berny, Antony, France

DAVID W. ECKHOFF (219), H. F. Ludwig and Associates, New York, New York

D. W. FUERSTENAU (91), Department of Materials Science and Engineering, University of California, Berkeley, California

ROBERT B. GRIEVES (175, 183, 191), Department of Chemical Engineering, University of Kentucky, Lexington, Kentucky

P. A. HAAS (279), Chemical Technology Division, Oak Ridge National Laboratory, Oak Ridge, Tennessee

T. W. HEALY* (91), Department of Materials Science and Engineering, University of California, Berkeley, California

C. JACOBELLI-TURI (265), Laboratorio di Metodologie Avanzate Inorganiche del Consiglio Nazionale delle Ricerche, Rome, Italy

DAVID JENKINS (219), Department of Sanitary Engineering, University of California, Berkeley, California

BARRY L. KARGER (145), Department of Chemistry, Northeastern University, Boston, Massachusetts

ROBERT LEMLICH (1, 33, 133), Department of Chemical and Nuclear Engineering, University of Cincinnati, Cincinnati, Ohio

F. MARACCI (265), Laboratorio di Metodologie Avanzate Inorganiche del Consiglio Nazionale delle Ricerche, Rome, Italy

A. MARGANI (265), Laboratorio di Metodologie Avanzate Inorganiche del Consiglio Nazionale delle Ricerche, Rome, Italy

M. PALMERA (265), Laboratorio di Metodologie Avanzate Inorganiche del Consiglio Nazionale delle Ricerche, Rome, Italy

T. A. PINFOLD (53, 75), Department of Chemistry including Biochemistry, University of the Witwatersrand, Johannesburg, South Africa

V. V. PUSHKAREV (299), The Urals Kirov Polytechnical Institute, Sverdlovsk, U.S.S.R.

*Present address: Department of Physical Chemistry, University of Melbourne, Parkville, Victoria, Australia.

ALAN J. RUBIN (199), Water Resources Center, College of Engineering, Ohio State University, Columbus, Ohio

ELIEZER RUBIN (249), Department of Chemical Engineering, Technion—Israel Institute of Technology, Haifa, Israel

T. SASAKI (273), Department of Chemistry, Faculty of Science, Tokyo Metropolitan University, Setagaya, Tokyo, Japan

JAN SCHERFIG (219), Department of Sanitary Engineering, University of California, Irvine, California

Preface

This is the first comprehensive book to cover the various adsorptive bubble separation techniques. It is a contributed volume, with various authors for its twenty chapters.

The editor is responsible for defining the scope of the work, and in general for selecting the topic for each chapter. The authors of the chapters were also selected by the editor and invited to write on the basis of their specialized knowledge in their respective areas. However, each author was given wide latitude to treat his material as he saw fit. The overall result is a highly authoritative compilation which, it is hoped, will prove to be both informative and readable.

Chapter 1 introduces the various adsorptive bubble separation techniques. Chapter 2 deals with certain pertinent properties of foam which are common to many of them. Then, Chapters 3 through 8 individually discuss several of these techniques; namely, foam fractionation, ion flotation, precipitate flotation, mineral flotation, bubble fractionation, and solvent sublation. The remaining chapters, 9 through 20, summarize the results of numerous separations, as well as the results of additional investigations into the mechanisms of the various techniques. As a special feature of interest, the final six chapters (arranged in alphabetical order by country) comprise a summary of work, dealing principally with the separation of surfactants and metallic ions, at several places around the world.

The editor expresses his thanks to the contributors and to the staff of Academic Press for making this volume possible.

ROBERT LEMLICH

Contents

Chapter 20   Separation of Surfactants and Ions from Solutions by Foaming: Studies in the U.S.S.R.

V. V. Pushkarev

adsorptive bubble separation techniques

CHAPTER 1

INTRODUCTION

Robert Lemlich
   Department of Chemical and Nuclear Engineering
   University of Cincinnati
   Cincinnati, Ohio

I. OVERVIEW

All techniques or methods of separation, whether physical or chemical, are based on differences in properties. For example, among the more familiar techniques, distillation is based on differences in volatility, and liquid extraction is based on differences in solubility.

The *adsorptive bubble separation* techniques are among the less familiar methods. This generic name was first proposed by Lemlich (1966), with *adsubble* techniques as the convenient contraction. The full generic name has since been accepted by common consent (Karger *et al.*, 1967).

The adsubble techniques are based on differences in surface activity. Material, which may be molecular, colloidal, or macroparticulate in size, is selectively adsorbed or attached at the surfaces of bubbles rising through the liquid, and is thereby concentrated or separated. A substance which is not surface active itself can often be made effectively surface active through union with or adherence to a surface active *collector*. The substance so removed is termed the *colligend* (Sebba, 1962).

Adsubble processes can be found in nature: in sea foam and bubbling marshes. Among human endeavors, the earliest occurrence is probably among the culinary arts in such phenomena as the slightly bubble-aided floating of some constituents in certain boiling soups and other preparations. Another early example is in the pouring of beer. Certain components of the beer can concentrate in the foam to a sufficient degree to alter the flavor (Nissen and Estes, 1940).

In 1878 Gibbs derived the celebrated adsorption equation that bears his name (Gibbs, republished 1928). About the turn of the century, attempts were begun to test this equation in the laboratory by indirectly measuring the extent of adsorption of solute on rising bubbles (von Zawidzki, 1900). Some years later, but much further along the spectrum of particle sizes, the technique of mineral flotation by air became commercial. Since then, various adsubble techniques have been employed in a number of industrial and laboratory separations.

II. CLASSIFICATION OF TECHNIQUES

There are a number of individual adsubble techniques. Figure 1 shows the accepted scheme of classification (Karger et al., 1967). It is a compromise between rational systemization and actual usage of the terms by various

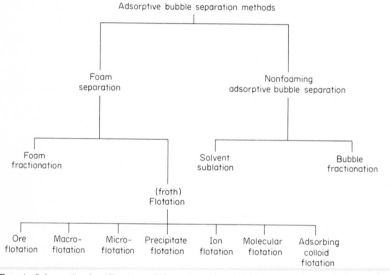

FIG. 1. Schematic classification of the adsorptive bubble separation techniques. [From Karger et al. (1967).]

writers. Accordingly, the reader should not be surprised at the overlap in the definitions, or when he encounters inconsistencies in terminology between the works of one author and another—or even within the work of a single author.

As indicated in Fig. 1, the adsubble techniques (or methods) are divided unequally into two main groups: The larger, called *foam separation*, requires the generation of a foam or froth to carry off material. The smaller, which is termed *nonfoaming adsorptive bubble separation*, does not.

This smaller division is further divided. *Bubble fractionation* (Dorman and Lemlich, 1965) is the transfer of material within a liquid by bubble adsorption or attachment, followed by deposition at the top of the liquid as the bubbles exit. *Solvent sublation* (Sebba, 1962) is the similar transfer to, or to either interface of, an immiscible liquid placed atop the main liquid.

The larger division is also subdivided. *Foam fractionation* is the foaming off of dissolved material from a solution via adsorption at the bubble surfaces. *Froth flotation*, or simply *flotation* (Gaudin, 1957), is the removal of particulate material by frothing (foaming).

Froth flotation, in turn, has many subdivisions. *Ore flotation* (Gaudin, 1957) is the separation of minerals. *Macroflotation* is the separation of macroscopic particles. *Microflotation* (Dognon and Dumontet, 1941) is the separation of microscopic particles, especially colloids or microorganisms. (Under certain conditions, the separation of colloids may sometimes be termed *colloid flotation*.)

In *precipitate flotation* (Baarson and Ray, 1963), a precipitate is formed and then foamed off. *Ion flotation* (Sebba, 1959) is the separation of surface inactive ions by foaming with a collector which yields an insoluble product, particularly if the product is removed as a scum. Similarly, *molecular flotation* is the separation of surface inactive molecules by foaming with a collector which gives an insoluble product. Finally, *adsorbing colloid flotation* is the separation of a solute through adsorption on colloidal particles which are then removed by flotation.

The overall classification scheme also lends itself to the incorporation of newly proposed adsubble techniques and others yet to come. For example, *laminae column foaming* (Maas, 1969a) is a subdivision of foam fractionation that utilizes relatively large wall-to-wall bubbles rising up the column. *Booster bubble fractionation* (Maas, 1969b) can be viewed as a subdivision of bubble fractionation that involves the use of certain volatile organic compounds in the gas bubbles in order to improve selectivity of separation.

III. DROPLET ANALOGS

Before the conclusion of this introductory chapter, it is of interest to point out that there exist droplet analogs to the adsorptive bubble separation techniques. These analogs involve adsorption or attachment at a liquid–liquid interface rather than at a liquid–gas interface. Accordingly, they have been called, by analogy, the *adsorptive droplet separation* techniques (Lemlich, 1968), with *adsoplet* techniques as the corresponding contraction.

The occurrence of adsoplet phenomena also appears to have a long history. Adsorption on the surfaces of the colloidal particles that rise in milk to form cream comes to mind. *Emulsion fractionation* is the analog of foam fractionation, and Eldib (1963) discusses several separations.

Emulsion fractionation, as well as adsorption of solute on a continuous stream of immiscible droplets, was employed early in the century (Lewis, 1908) in another preliminary attempt to verify the Gibbs adsorption equation. The latter technique, involving the stream of droplets, has been termed *droplet fractionation* (Lemlich, 1968) by analogy with bubble fractionation.

The most significant adsoplet techniques were undoubtedly to be found in extractive metallurgy. A forerunner is described by T. J. Hoover in his introduction to the first edition of Gaudin (1932); namely, the report of Herodotus that the maidens of ancient Gyzantia used to extract gold dust from mud by means of bird feathers smeared with pitch. Much later, the Persians of the fifteenth century employed procedures whereby water-wetted particles were freed under water from an oiled mass (Gaudin, 1957). During the last century, *oil flotation* (which may be viewed as the adsoplet analog of modern ore flotation) was employed in the mineral industries (Taggart, 1927). However, due to its high oil consumption, oil flotation was supplanted early in the present century by ore flotation which uses air.

As a whole, the adsoplet techniques have not attracted the measure of attention accorded to the adsubble techniques. However, the interested reader may pursue the subject further in the aforementioned references, as well as in some representative patents and other reports (Elmore, 1902; Lewis, 1909; Rideout, 1925; Schutte, 1947; Denekas *et al.*, 1951; Strain, 1953; Dunning *et al.*, 1954; Winkler and Kaulakis, 1955; Hardy, 1957).

Further discussion of the adsoplet techniques is beyond the present scope. The remainder of this book is devoted to the adsubble techniques.[1]

REFERENCES

Baarson, R. E., and Ray, C. L. (1963). Precipitate Flotation, a New Metal Extraction and Concentration Technique, paper presented at the Amer. Inst. Mining, Metall., Petrol. Eng. Symp. Dallas, Texas.

Denekas, M. O., Carlson, F. T., Moore, J. W., and Dodd, C. G. (1951). *Ind. Eng. Chem.* **43,** 1165.

Dognon, A., and Dumontet, H. (1941). *C. R. Acad. Sci., Paris* **135,** 884.

Dorman, D. C., and Lemlich, R. (1965). *Nature* **207,** 145.

Dunning, H. N., Moore, J. W., and Myers, A. T. (1954). *Ind. Eng. Chem.* **46,** 2000.

Eldib, I. A. (1963). *In* "Advances in Petroleum Chemistry and Refining" (K. A. Kobe and J. J. McKetta, Jr., eds.), Vol. 7. Wiley (Interscience), New York.

Elmore, E. (1902). U.S. Pat. 692,643.

[1] Certain material of a general nature dealing with the pertinent properties of foam is presented in the next chapter. The discussion of the separation techniques themselves commences with Chapter 3.

Gaudin, A. M. (1932). "Flotation," 1st ed. McGraw-Hill, New York.

Gaudin, A. M. (1957). "Flotation," 2nd ed. McGraw-Hill, New York.

Gibbs, J. W. (1928). "Collected Works." Longmans, Green, New York.

Hardy, R. L. (1957). U.S. Pat. 2,789,083.

Karger, B. L., Grieves, R. B., Lemlich, R., Rubin, A. J., and Sebba, F. (1967). *Separ. Sci.* **2**, 401.

Lemlich, R. (1966). *Chem. Eng.* **73**(21), 7.

Lemlich, R. (1968). *Ind. Eng. Chem.* **60**(10), 16.

Lewis, W. C. M. (1908). *Phil. Mag.* **6**(15), 499.

Lewis, W. C. M. (1909). *Phil. Mag.* **6**(17), 466.

Maas, K. (1969a). *Separ. Sci.* **4**, 69.

Maas, K. (1969b). *Separ. Sci.* **4**, 457.

Nissen, B. H., and Estes, C. (1940). *Amer. Soc. Brewing Chem.* 23.

Rideout, W. H. (1925). U.S. Pat. 1,562,125.

Schutte, A. H. (1947). U.S. Pat. 2,429,430.

Sebba, F. (1959). *Nature* **184**, 1062.

Sebba, F. (1962). "Ion Flotation." American Elsevier, New York.

Strain, H. H. (1953). *J. Phys. Chem.* **57**, 638.

Taggart, A. F. (1927). "Handbook of Ore Dressing," Wiley, New York.

von Zawidzki, J. (1900). *Z. Phys. Chem.* **35**, 77.

Winkler, R. W., and Kaulakis, A. F. (1955). U.S. Pat. 2,728,714.

Albert J. de Vries
Pechiney-Saint-Gobain
Centre de Recherches de la Croix-de-Berny
Antony, France

I. INTRODUCTION

Gas bubbles uniformly dispersed in a given amount of liquid may give rise to the formation of a foam under a wide range of conditions in spite of the concomitant increase of surface free energy associated with an extended gas–liquid interface. Because of this free-energy excess, foams, like other dispersed systems, are fundamentally unstable in the thermodynamical sense. The spontaneous decrease of interfacial area, which should ultimately result in a complete separation of the gaseous and the liquid phases, often progresses at a relatively low rate in many foams. The persistence of foam structure during a measurable span of time is, obviously, a prerequisite for many of the separation techniques discussed in this book. On the other hand, a more or less rapid collapse of the foam structure may be desirable or even necessary at a certain stage of the separation process. The present chapter is concerned with the actual state of knowledge regarding the various irreversible processes occurring in collapsing foams, and with the experimental means of investigation liable to provide quantitative information on the fundamental parameters which appear in the corresponding rate equations.

As indicated in the title of this chapter, the discussion will be limited to processes associated with bubble coalescence. Foam drainage, one of the fundamental irreversible processes occurring in all foams, will not be considered explicitly, since it does not directly produce any change in the number of gas bubbles nor in the bubble size distribution. The latter statement is strictly true only for foams of definitely established morphology, in particular for foams containing exclusively bubbles of either spherical or polyhedral shape. Relatively "wet" foams may be subject to a rapid draining of liquid from between the gas bubbles accompanied by local changes in foam density and bubble shape. In such foams the occurrence of liquid flow associated with rapid deformations of the bubble walls might affect appreciably the rate of bubble coalescence.

II. MORPHOLOGY AND STRUCTURE OF FOAMS

A. Gas-to-Liquid Ratio and Its Effect on Bubble Shape

As a consequence of the positive surface tension existing at any gas–liquid interface, gas bubbles dispersed in a volume of liquid will spontaneously tend to adopt a spherical shape independently of the particular method used for obtaining the dispersion. Spargers or spinnerets with well-defined orifices are usually preferred if bubbles of fairly uniform size are to be generated. In those cases, only foams of relative high density should be obtained, unless the bubbles change shape. Low-density foams, in which more than 74% of the total volume is occupied by nearly spherical gas bubbles, can nevertheless exist if the bubbles are of unequal size, which follows immediately from space-filling requirements. Foams of the latter type may persist during a considerable period of time (more than 15–20 min) without any change of bubble shape if the viscosity of the liquid is sufficiently high, as shown by this author (de Vries, 1958a).

The effect of viscosity may be easily explained by realizing that the spontaneous change in gas-to-liquid ratio in the first stages of foam formation, and which should ultimately lead to a change in foam morphology, is due to the drainage of liquid under the influence of the density difference between the gas bubbles and the surrounding liquid phase. The rate of this laminar flow process is inversely proportional to the viscosity of the liquid. Therefore, the time interval during which the bubbles maintain their initial spherical shape will be longer the higher the viscosity of the liquid. In order to maintain the spherical shape for periods of time longer than 15 min in the above foams (and in which the volumetric gas to liquid ratio was higher than 10), the viscosity of the liquid phase had to be at least equal to 0.5 P. It follows that foams generated from dilute aqueous solutions having a viscosity in the order of 0.01 P, in general, will rapidly (within a few minutes) show a change in morphology; liquid is squeezed out from between the gas bubbles, whose shape changes progressively from spherical to polyhedral.

In the case of bubbles of uniform size, this change in morphology should appear for relatively low values of the volumetric gas-to-liquid ratio, equal to at most 2.85, corresponding to the lowest possible void volume in a packed bed of spheres. However, as mentioned before, one may generate foams of much higher gas content that contain only approximately spherical bubbles if the bubble size distribution is sufficiently broad. This type of foam is most readily generated by means of mechanical foaming methods liable to furnish the energy required to break down larger bubbles into smaller ones in order to attain a high degree of dispersion.

An important parameter for separation techniques based on selective adsorption at a gas–liquid interface is the total interfacial area A generated in a given volume of liquid V_l. If the size of each bubble can be characterized by the value of its diameter d_i, the total interfacial area in a foam per unit volume of liquid is determined by

$$A/V_l = f(\sum n_i d_i^2 / \sum n_i d_i^3)(V_g/V_l). \qquad (1)$$

The volumetric gas-to-liquid ratio is given by V_g/V_l, and f is a shape factor equal to or larger than 6. If bubbles of various shapes are present in the foam, the value of f occurring in Eq. (1) should be considered as an average value.

Neglecting any other changes which may occur simultaneously, in particular, bubble coalescence, the spontaneous modification of bubble shape as a result of foam drainage should be beneficial since all three factors on the right hand side of Eq. (1), i.e., shape factor, average reciprocal bubble diameter, and gas–liquid ratio, would be, in theory, increasing functions of time during this stage of foam formation. In real foams, however, bubble coalescence will always proceed more or less rapidly, involving a simultaneous decrease of interfacial area. Nevertheless, considerable amounts of interfacial area may be generated by foaming even if the bubble shape remains approximately spherical. In the foams of various composition, investigated by this author (de Vries, 1958c) and produced by means of a high-speed mixer, values of A/V_l amounting to 5000 cm^2/cm^3 were readily obtained for gas-to-liquid ratios of the order of 10. The liquid phase in this type of foams is, as a first approximation, rather uniformly distributed between the spherical bubbles, in which case the total interfacial area per unit volume of liquid may be assumed equal to twice the reciprocal average thickness of the liquid lamellae separating the bubbles. This is obvious from simple geometrical considerations as shown by Eq. (2),

$$V_l = \tfrac{1}{2}\theta A, \tag{2}$$

where θ is the average bubble wall thickness. Wall thicknesses calculated by means of Eq. (2) from observed values of liquid content and total interfacial area may be expected, in general, to be somewhat overestimated and are only meaningful in the case of bubbles of approximately spherical shape. In foams containing mainly bubbles of polyhedral shape, the distribution of liquid between the bubbles is highly nonuniform, as will be discussed further in the next section.

B. Shape and Structure of Bubble Walls

As a result of the outflow of liquid, the original spherical bubbles start to be distorted after a period of time that will be longer the higher the viscosity of the liquid and the greater the degree of polydispersity of the bubbles. The distortion of a spherical bubble entails a change in its surface curvature that will no longer remain uniform along the gas–liquid interface but, instead, will vary continuously as a function of the degree of distortion. Because of this continuously varying surface curvature, pressure gradients will be produced in the liquid which will affect the flow phenomenon occurring in the bubble walls. The capillary suction created in this way tends to drain the liquid towards the regions of greatest curvature; this means to the edges where three or more bubbles meet. It is well known since the classical works of Plateau (1873) that an assembly of bubbles of polyhedral shape can only be in mechanical equilibrium if a certain number of geometrical conditions are met. Pressure differences across the strongly curved interfaces, which are due to the existence of surface tension, will have to cancel mutually at the locations where several bubbles meet if a more or less stable configuration is to be obtained. The resulting foam structure, as has been amply verified experimentally, is characterized by the presence of dihedral angles of 120° between three intersecting bubble walls, whatever the degree of polydispersity of the bubble sizes may be. The apparently negligible effect of bubble size or bubble polydispersity can be explained by the fact that the pressure difference across a curved interface is mainly determined by the *smallest* radius of curvature, as follows from the fundamental relationship between pressure difference Δp and surface tension γ derived in the beginning of the nineteenth century by Young and Laplace:

$$\Delta p = \gamma(1/R_1 + 1/R_2),\qquad(3)$$

where R_1 and R_2 are the principal radii of curvature of the interface. For a spherical bubble, $R_1 = R_2$; for a distorted bubble the difference between R_1 and R_2 grows with increasing degree of distortion and, ultimately, after the transition to polyhedral shape has been virtually completed, R_2 may approach infinity whereas R_1 will adopt a value, independent of bubble size, and equal for any set of three bubbles meeting along a common edge (Fig. 1).

Attainment of mechanical equilibrium also requires that no more than four edges come together in a single point, each pair meeting at an angle of approximately 109° (the tetrahedral angle). It may be concluded then that the most probable shape for a foam bubble is the pentagonal dodecahedron (figure bounded by 12 equilateral pentagons) and this ideal shape is, in fact, frequently approximated in real foams containing only bubbles of nearly uniform size. In polydisperse foams, however, bubbles of quite different and less regular shape are also observed but the general condition concerning the dihedral angle of 120° between intersecting bubble walls appears to be

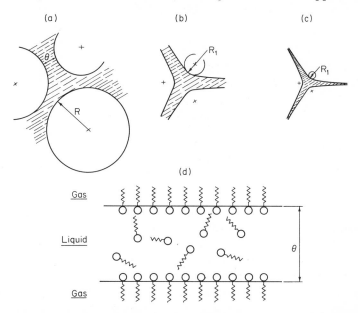

FIG. 1. (a), (b), and (c) Schematic representation of successive stages of the formation of a Plateau border with the radius of curvature R_1. (d) Schematic representation of the sandwichlike structure of a bubble wall with surface layers containing a surfactant with a lyophilic part (O) and a lyophobic part (∿∿).

fairly well respected in all foams of polyhedral structure after an initial period of transition during which unstable bubble configurations progressively disappear.

The strong curvature of the edges (usually called Plateau borders) is responsible for the capillary suction exerted on the liquid in the bubble walls, as a result of which the thickness of the latter is rapidly reduced. The combined action of capillary suction and gravity may produce quite complex flow patterns in the interior of the bubble walls, which have been the object of detailed experimental studies in the case of single isolated lamellae (Mysels *et al.*, 1959; Overbeek, 1967). Similar phenomena observed in real foams have also been reported in the literature (Leonard and Lemlich, 1965).

The greater part of the liquid content of polyhedral foams is concentrated in the Plateau borders and subject to laminar channel flow under the influence of gravity. The interstitial flow is the main source of foam drainage in low-density foams; Leonard and Lemlich (1965) have given a comprehensive theoretical treatment of this flow process and obtained results that are in good agreement with experimental data (Shih and Lemlich, 1967).

Their treatment is based on a model assuming dodecahedral bubbles and symmetrical Plateau borders of cross-sectional shape as indicated in Fig. 1.

The cross-sectional area of the Plateau borders[1] A_{Pb} is given by

$$A_{Pb} = (\sqrt{3} - \pi/2)R_1{}^2 + R_1\theta\sqrt{3} + (\sqrt{3}/4)\theta^2. \qquad (4)$$

The relationship between the gas-to-liquid ratio and the various geometric parameters of this foam model may be described by the following expression:

$$\frac{V_l}{V_g} = \frac{60A_{Pb}}{2.445\pi} \frac{\sum n_i d_i}{\sum n_i d_i{}^3} + 3\theta \frac{\sum n_i d_i{}^2}{\sum n_i d_i{}^3}. \qquad (5)$$

The value of d_i is that of the diameter of an individual spherical bubble before it has become distorted into a dodecahedron; it may be assumed to be only slightly different from the actual diameter of the polyhedral foam bubble.

The second term on the right-hand side of Eq. (5) represents the liquid content in the bubble walls. In a foam of undistorted spherical bubbles this term represents the only contribution to the liquid-to-gas ratio; the corresponding simplified expression can be derived by combining Eqs. (1) and (2) after inserting the value 6 for the shape factor f of a spherical bubble.

The first term on the right-hand side of Eq. (5) takes account of the liquid content in the Plateau borders and will, therefore, be the most important one in the case of low-density foams with polyhedral bubbles of small wall thickness θ.

[1] Editor's note: The cross section of the Plateau border is shown in detail in Fig. 5 of Chapter 3. (A Plateau border is also called a capillary.)

As a typical example one may cite the foams investigated by Leonard and Lemlich (1965): for average bubble diameters of 0.4–0.5 cm, wall thicknesses between 0.2 and 2×10^{-4} cm were observed by these authors. According to Eqs. (4) and (5), when the gas-to-liquid ratio in these foams varied between 100 and 250, between 80 and 95 % of the total liquid content was located in the Plateau borders.

Whatever the particular morphology may be, more or less persistent foams can only exist if the liquid phase contains at least one component which is preferentially adsorbed at the gas–liquid interface. This implies that, on a molecular scale, the structure of bubble walls and Plateau borders is never homogeneous but characterized by the presence of surface layers whose composition is different from the inner core separating them. This fundamental property, which forms the basis of all adsorptive bubble separation techniques, is also of essential importance for understanding the mechanism of the various processes leading to bubble coalescence. The sandwichlike structure of the liquid phase in a low-density foam is schematically indicated in Fig. 1(d).

C. Bubble Size Distribution and Thickness of Bubble Walls

Generally speaking, it is not feasible to observe directly the complete distribution of bubble sizes in a given volume of foam. However, it is possible to measure the size distribution of bubbles in a plane cross section cut through the foam. One method consists of rapidly freezing the foam, e.g., by means of liquid oxygen, and then to cut through the frozen foam with a razor blade at different depths and measure the observed bubble size distribution in the various sections that have become accessible in this way. Freezing and subsequent thawing of the foam does not seem to affect significantly the bubble size frequency distribution, as may be inferred from the results reported by Savitskaya (1951) and by Chang et al. (1956). It has also been concluded by both groups of authors that the bubble size distribution in an outer surface layer of the foam is identical with (or only slightly different from) the frequency distribution measured in any other section within the foam mass. New (1967) has measured the bubble size distribution in the top layer of various foams and concluded that the size distribution in that layer did not deviate appreciably from that observed in layers situated a few millimeters below the surface. A most convenient method, which may be expected therefore to yield representative information on the size frequency distribution, is to observe the bubbles in the plane of contact between the foam and a flat glass wall, neglecting the slight degree of distortion of the bubbles which might result from the presence of the glass wall.

The size frequency distribution measured in any plane cross section will be different, in general, from the real size distribution because the number of bubbles observable in a plane will vary as a function of the bubble size. If N is the total number of bubbles per unit volume of foam and $F(r)$ is the frequency distribution function of their radii, $NF(r)\,dr$ represents the number of bubbles per unit volume of foam with radii between r and $r + dr$. Let us call n_s the total number of bubbles counted per unit of surface area in any arbitrary plane cut through the same foam, and let $f(r)$ represent the size frequency distribution function of the latter bubbles. If all bubbles are randomly distributed throughout the foam volume regardless of their respective sizes, the number of bubbles of a given diameter $2r$ per unit area of surface may be equated to the number of bubbles of the same diameter contained in a volume of foam equal to unit area times the bubble diameter. More generally, it might be assumed (de Vries, 1958b) that these numbers are not strictly identical but proportional one to the other. Supposing that the proportionality constant C_s is independent of bubble size, we may express the sought relationship by means of

$$n_s f(r)\,dr = 2rC_s NF(r)\,dr. \tag{6}$$

By integration of Eq. (6) between $r = 0$ and $r = \infty$, we obtain

$$n_s = 2C_s N \langle r \rangle, \tag{7}$$

where $\langle r \rangle$ is the average bubble radius.

The relationship between the two frequency distribution functions is found after substitution of Eq. (7) into Eq. (6),

$$f(r) = (r/\langle r \rangle)F(r). \tag{8}$$

Equation (8) does not contain C_s and illustrates the fact that bubbles larger than the average size will occur more frequently in a cross section than might have been inferred from the real distribution function $F(r)$. The inverse is true for bubbles smaller than average size. In order to calculate the distribution function $F(r)$ from the observed one $f(r)$ the following relationships are useful. First, Eq. (8) may be written as

$$(1/r)f(r) = (1/\langle r \rangle)F(r). \tag{8'}$$

The general relationship between the successive moments of both distribution functions is given, therefore, by

$$\int_0^\infty r^{i-1} f(r)\,dr = (1/\langle r \rangle)\int_0^\infty r^i F(r)\,dr. \tag{9}$$

For $i = 0$, we find

$$\langle r_s^{-1} \rangle = \langle r \rangle^{-1}. \tag{10}$$

The index s indicates that we refer to a directly measured average in a layer of foam bubbles observable in a cross section. Higher averages can also be obtained from Eq. (9) after substitution of the appropriate values of i and by making use of Eq. (10). Of particular interest are the average square radius $\langle r^2 \rangle$, and the average cube radius $\langle r^3 \rangle$ corresponding to $i = 2$ and $i = 3$, respectively:

$$\langle r^2 \rangle = \langle r_s \rangle / \langle r_s^{-1} \rangle, \tag{11}$$

$$\langle r^3 \rangle = \langle r_s^2 \rangle / \langle r_s^{-1} \rangle. \tag{12}$$

Equations (10)–(12) are particular examples of the general relationship given by

$$\langle r^i \rangle = \langle r_s^{i-1} \rangle / \langle r_s^{-1} \rangle. \tag{13}$$

In studies of the kinetics of bubble coalescence it may be sometimes sufficient to know the change in the number of bubbles per unit volume of foam as a function of time. This number N is related to the observed number of bubbles per unit area of cross section n_s by means of the expression given in Eq. (7).

The latter expression is, in general, time dependent because of the continuous increase of average bubble size $\langle r \rangle$. As to the proportionality constant C_s, its value as a function of time may be determined, in principle, with the aid of the following equations:

$$n_s = 2C_s N \langle r \rangle = \frac{3C_s}{2\pi} \left(\frac{V_g}{V_g + V_l} \right) \frac{\langle r \rangle}{\langle r^3 \rangle} \tag{14}$$

if the average bubble volume is equal to $(4\pi/3)\langle r^3 \rangle$.

Taking account of the general relationship expressed by Eq. (13), the following equation may be obtained:

$$C_s = \left(\frac{2\pi n_s \langle r_s^2 \rangle}{3} \right) \left(\frac{V_g + V_l}{V_g} \right). \tag{15}$$

Application of Eq. (15) to the experimental data published by various authors shows that the value of C_s is always smaller than unity and varies, in general, between 0.4 and 0.7, approximately. During bubble coalescence the calculated values of C_s may vary somewhat as a function of time, but this dependence appears to be only slight and may probably be neglected in many cases without introducing important errors. Use of Eq. (15) allows those errors to be estimated if the bubble size distribution in the cross section is known, as well as the gas-to-liquid ratio in the foam.

It is obvious from Eq. (15) that calculated values of C_s smaller than unity do not necessarily imply a nonrandom distribution of gas bubbles, as has been assumed in deriving Eq. (6). They may also be imputed to errors in the measured values of r_s due to distortion of the bubbles, or limited transparency of the liquid phase. Irrespective of the exact interpretation of C_s, kinetic studies of bubble coalescence based on a simple counting of bubbles in a cross section or a surface layer are only justified if it has been established beforehand that the time dependence of C_s is negligible.

The average bubble wall thickness θ is an important parameter in all coalescence processes. In foams containing only bubbles of spherical shape, its value may be calculated from the bubble size frequency distribution and the gas-to-liquid ratio, as has already been mentioned before. In terms of measured bubble radii in a cross section, the required relationship is given by

$$\theta = 2V_l \langle r_s^2 \rangle / 3V_g \langle r_s \rangle. \tag{16}$$

The more general relationship valid for polyhedral foams, given by Eq. (5), may be expressed in the same parameters as follows:

$$V_l/V_g = (15A_{Pb}/2.445\pi \langle r_s^2 \rangle) + (3\theta \langle r_s \rangle / 2 \langle r_s^2 \rangle). \tag{17}$$

Although the average cross-sectional area of the Plateau borders A_{Pb} might be estimated more or less accurately from photomicrographs, Eq. (17) is not recommended for use in obtaining reliable values of θ since the second term on the right-hand side of Eq. (17) is, in general, much smaller than the first one (see Section II,B). If, however, the value of θ is known by direct observation, Eq. (17) may be used to calculate an average value for the cross-sectional area of the Plateau borders, as has been shown by Shih and Lemlich (1967) for the case of foams with polyhedral bubbles of uniform size.

D. Experimental Techniques

Bubble size frequency distributions are most readily measured by means of photomicrography using a low-power microscope. Linear magnifications between 20 and 40 are usually sufficient. The layer of foam bubbles observed may be either a cross section obtained by slicing a frozen foam, or the plane of contact between the foam and the transparent wall of a container or column, or, finally, one of the upper layers of a strongly illuminated foam. Bubble diameters measured on the photomicrographs with the aid of a comparator are assigned to a number of successive size groups having an interval of 25 or 50×10^{-4} cm. The various averages then can be calculated by counting the number of bubbles in each size group and by assigning the same average value of diameter to all bubbles in one group. If required, integral and differential frequency distribution functions may be constructed from the same experimental data, but it should be kept in mind that the accuracy of the measured bubble diameters is often limited as a result of bubble distortion and insufficient sharpness of the outer contours.

In the case of low-density foams with polyhedral bubbles the same photo-micrographic technique may be used in order to estimate an average value of the cross section of the Plateau borders.

Bubble wall thicknesses in polyhedral foams may be estimated more or less with accuracy from their color when observed in white light. The colors are due to the interference of light reflected at the two gas–liquid interfaces of a bubble wall. The intensity of this reflected light varies as a periodic function of the ratio of wall thickness to wavelength, according to the well-known expression derived by Lord Rayleigh (1936). This method of thickness estimation has been extensively used in the study of single isolated liquid lamellae ("soap films") and should give reasonably accurate values for thicknesses between about 10^{-5} and 10^{-4} cm (Overbeek, 1967). More accurate values can be obtained by using monochromatic light and by comparing the intensities of the incident and the reflected light. A detailed review of the various experimental methods used for measuring the thickness of liquid lamellae has been given by Sheludko (1967).

In the case of very curved bubble walls the local values of thickness may be evaluated by observing interference effects in transmitted light, as has been demonstrated by Princen and Mason (1965b) who tabulated the colors observed in transmitted daylight as a function of the wall thickness of a symmetrical bubble having an axis of revolution normal to the light beam. It is not obvious, however, that the latter method is applicable to an assembly of bubbles as found in a real foam.

The attenuation of light transmitted through a layer of foam bubbles has been measured by Clark and Blackman (1948b), who found that the ratio of transmitted to incident light intensity decreased with decreasing average bubble size, for a foam layer of given thickness. The theoretical interpretation of these results, presented by the same authors, has been criticized (Bikerman, 1953; Chang et al., 1956). Indeed, one might expect that the general relation-ship between light transmission and degree of dispersion of the gas bubbles would be a rather complex function of various structural parameters (thick-ness of bubble walls and their orientation in space, for instance) which have been neglected in the simplified theoretical analysis published by Clark and Blackman. According to Chang et al. (1956), a modified light transmission technique has been investigated in an unpublished study by Stenuf (1953) but its results, apparently, have not given rise to any further development of this experimental method.

The electrical properties of liquid lamellae have been discussed by Bikerman (1953) and Sheludko (1967). The electrical conductivity of aqueous foams is directly related to the liquid content as first shown by Miles et al. (1945). The results obtained by Clark (1948) for aqueous foams of various composi-tions can be represented by means of a unique relationship between the ratio of electrical conductivity of the foam to that of the solution on the one hand, and the liquid content of the foam on the other hand, irrespective of the bubble size distribution. After a suitable calibration, electrical conductivity measurements are particularly appropriate for determinations of the local liquid content in a moving foam column (Fanlo and Lemlich, 1965).

Finally, the use of cinephotographic methods in the study of bubble coalescence (de Vries, 1958d; New, 1967) should also be mentioned.

III. GAS DIFFUSION IN FOAMS

A. Pressure Distribution in the Gas Phase

The nonuniform distribution of pressure in the interior of a foam is a consequence of the free energy excess due to the existence of a positive interfacial tension. In foams with polyhedral bubbles the curvature of the greater part of the bubble walls is nearly zero, but increases rapidly towards the edges. The pressure of the gas in a given bubble will be the same throughout the volume of the bubble, but since the pressure difference across the interface will vary with the curvature, as indicated by Eq. (3), the liquid phase surrounding the bubbles will be the seat of pressure gradients, the importance of which has already been discussed in Section II,B.

If, however, the curvature of the bubble surface is uniform, as is the case, in particular, for spherical bubbles, the pressure of the gas inside a bubble will exceed the average hydrostatic pressure existing in the surrounding liquid phase. This pressure excess Δp will be only dependent on the bubble size and the interfacial tension. In such a foam, important pressure differences may exist between adjacent bubbles of different sizes, which gives rise to gas transfer from smaller to larger bubbles as a result of diffusion through the bubble walls.

The excess pressure inside the bubble may be considered to be a direct consequence of the tendency of the bubble surface to contract spontaneously under the influence of the interfacial tension. Equating the work done by compressing the gas inside the bubble with the surface free energy liberated upon compression, leads to

$$\Delta p \, dv_i = \gamma \, da_i. \tag{18}$$

The variation of surface area, da_i, with respect to an isotropic volume change, dv_i, irrespective of the exact bubble shape, is given by

$$da_i/dv_i = 2a_i/3v_i. \tag{19}$$

Accordingly, the excess pressure Δp may be written as follows:

$$\Delta p = (2a_i/3v_i)\gamma. \tag{20}$$

Ross (1969) has recently drawn attention to the fact that application of the ideal gas law to all bubbles obeying Eq. (20) would lead to the definition of an equation of state for the gaseous phase in a foam. If the foam contains n_m moles of gas, and if the average hydrostatic pressure in the liquid phase is equated, as a first approximation, to the ambient pressure p_a of the atmosphere in contact with the foam, we may write

$$n_m RT = \sum_i (p_a + \Delta p)v_i, \tag{21}$$

where R is the gas constant and T the absolute temperature. After substitution of Eq. (20), we obtain

$$n_m RT = p_a V_g + \tfrac{2}{3}\gamma A. \tag{22}$$

The average pressure of the gaseous phase in the foam is given by

$$\langle p \rangle = n_m RT/V_g = p_a + \tfrac{2}{3}\gamma A/V_g. \tag{23}$$

An equation of state similar to Eq. (22) has been used by Derjaguin (1933) in order to deduce an expression for the compressibility of foams. The second term on the right-hand side of Eq. (23) represents the average excess pressure with respect to the ambient atmosphere; it is directly proportional to the total interfacial area per unit volume of gas, and amounts for most aqueous foams to only a small fraction of the atmospheric pressure (of the order of 10^{-2}). Bubble coalescence will result in a decrease of total interfacial area and should be accompanied by a small volume expansion, the value of which can be calculated from Eq. (22). Consequently, the decrease of interfacial area resulting from bubble coalescence can be followed, in principle, by measuring the volume expansion or the change in foam density as a function of time. The feasibility of such an experimental technique, suggested by Ross (1969) apparently has never been tested and would require a very accurate measurement of foam density.

Pressure measurements in the gas liberated from collapsing foams have been reported by Alejnikoff (1931) and confirmed the existence of a pressure excess, but the development of a reliable experimental technique based on the application of Eq. (23) does not seem to have been attempted so far.

The transfer of gas from smaller to larger bubbles as a result of differences in excess pressure has been the object of detailed studies in the case of foams with bubbles of approximately spherical shape. In polydisperse foams the diffusion process may lead to a relatively rapid decrease in interfacial area due to the progressive shrinkage and ultimate disappearance of the smallest bubbles.

The mechanism and kinetics of this transfer process will be considered in the following sections.

B. Mechanism of Gas Transfer between Foam Bubbles

The pressure difference between the gas in a spherical bubble of radius r and that in a larger bubble of radius R is equal to

$$\Delta p_{12} = 2\gamma(1/r - 1/R). \tag{24}$$

If these bubbles are separated by a liquid layer of average thickness θ, the flux of gas across this layer is directed towards the larger bubble and may be expressed as

$$dn/dt = PA_{12}\,\Delta p_{12}, \tag{25}$$

where A_{12} is the area of the liquid layer perpendicular to the direction of the flux, and P is the "permeability" of the liquid layer. Equation (25) is only valid, of course, for a steady state of permeation but such a state may be assumed to be attained very rapidly. For a relatively thick layer the rate of permeation is mainly determined by the rate of diffusion of the gas across the liquid layer under the influence of a constant concentration gradient $\Delta C_0/\theta$.

The value of ΔC_0 can be derived from Henry's law

$$\Delta C_0 = S\,\Delta p_{12}, \tag{26}$$

where S is the solubility coefficient of the gas in the liquid, expressed as an equilibrium concentration in the liquid phase per unit of gas pressure.

In general, however, the rate of permeation may be partly determined by the rate of transfer across the interfaces (gas–liquid and liquid–gas) at both sides of the liquid layer. In foams these interfaces are usually composed of a monolayer of surface-active compound which may offer a nonnegligible resistance to mass transfer. A general expression for the permeability P taking account of interfacial resistance to mass transfer, has been derived by Van Amerongen (1943) in order to explain the permeation of gases through rubber membranes. It can be written as

$$P = DS/(\theta + 2D/k_i), \tag{27}$$

where D is the diffusion coefficient of the gas in the liquid phase and k_i is the rate constant of the mass transfer process across the gas–liquid interface, defined by

$$dn/dt = k_i A_{12}(C_0 - C), \tag{28}$$

where C is the actual concentration of dissolved gas at the phase boundary and C_0 is the equilibrium concentration given by Henry's law.

Two limiting cases of interest are described by Eqs. (29) and (30):

$$P = DS/\theta \quad \text{if} \quad 2D/k_i \ll \theta, \tag{29}$$

and

$$P = \tfrac{1}{2}k_iS \quad \text{if} \quad 2D/k_i \gg \theta. \tag{30}$$

Princen and Mason (1965a) succeeded in obtaining experimental values for k_i by means of permeability measurements on extremely thin equilibrated soap films surrounding single gas bubbles resting at a gas–liquid interface. After a period of draining, these films attained equilibrium thicknesses equal to twice the length of the surfactant molecule (hexadecyltrimethyl-ammonium bromide) plus a central aqueous layer of thickness θ approximately equal to 14×10^{-8} cm (Princen et al., 1967). In that case the resistance to mass transport offered by the two monolayers with total thickness of about 4×10^{-7} cm is rate determining for the permeation process and Eq. (30) may be used as a good approximation, leading to the following expression for the flux of gas molecules:

$$dn/dt = \tfrac{1}{2}k_iA_{12}S\,\Delta p_{12}. \tag{31}$$

The experimental values of k_i determined in this way by Princen and Mason for eight different gases vary between 13 and 90 cm/sec, but the value of $2D/k_i$ does not vary appreciably with the nature of the gas, its lowest value being 1.3×10^{-6} cm (for He) and its highest being 2.9×10^{-6} cm (for N_2O). For nitrogen, oxygen, and air the value of $2D/k_i$ is of the order of 2×10^{-6} cm. This value may be interpreted as the equivalent thickness of an aqueous layer having the same permeability as the two monolayers together. It may be concluded, then, that the use of Eq. (29) instead of the general expression given by Eq. (27) will not lead to any appreciable error if the liquid layer has a thickness of the order of 10^{-4} cm, as is the case for many foams. For bubble wall thicknesses of the order of 10^{-5} cm, which may be encountered in foams of very high gas-to-liquid ratios, the effect of the surfactant monolayers on the permeability of the bubble wall is no longer negligible.

Princen et al. (1967) have also provided evidence that the transport of gas through the soluble monolayer may be explained satisfactorily by assuming that a mechanism of simple Fickean diffusion takes place, probably within aqueous pores between the surfactant molecules. Indeed, the area available per surfactant molecule in these layers is large enough to leave room for pores of sufficient width. This may not be true, however, for rigid or insoluble condensed monolayers, and the permeation process in those cases, therefore, is preferably described by means of the energy-barrier theory (Blank, 1962, 1964; La Mer, 1962). The permeability of insoluble monolayers may be several orders of magnitude lower than that of soluble ones and depends strongly on the surface concentration of the surfactant (Hawke and Parts,

1964). If this type of monolayer is present at the interface of foam bubbles, the use of Eq. (29) is obviously not justified since the contribution of the monolayers to the resistance to gas transfer cannot be neglected. Following Blank (1964), the total resistance to gas transfer may be defined as the pressure difference required for obtaining a unit flux of gas across a unit of surface area. Its value can be derived from Eqs. (25) and (27), and is equal to

$$P^{-1} = (\theta/DS) + (2/k_iS). \tag{32}$$

Equation (32) shows quite clearly that the total resistance to the permeation of a bubble wall is the sum of the resistance due to diffusion through the central liquid layer and the resistances of the two interfacial layers, each of the latter contributing an amount $1/k_iS$. The additivity of the relationship expressed by Eq. (32) is generally valid irrespective of the particular mechanism of interfacial resistance, provided that it obeys Eq. (28).

C. Kinetics of Gas Transfer and Its Effect on Bubble Size Distribution

A polydisperse foam with bubbles of approximately spherical shape always contains bubbles of widely varying sizes if the gas-to-liquid ratio is relatively important. Each large bubble in such a foam, in general, will be surrounded by several smaller bubbles which more or less rapidly disappear as a result of gas transfer to the adjacent larger bubbles. Therefore, in a first approximation, one may identify the total surface area of a small bubble with the escape area A_{12} through which the gas permeates. Moreover, if the difference in size between the small and the surrounding larger bubbles is relatively important, the pressure difference Δp_{12} may be approximated by the excess pressure $2\gamma/r$ in the smaller bubble, and the second term in the right-hand side of Eq. (24) may be neglected. Since the smaller bubble shrinks during the process of gas transfer whereas the larger bubbles grow, the latter approximation will be a particularly good one during the last stage of the shrinkage process. The rate of shrinkage of a small bubble is now only dependent on its own radius and can be derived from Eqs. (24) and (25):

$$-r(dr/dt) = 2\gamma P(RT/p_a). \tag{33}$$

In deriving Eq. (33) it has been assumed that the ideal gas law holds and that the excess pressure in the bubble can be neglected with respect to the ambient atmospheric pressure. The decrease of radius as a function of time is obtained by integration of Eq. (33):

$$r^2(t = 0) - r^2(t) = (4RT/p_a)\gamma Pt. \tag{34}$$

According to Eq. (34) the surface area of a shrinking bubble decreases as a linear function of time, and the lifetime of the bubble may be expressed as a function of its initial radius r_0 by

$$t(r_0) = (4RT\gamma P/p_a)^{-1} r_0^2. \tag{35}$$

Equation (34) has been applied (de Vries, 1958b) to the observed rate of shrinkage of foam bubbles with the aid of a photomicrographic technique. The experimental results, as well as those reported by Clark and Blackman (1948a), could be adequately described by Eq. (34), and allowed the calculation of permeability values which were in reasonable agreement with the expected values based on the application of Eq. (29) if θ was identified with the average bubble wall thickness estimated from the bubble size frequency distribution and the gas-to-liquid ratio in the various investigated foams. Apparently, the effect of interfacial resistance to gas permeation was small in these foams, which is not surprising in view of the relatively large values of bubble wall thickness (between about 10^{-4} and 10^{-3} cm).

The slope of the straight lines representing the square of the radius as a function of time was, within experimental error, independent of the initial value of the radius for the foams investigated by de Vries. This seems to indicate that the liquid phase was homogeneously distributed throughout the foam volume, resulting in nearly constant values for bubble wall thickness and permeability in a given foam. Application of Eq. (34) to the results of Clark and Blackman reveals slight but significant variations in bubble wall permeability which may be due, although not necessarily, to nonuniformity of either the bubble wall thickness or the interfacial resistance to permeation. If the bubble wall permeability is practically the same for all bubble walls in a given foam volume, the change in bubble size distribution as a result of gas transfer might be calculated, in principle, by means of Eq. (34) if appropriate assumptions are made regarding the distribution of the permeated gas over the growing larger bubbles surrounding a small one. For example, New (1967) has assumed that the permeated gas escaping from the smaller bubbles is distributed over the larger bubbles in proportion to their surface area. In general, however, it will be impossible to predict accurately the changes occurring in bubble size distribution without the aid of other specific parameters characteristic of the foam structure.

Shrinkage and subsequent disappearance of the smaller bubbles will result in a rapid change in the total number of bubbles per unit volume of foam. The total number per unit volume at any time t is related to the initial number N_0 by means of

$$N_t = N_0 \left(1 - \int_0^{r_{0,t}} F(r_0) \, dr_0\right), \tag{36}$$

where $F(r_0)$ is the bubble size frequency distribution function at $t = 0$, and the upper integration limit $r_{0,t}$ represents the initial radius of a bubble with lifetime t. All bubbles having an initial radius equal to or smaller than $r_{0,t}$ will obviously have vanished during the time interval t. The relationship between $r_{0,t}$ and t is given by Eq. (35).

Equation (36) shows that knowledge of the initial bubble size distribution is sufficient for an exact calculation of the change in the number of bubbles per unit volume as a function of time. Since the total volume of gas should remain constant if we neglect the very small volume change due to the decrease of average excess pressure, Eq. (36) also allows an exact calculation of the average bubble volume at any time t. The latter volume is equal to the gas content $V_g/(V_g + V_l)$ divided by the number of bubbles per unit volume, N_t.

Equation (36) has been used by the present author (de Vries, 1958c) in a study of gas diffusion in foams of various compositions. The initial bubble size frequency distribution function in these foams, mechanically generated with the aid of a high-speed mixer, was fairly well represented by

$$F(r_0) = 6\alpha r_0/(1 + \alpha r_0^2)^4, \tag{37}$$

where α is an adjustable parameter of the distribution function.

After substitution of Eqs. (35) and (37) into Eq. (36) and subsequent integration, one obtains

$$N_t = N_0/(1 + k_d t)^3, \tag{38}$$

where the rate constant k_d is given by

$$k_d = (4RT/p_a)\gamma P\alpha. \tag{39}$$

Equation (38) was found to be in reasonable agreement with experimental data if the permeability P was assumed to be given by Eq. (29). The experimental values for k_d in these foams varied between 1.0 and 3.3 × 10^{-3} sec^{-1}, depending on the composition of the liquid and gaseous phases of the foam. They are in fair agreement with values calculated from the observed rate of shrinkage of individual bubbles in the same foams, and their dependence on diffusion and solubility coefficients also appeared essentially correct, within the limits of experimental accuracy.

It may be concluded then, that in polydisperse foams with permeable bubble walls, the smallest bubbles of approximately spherical shape rapidly disappear as a result of gas transfer. This process also involves a spontaneous decrease of total interfacial area, the importance of which depends on the initial bubble size distribution and its subsequent variation with time. For the particular foams investigated by this author and referred to above, the decrease in interfacial area A_t as a function of time could be represented by

$$A_t = A_0/(1 + k_d t). \tag{40}$$

Comparison of Eqs. (38) and (40) shows that in this case the relative decrease in interfacial area was three times slower than the relative decrease in the number of bubbles. The latter finding may be expected to have some general qualitative value, but it should be remembered that it is not possible to derive simple general relationships for the variation of interfacial area comparable to Eq. (36).

IV. THINNING AND RUPTURE OF BUBBLE WALLS

A. Bubble Wall Thickness and Probability of Rupture

Rupture of a bubble wall followed by coalescence of the two bubbles originally separated by the same lamella is one of the main causes of spontaneous decrease of interfacial area in foams. It is well known that liquid lamellae are more liable to rupture the smaller their thickness; the probability of film rupture and subsequent coalescence therefore increases with increasing gas-to-liquid ratio in a foam with a given bubble size distribution. The lifetime of a bubble wall is determined by its rate of thinning and by the probability of rupture as a function of thickness. Both processes, thinning and rupture, are dependent in general on a number of various parameters whose quantitative effects are still incompletely known. This is particularly true for the rupture process. The thinning of liquid lamellae due to the combined effects of gravity and capillary suction is understood relatively well and has been the object of detailed studies by Mysels et al. (1959), as well as by other authors (Sheludko, 1967). Unlike the channel flow in the Plateau borders which is responsible for foam drainage, the hydrodynamics of the thinning process taking place in the bubble walls are often quite complex and cannot be treated as a simple problem of viscous flow. Relatively simple flow patterns are observed only in the case where lamellae are stabilized by rigid monolayers. Such lamellae thin rather slowly, and the rate of thinning is inversely proportional to the viscosity and directly proportional to the square of the thickness. More often, however, foam bubble walls are of the "mobile" type and do not thin in a regular way; patches of much smaller thickness than average occur locally near the edges and, in particular, at the top of the bubble walls. The spontaneous formation of these "black spots" is a characteristic feature of relatively persistent bubble walls. Unstable liquid lamellae apparently do not allow the formation of black spots but thin and collapse very rapidly after having attained a "critical" thickness of the order of 10^{-5} cm or less. Therefore, the lifetime of inherently unstable foam bubble walls is only determined by the rate of thinning and is as a first approximation, proportional to the viscosity of the liquid.

In metastable foams attainment of the critical bubble wall thickness does not entail immediate rupture but leads to the appearance of discontinuous changes in thickness, followed by a progressive increase of the area of smallest thickness. Ultimately the greater part of the bubble wall may attain this very low "equilibrium" value of thickness without rupturing, but more often collapse occurs during rapid growth of the area of the thinnest part of the lamella (de Vries, 1958d). Single liquid lamellae or soap bubbles having drained to very low equilibrium thicknesses of the order of 10^{-6} cm may, nevertheless, persist over extended periods of time provided they are kept free from any external perturbations. Obviously, for these metastable lamellae no simple relationship exists between thickness and lifetime.

B. Initiation, Propagation, and Activation Energy of Rupture

The rupture of a bubble wall, initiated by the formation of a small hole, and its ultimate disappearance are associated with a decrease in interfacial area. The surface free energy released during growth of the hole is converted into kinetic energy of the receding liquid at the rim of the hole. In a liquid film of uniform thickness the velocity of propagation is constant, as has been shown by various authors (de Vries, 1958d; Ranz, 1959; McEntee and Mysels, 1969). This result had been predicted by Dupré (1867) on the basis of a simple energy balance. A more elaborate treatment proposed by Culick (1960) and discussed further by Frankel and Mysels (1969) leads to a slightly smaller calculated velocity, as given by

$$u = (2\gamma/\theta\rho)^{1/2}, \qquad (41)$$

where ρ is the density of the liquid and u is the linear velocity of propagation. Equation (41) differs from the original Dupré relationship only by a factor of $\sqrt{2}$. For a lamella thickness of the order of 10^{-4} cm, velocities approaching 10^3 cm/sec are found, which means that rupture of a foam bubble wall is completed within some milliseconds after its initiation, in agreement with reported cinephotographic observations (de Vries, 1958d). It follows that initiation is the rate-determining step in the rupture process.

Although nothing is known about the very first stages of the rupture process, the formation of a hole or flaw seems to be a necessary condition for rupture to occur. Hole formation is always associated with a temporary increase of interfacial area (Fig. 2), and the subsequent decrease of total interfacial area during growth of the hole is only realized if the diameter of the hole becomes of the same order of magnitude as the film thickness. Various hypotheses may be formulated concerning the causes of hole formation, such as surface deformation and corresponding fluctuations in the local film

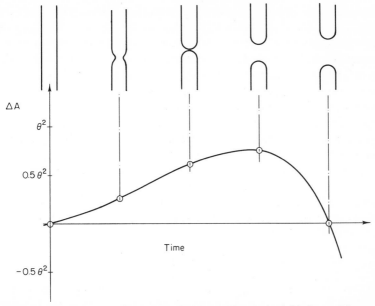

FIG. 2. Time-dependent variation of interfacial area (ΔA) during initiation and first stages of propagation of rupture: ΔA units, (local lamella thickness)^2; time units, arbitrary. The adopted model yields a maximum ΔA of $0.73\theta^2$; other models (which might be more realistic) yield still higher values for the activation energy of rupture.

thickness (de Vries, 1958e) or cavitation in the interior of the film (Gleim and Chelomov, 1959). Both hypotheses lead to the conclusion that the temporary increase of interfacial area is approximately equal to the square of the local film thickness. The corresponding increase of interfacial energy is sufficiently high to exclude the possibility of spontaneous rupture of a foam lamella with a thickness of the order of 10^{-6} cm in the absence of an external source of energy. Since the activation energy decreases with the square of the local film thickness, spontaneous random energy fluctuations may be expected to induce rupture only if the thickness is a few 10^{-7} cm.

Foam bubble walls frequently break at a much larger average thickness. Thus far, no comprehensive and adequate theoretical analysis has been proposed to explain the rupture process in liquid lamellae stabilized against ultimate thinning by the presence of either ionized monolayers (responsible for long-range electric double layer repulsion) or nonionic surfactant molecules with long hydrophilic chains. The rapid thinning after attainment of a critical thickness which, in the case of unstable lamellae, leads to immediate rupture, can be adequately explained by the action of inter-molecular forces (van der Waals forces) which will always tend to drive the liquid from thinner to thicker parts of the lamella (de Vries, 1958d, e, 1960).

In the absence of important long-range repulsive forces, the effect of the van der Waals forces is sufficiently strong to induce rapid thinning at a critical thickness of the order of 5×10^{-6} cm, in agreement with numerous experimental observations. The exact value of the critical thickness depends on the magnitude of various parameters, as well as on geometric characteristics of the thinning process. Explicit quantitative relationships may be obtained by application of the theory of fluctuations, as recently shown by Vrij (1966). Due to thermal motion, both surfaces of a liquid lamella are subject to periodic deformations which can be determined by means of light scattering measurements (Vrij, 1964). The corresponding thickness fluctuations will grow spontaneously in amplitude if the associated decrease in free energy, due to the van der Waals forces, is sufficient to compensate for the increase in interfacial free energy. This is only possible, according to Vrij (1966) and Vrij and Overbeek (1968), when the wavelength λ of the fluctuation is larger than a critical value λ_{crit} :

$$\lambda > \lambda_{\text{crit}} = [-2\pi^2\gamma/(d^2G/d\theta^2)]^{1/2}, \tag{42}$$

where G is the total free energy of interaction per unit area (including attractive as well as repulsive forces, in the general case), whereas the interfacial tension γ is supposed to be uniform and constant.

Fluctuations with a wavelength smaller than λ_{crit} are damped, but for $\lambda > \lambda_{\text{crit}}$ the amplitude increases exponentially with time. An accurate description of the fluctuation kinetics requires a model of the mechanism of liquid transport in the lamella. The simple model, based on laminar flow parallel to the surfaces without slip at the interface, which was used by Vrij, yields the following expressions for the characteristic time constant τ :

$$\tau = -(3\eta\lambda^2/\pi^2\theta_0^3)[(2\pi^2\gamma/\lambda^2) + (d^2G/d\theta^2)_{\theta=\theta_0}]^{-1}, \tag{43}$$

and

$$\tau_{\text{min}} = (24\gamma\eta/\theta_0^3)(d^2G/d\theta^2)_{\theta=\theta_0}^{-2}, \tag{44}$$

where θ_0 is the initial lamella thickness and τ_{min} represents the characteristic time constant of the most rapid growing fluctuation. According to Eq. (44), the characteristic time for a lamella in which the repulsive forces may be neglected would be proportional to the fifth power of the thickness, provided the usual Hamaker approximation for the van der Waals energy is assumed to be valid. Substitution of reasonable values for the parameters in Eq. (44) leads to the conclusion that an unstable lamella would break within a second after it has attained a critical thickness of the order of 5×10^{-6} cm. Detailed experimental data on the critical thickness of regularly draining microscopic lamellae, as reported by Sheludko and Manev (1968), are in fair agreement with the theoretical predictions, in particular concerning the independence of critical thickness with viscosity and the rather weak influence of surface tension.

C. Rupture of Metastable Bubble Walls and Kinetics of Coalescence

The theory of spontaneous growth of thermal fluctuations, discussed in the preceding section, may explain the formation of black spots in metastable lamellae. However, it does not yield any prediction regarding the mechanism of rupture in this type of lamellae which is characterized by the presence of electric double layer repulsion. Because of this or any other repulsive effect, spontaneous fluctuations are rapidly damped out, according to Eqs. (42) and (43), in any part of the lamella having attained its equilibrium thickness. Moreover, damping of surface waves is also favored by the viscoelastic resistance to the local increase of interfacial area (Hansen and Mann, 1964; Van den Tempel and Van de Riet, 1965).

The inherent resistance of lamellae to local extension, which is the basis of the Gibbs definition of surface elasticity, is caused by the time-dependent variations in the local surface concentration of adsorbed surfactant and the corresponding surface tension gradient. In the particular case considered by Gibbs (1876) and others, the induced surface tension gradient always tends to restore the initial equilibrium. Surface tension gradients may, however, also cause local thinning of a lamella with subsequent rupture, as first pointed out by Ewers and Sutherland (1952). If, for any reason, the surface tension in a small element of the lamella is lower than in adjacent volume elements, surface free energy may be gained by an extension of the area of lower tension associated with a contraction of the surrounding area of higher tension. Liquid motion under the influence of the transient surface tension gradient will then result in local thinning of the lamella, leading to ultimate rupture if the rate of thinning is higher than the rate of annihilation of the surface tension gradient. This general mechanism of rupture might well explain the observed effect of a temperature gradient (de Vries, 1958e) as well as the action of foam breakers (Ewers and Sutherland, 1952).

In general, any local fluctuation in the composition or concentration of the adsorbed monolayer at the interface will give rise to transient surface tension gradients affecting the stability of the lamella. This type of fluctuation will occur quite frequently under dynamic conditions (such as foam formation, geometric rearrangements, rapid changes in bubble size distribution due to gas transfer, etc.) and may be assumed to be normally attenuated by diffusion of the surface-active species. Occasionally, however, diffusion may not be rapid enough to restore equilibrium before local thinning has led to rupture.

In spite of still differing opinions regarding the relative importance of the various parameters responsible for the strength of metastable lamellae (namely, double-layer repulsion, surface elasticity or viscosity, etc.), the most challenging problem remaining to be solved is not so much the stability as it is the exact mechanism of breakdown.

In view of the random character of lamella breakdown, the coalescence process is supposed to be mostly of first order, as indicated by

$$-dN/dt = k_r N, \tag{45}$$

where k_r is a characteristic rate constant of the rupture process. Experimental verification of Eq. (45) requires appropriate corrections to be made for the decrease in the number of bubbles due to other processes, in particular, gas diffusion, unless the number of collapses as a function of time can be directly measured by means of cinephotographic methods.

New (1967) has reported some results for the value of k_r for several foams generated with the aid of a high-speed mixer, but sufficient experimental data are not yet available to allow any quantitative conclusions to be drawn regarding the influence of parameters such as bubble size distribution, gas-to-liquid ratio, etc. The more systematic use of photographic and cinephotographic methods is desirable in order to obtain reliable experimental data for the kinetics of bubble wall rupture and bubble coalescence in real foams under the widely varying conditions of practical interest.

REFERENCES

Alejnikoff, N. A. (1931). *Tsvet. Metal.* **6**, 1546.
Bikerman, J. J. (1953). "Foams: Theory and Industrial Applications." Van Nostrand-Reinhold, Princeton, New Jersey.
Blank, M. (1962). *J. Phys. Chem.* **66**, 1911.
Blank, M. (1964). *J. Phys. Chem.* **68**, 2793.
Chang, R. C., Schoen, H. M., and Grove, C. S., Jr. (1956). *Ind. Eng. Chem.* **48**, 2035.
Clark, N. O. (1948). *Trans. Faraday Soc.* **44**, 13.
Clark, N. O., and Blackman, M. (1948a) *Trans. Faraday Soc.* **44**, 1.
Clark, N. O., and Blackman, M. (1948b). *Trans. Faraday Soc.* **44**, 7.
Culick, F. E. C. (1960). *J. Appl. Phys.* **31**, 1128.
Derjaguin, B. (1933). *Kolloid Z.* **64**, 1.
de Vries, A. J. (1958a). *Rec. Trav. Chim.* **77**, 81.
de Vries, A. J. (1958b). *Rec. Trav. Chim.* **77**, 209.
de Vries, A. J. (1958c). *Rec. Trav. Chim.* **77**, 283.
de Vries, A. J. (1958d). *Rec. Trav. Chim.* **77**, 383.
de Vries, A. J. (1958e). *Rec. Trav. Chim.* **77**, 441.
de Vries, A. J. (1960). *Proc. Int. Congr. Surface Active Substances, 3rd, Cologne* **2**, 566.
Dupré, A. (1867). *Ann. Chim. Phys.* **11**, 194.
Ewers, W. E., and Sutherland, K. L. (1952). *Aust. J. Sci. Res.* **A5**, 697.
Fanlo, S., and Lemlich, R. (1965). *Amer. Inst. Chem. Eng. Inst. Chem. Eng. Symp. Ser.* **9**, 75.
Frankel, S., and Mysels, K. J. (1969). *J. Phys. Chem.* **73**, 3028.

Gibbs, J. W. (1876). *Trans. Connecticut Acad.* **3**, 108.

Gleim, V. G., and Chelomov, I. K. (1959). *J. Appl. Chem. (URSS)* **32**, 778.

Hansen, R. S., and Mann, J. A. (1964). *J. Appl. Phys.* **35**, 152.

Hawke, J. G., and Parts, A. G. (1964). *J. Colloid Sci.* **19**, 448.

La Mer, V. K. (1962). "Retardation of Evaporation by Monolayers: Transport Processes." Academic Press, New York.

Leonard, R. A., and Lemlich, R. (1965). *A.I. Ch. E. J.* **11**, 18, 25.

McEntee, W. R., and Mysels, K. J. (1969). *J. Phys. Chem.* **73**, 3018.

Miles, G. D., Shedlovsky, L., and Ross, J. (1945). *J. Phys. Chem.* **49**, 93.

Mysels, K. J., Shinoda, K., and Frankel, S. (1959). "Soap Films: Studies of Their Thinning." Pergamon, Oxford.

New, G. E. (1967). *Proc. Int. Congr. Surface Active Substances, 4th, Brussels, 1964* **2**, 1167.

Overbeek, J. Th. G. (1967) *Proc. Int. Congr. Surface Active Substances, 4th, Brussels, 1964* **2**, 19.

Plateau, J. (1873). "Statique expérimentale et théorique des liquides soumis aux seules forces moléculaires." Gauthier-Villars, Paris.

Princen, H. M., and Mason, S. G. (1965a). *J. Colloid Sci.* **20**, 353.

Princen, H. M., and Mason, S. G. (1965b). *J. Colloid Sci.* **20**, 453.

Princen, H. M., Overbeek, J. Th. G., and Mason, S. G. (1967). *J. Colloid Interface Sci.* **24**, 125.

Ranz, W. E. (1959). *J. Appl. Phys.* **30**, 1950.

Rayleigh, Lord (1936). *Proc. Roy. Soc. (London)* **A156**, 343.

Ross, S. (1969). *Ind. Eng. Chem.* **61**, 48.

Savitskaya, E. M. (1951). *Kolloid Zh.* **13**, 309.

Sheludko, A. (1967). *Advan. Colloid Interface Sci.* **1**, 391.

Sheludko, A., and Manev, E. (1968). *Trans. Faraday Soc.* **64**, 1123.

Shih, F. S., and Lemlich, R. (1967). *A.I. Ch. E. J.* **13**, 751.

Stenuf, T. J. (1953). Ph.D. Thesis, Syracuse Univ.

Van Amerongen, G. J. (1943). Doctoral Thesis, Delft.

Van den Tempel, M., and Van de Riet, R. P. (1965). *J. Chem. Phys.* **42**, 2769.

Vrij, A. (1964). *J. Colloid Sci.* **19**, 1.

Vrij, A. (1966). *Discuss. Faraday Soc.* **42**, 23.

Vrij, A., and Overbeek, J. Th. G. (1968). *J. Amer. Chem. Soc.* **90**, 3074.

CHAPTER **3**   PRINCIPLES OF FOAM FRACTIONATION
AND DRAINAGE

Robert Lemlich
Department of Chemical and Nuclear Engineering
University of Cincinnati
Cincinnati, Ohio

I. INTRODUCTION

Foam fractionation is the name given to the partial separation of a dissolved (or sometimes colloidal) substance from a liquid by adsorption on the surfaces of bubbles which ascend through the liquid to form a foam which carries the substance off. As indicated in Chapter 1, a substance which is not surface active itself can sometimes be made effectively surface active through the deliberate addition, or presence otherwise, of a suitable surfactant (termed the collector) which will combine with the substance in question (termed the colligend) so that it may be adsorbed (Schnepf et al., 1959, Sebba, 1959). This union between collector and colligend may be by chelation, counterionic attraction, or some other mechanism.

When foam fractionation is deliberate, the gas which forms the bubbles is usually injected at the bottom of the liquid pool. However, in the simple "ring test" for traces of surfactant in water (Crits, 1961), air bubbles are produced by shaking. This test has been shown by Lemlich (1968a) to operate by virtue of transient foam fractionation (or, at very low concentrations, by bubble fractionation).[1] This is discussed by the author (Lemlich, 1971a, in press).

Bubbles can also be formed by electrochemical or chemical reaction in the liquid, or by the liberation of dissolved gas. The separation that occurs in beer foam, mentioned in Chapter 1, is an example of the latter.

II. ADSORPTION

A. Effect of Concentration

The degree of adsorption of component i at the bubble surface is expressed by the surface excess, Γ_i, in such units as gm mole/cm^2. For present purposes, Γ_i can be considered to be simply the concentration of component i at the surface.

Theoretically, Γ_i at equilibrium can be found from the Gibbs adsorption equation (Gibbs, republished 1928; Osipow, 1962),

$$d\gamma = -\mathbf{R}T \sum \Gamma_i \, d \ln a_i, \tag{1}$$

where γ is the surface tension, $\mathbf{R}$ is the gas constant, T is the absolute temperature, and a_i is the activity of component i. In practice, the difficulties in measuring small changes in γ accurately and the uncertainties in identifying the species and evaluating their activity coefficients has severely limited the utility of Eq. (1) as a quantitative tool. However, one useful exception is the case of a nonionic surfactant in pure water at concentrations below the critical micelle concentration. For this case Eq. (1) simplifies to Eq. (2), where C represents concentration in the liquid and subscript s refers to the surfactant,

$$\Gamma_{\mathrm{s}} = -\frac{1}{\mathbf{R}T} \frac{d\gamma}{d \ln C_{\mathrm{s}}}. \tag{2}$$

Equation (2) also applies to ionic surfactants under certain conditions (Pethica, 1954; Davies and Rideal, 1963; Lemlich, 1968b).

Figure 1 illustrates typical variation of Γ_i with C_i. The curve can often be approximated by a Langmuir type of isotherm (Davies and Rideal, 1963). At sufficiently low concentrations, the linear isotherm, $\Gamma_i = K_i C_i$, applies, where K_i is the equilibrium constant. At sufficiently high concentration, Γ_i may approach a constant maximum which corresponds to saturation of the

[1] When shaking is stopped, the coalescing bubbles momentarily enrich the liquid meniscus, causing a sudden drop in surface tension which produces a circular ring or pulse that is faintly visible as it quickly rises up the free walls of the cylindrical glass container to a height which is very roughly indicative of surfactant concentration.

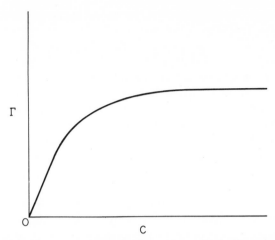

FIG. 1. Variation of surface excess (concentration at the surface) with concentration in the liquid at equilibrium.

surface with an adsorbed monolayer. This appears to be the usual situation for the Γ_s of the major surfactant in a foam.

The surface excess for a monolayer can be roughly estimated from the probable packing and size of the molecules as determined from inter-molecular forces and interatomic distances (Weast and Selby, 1967). Values on the order of 10^{-10} to 10^{-9} gm mole/cm^2 are common for the Γ_s of a surfactant monolayer.

Experimental methods for finding the surface excess include skimming a stationary pool (McBain and Swain, 1936), skimming a liquid jet (Snavely et al., 1962), surface compression combined with interferometry (McBain et al., 1940), surface radioactivity measurements (Dixon et al., 1949 ; Hutchinson, 1949), and foam fractionation, to be discussed presently. Where applicable, this last method offers the inherent advantages of simplicity as well as a partial cancellation of errors with respect to subsequent applications to foam fractionation.

B. Effect of Collector

Insufficient collector can reduce K_i for the colligend. Excess collector can do the same, either by competing against the colligend–collector complex for the available surface (Schnepf et al., 1959), or by forming micelles in the liquid which compete against the collector at the surface for the adsorption of colligend (Sebba, 1962). If E, the relative effectiveness (in adsorbing colligend) of the surface collector compared to the micellar collector, is known, then the effect of micelles on the apparent K_i for the colligend can be found theoretically by (Lemlich, 1968b)

$$1/K_2 = 1/K_1 + (C_s - C_{sc})/\Gamma_s E \tag{3}$$

where K_1 is K_i just below the critical micelle concentration of the collector C_{sc}, and K_2 is K_i at C_s which is some higher concentration of collector. E is unity if the collector molecules (or ions) at the surface and in the micelles are equally effective at adsorbing colligend.

Multiplying K_2 from Eq. (3) by the bulk liquid concentration of colligend in the presence of the micelles will give the corresponding surface excess of colligend in the presence of these micelles. Alternatively, but perhaps less conveniently, one could obtain this same surface excess by multiplying a a modified K_2 by the original colligend concentration in the liquid before the addition of the excess collector. This modified K_2 is obtainable from Eq. (3) by dividing the last term of Eq. (3) by the quantity $(1 + a' K_1)$, where a' is the ratio of surface to liquid volume (Lemlich, 1968c, d). In a quiescent pool (as opposed to a foam or a highly bubbled pool) a' approaches zero, which makes the aforementioned two colligend concentrations in the liquid the same and the two values of K_2 identical.

Further discussion of adsorption can be found in subsequent chapters.

III. COLUMN OPERATION

A. Modes of Operation

Figure 2(a) shows batchwise foam fractionation in the simple mode. Figure 2(b) shows simple continuous operation. Each is approximately one theoretical (perfect) stage, provided the pool is well mixed, sparger submergence is sufficient,[2] and there is no coalescence in the rising foam. Such coalescence releases adsorbed solute which runs back down through the rising foam as internal reflux, thus enriching the foam overflow beyond that of a single stage of separation. Coalescence of bubbles in the liquid just before they enter the foam can also make for a richer foam. Furthermore, if the pool is elongated vertically and is not well mixed, additional separation is possible through the action of bubble fractionation (Dorman and Lemlich, 1965). This is discussed by the author in Chapter 7.

Some[3] or all of the foam overflow can be deliberately collapsed and returned to the column (a short distance below the start of the overhead bend)

[2] A submergence of 30 cm should be more than ample in most cases.

[3] If the foam is only partially collapsed, the foam breaker acts as a dephlegmator and theoretically furnishes one additional perfect stage of separation.

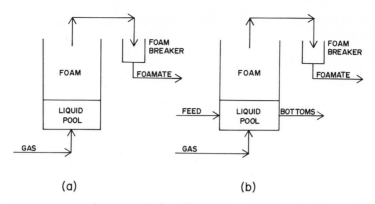

FIG. 2. Foam fractionation in the simple mode, together with a foam breaker to collapse the overflow: (a) batchwise operation, (b) continuous flow operation.

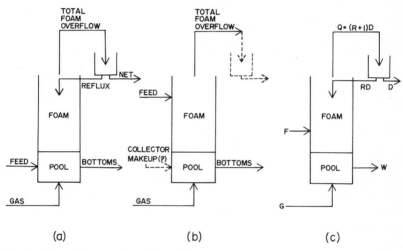

FIG. 3. Three higher modes of foam fractionation involving counterflow: (a) enriching operation; (b) stripping operation, with optional foam breaker and possible collector makeup; (c) combined operation, showing symbols.

as external reflux to trickle down through the rising foam, as shown in Fig. 3(a). The enriching effect so obtained for the foamate (collapsed overflowing foam) by virtue of this counterflow and concomitant mass transfer has been well verified experimentally for solutes that completely saturate the surface (Lemlich and Lavi, 1961), as well as for those that do not (Schonfeld and Kibbey, 1967). As would be expected, it has also been shown experimentally (Hastings, 1967; Jashnani, 1970) that increasing the height of a refluxing column increases the degree of separation obtained.

This was further confirmed (Jashnani, 1970) by some recent experiments with reflux in a 3-ft high, 2-in. i.d. glass column with aqueous solutions of the dye patent blue in the presence of the surfactant hexadecyltrimethyl-ammonium bromide. With appropriate steady-state conditions, a prominent color gradient (from white to blue) could be made to appear anywhere along the entire length of the foam column. This indicates that the entire length of column was effective in carrying out the enrichment. Therefore, for fixed conditions, the greater the height, the greater is the enrichment.

It is worth mentioning at this point that a refluxing foam column can take a surprisingly long time to reach steady-state conditions. Data or observations taken too soon can prove very misleading.

Stripping action can be obtained by injecting the feed into the foam some distance above the pool, as shown in Fig. 3(b). As the feed trickles down through the rising foam, the resulting counterflow and mass transfer result in a lower solute concentration in the bottoms stream. When stripping off a colligend, it may be necessary to add some additional collector directly to the pool in order to make up for the collector which is also stripped off.

Stripping and enriching can be combined as shown in Fig. 3(c).

B. Counterflow Theory

In a sense, foam fractionation is analogous to distillation *with entrainment*. The rising bubble surfaces correspond to the vapor, and the interstitial liquid carried up corresponds to the entrainment. The interstitial liquid flowing down through the foam corresponds to the downflow. Unfortunately, some workers have overlooked the rising interstitial liquid in their analyses as well as their analogies and have treated the upflow as consisting entirely of surface. This can be grossly incorrect since the amount of solute carried up by the rising interstitial liquid can be very important.

Equation (4) defines $\bar{C}$, which is the effective concentration of solute i in the rising stream at any level in the foam column (Lemlich, 1968b). The subscript i is omitted for convenience.

$$\bar{C} = C + GS\Gamma/U, \tag{4}$$

where G is the volumetric rate of gas flow and S is the ratio of bubble surface to bubble volume. Γ is taken in equilibrium with concentration C in the rising interstitial liquid. Also to simplify matters, the rate of interstitial liquid upflow U can be taken as equal to the rate of interstitial liquid overflow Q. A plot of $\bar{C}$ (now asterisked as $\bar{C}^*$) against C constitutes an effective equilibrium curve. Material balances around the entire column and around either or both ends yield the operating line or lines for steady continuous operation.

Strictly speaking, such balances require that liquid flow rates and concentrations be on a mass basis. However, since liquid volume is very nearly conserved, the more convenient volumetric units can be used with negligible error.

The number of theoretical stages in the foam can now be found by graphical steps, as illustrated in Fig. 4, or by other standard techniques (Smith, 1963). As

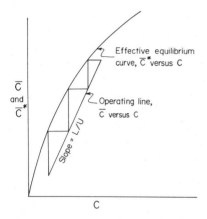

FIG. 4. A typical graphical stagewise calculation.

indicated earlier, the pool is generally considered to be one theoretical stage. (Its analogy to the reboiler in distillation is evident).

The number of transfer units in the foam based on the rising stream is expressed by Eq. (5), in which the lower integration limit $\overline{C_W}^*$ is the effective concentration $\overline{C}$ in equilibrium with C_W. C_Q is the concentration in the total foamate; that is, in the overflowing foam on a collapsed (gas-free) basis.

$$NTU = \int_{\overline{C_W}^*}^{C_Q} \frac{d\overline{C}}{\overline{C}^* - \overline{C}}. \tag{5}$$

A corresponding expression can be written based on the descending stream. Either expression can be integrated from the effective equilibrium curve and operating line (Lemlich, 1968b). Equation (6) illustrates the integration of Eq. (5) for an enriching column which is concentrating a surfactant of constant Γ while operating at a reflux ratio of R:

$$NTU = R \ln \frac{RGS\Gamma(F - D)}{(R + 1)GS\Gamma(F - D) - (R + 1)FD(C_D - C_F)}. \tag{6}$$

The height of a transfer unit (HTU) was found to be as low as several centimeters under proper conditions (Fanlo and Lemlich, 1965). For stripping and enriching operations with aqueous solutions of sodium lauryl sulfate in a column of 15.52 cm^2 cross section, Hastings (1967) correlated HTU against the flow number, $(L/U)[L/(G + L + U)]^{1/2}$, and found a minimum HTU of 2.54 cm at a flow number of 0.2905.

The HTU for a colligend may be larger (poorer) than the HTU for the collector in the same foam because the interstitial transport of the colligend involves edgewise entry into and exit from the bubble films, as well as surface adsorption and desorption, while a collector which saturates the bubble surfaces is involved in effective local transport chiefly within the capillaries. In similar vein, a high surface viscosity (see Section IV) could rigidify the bubble surfaces sufficiently so as to lessen the local transport and make for a larger HTU, especially for a colligend.

If a foam column is sufficiently tall, the concentrations of the counterflowing streams will approach each other in the foam (Brunner and Lemlich, 1963) at the start of the larger stream (Lemlich, 1965, 1968b). Theoretically, this concentration pinch requires infinite height, but in actuality the column could be almost any height, depending on circumstances.

For such sufficiently tall columns, the performance can be expressed more simply by Eqs. (7) and (8) if the column is an enricher,

$$C_D = C_W + GS\Gamma_W/D, \tag{7}$$

$$C_W = C_F - GS\Gamma_W/F, \tag{8}$$

and by Eqs. (9) and (10) if it is a combined column,

$$C_D = C_F + GS\Gamma_F/D, \tag{9}$$

$$C_W = C_F - GS\Gamma_F/W, \tag{10}$$

where F, W, and D are the flow rates of feed, bottoms, and net overflow, respectively. C_F, C_W, and C_D are the corresponding concentrations, and Γ_F and Γ_W are the corresponding surface excesses at equilibrium.

Of course, these equations must not be applied indiscriminately to mutually incompatible operating conditions, such as conditions that go *beyond* a pinch. For example, too high a reflux ratio may give an unreasonably low D for the C_W, yielding an absurdly high C_D from Eq. (7).

For short (finite) columns, Eqs. (7)–(10) give the limiting separations. Of course, calculation by transfer units is recommended for short columns if enabling information is available.

The applicability of the transfer unit approach has been demonstrated by experiments which show that under good operating conditions the HTU is approximately independent of countercurrent length (Haas and Johnson,[4] 1965; Hastings, 1967; Jashnani, 1970). This means, for example, that doubling the height of a well-operating stable foam column should approximately

[4]The HTU calculations in this particular study apparently did not include the rising interstitial liquid (entrainment).

double the NTU in the foam. Naturally, doubling the height of a sufficiently tall foam column which already approaches limiting operation, or doubling the height of a foam column which operates poorly with much channeling, may have very little effect on the overall separation. Twice infinity is no more than infinity, and twice zero is still zero.

Of course, there are circumstances under which the HTU does vary appreciably with the countercurrent length. Such variation can be found among some results obtained by Hastings (1967) for a stripping column operating with comparatively high ratios of downflow to upflow. A high rate of downflow tends to flood the capillaries.

Equations (9) and (10) also apply to a sufficiently tall stripper by replacing D with Q, and replacing C_D with C_Q. With these same replacements, Eqs. (7) and (8) also apply to operation in the simple mode regardless of column height. Thus, Eq. (7) or Eq. (8) can be used to find Γ by foaming experiments in the simple mode. But care should be exercised to minimize the afore-mentioned internal coalescence by employing a short foam column to lessen the residence time, and by using a high gas rate for the same reason and to keep the foam wet (Brunner and Lemlich, 1963). However, G should not be so high as to cause significant bubble coalescence in the upper portion of the liquid pool.

When measuring Γ via the simple mode, the collapsed foam (foamate) of Fig. 1(a) can be conveniently recycled to the pool. However, this may not be advisable when the Γ in question is for a colligend if interfering surfactant micelles that form in the foam breaker do not dissociate rapidly upon return to the pool.

The formation of micelles in external reflux can also reduce the separation of a colligend in finite enriching and combined columns. However, for sufficiently tall columns this effect of the micelles vanishes due to their virtually complete transfer from the reflux to the rising stream.

With spherical bubbles, $S = 6/d$. However, for polyhedral bubbles, it is better to employ $S = 6.3/d$ for stripping, enriching, and combined operation, where d is the diameter of the sphere equal in volume to the polyhedron (Lemlich, 1968d). For variation in bubble size, d is taken as $d_{32} = \Sigma\, n_i d_i^3 / \Sigma\, n_i d_i^2$. Fanlo and Lemlich (1965) discuss S further.

C. Additional Aspects

Channeling of liquid downflow or deviations from plug foam flow destroy the countercurrent pattern in the column and thus reduce the extra separation inherent in stripping, enriching, and combined operations. This loss of efficiency can be reduced by employing a low superficial gas velocity so as to produce a foam of low liquid content, and by using an appropriate sparger to produce bubbles of fairly uniform size. Both precautions make for foam bubbles of more polyhedral shape, and hence less interbubble slippage and more capillary suction to oppose channeling. Unfortunately, low G/A also makes for low production unless A is enlarged.

In an effort to circumvent this problem, Maas (1969a) employed rapidly rotating rods in a column to "stabilize" high throughput foam. Wace and Banfield (1966) investigated a bubble cap column, reporting plate efficiencies of up to 30 %. It is also possible to connect short, individual, high throughput columns in overall countercurrent array (Banfield et al., 1965).

Since channeling and bubble slippage are not intrinsically detrimental to the simple mode, a high throughput can be achieved in that mode by the direct use of a high gas rate (Wace et al., 1969), even to the extent of producing a highly mobile "gas emulsion" (Bikerman, 1965) rather than a genuine foam. The very wet overflow can be back-drained in a nearly horizontal section (Haas and Johnson, 1967) before it is collapsed externally.

Difficult multicomponent separations can sometimes be achieved by using tall columns in batch operation at very low gas rates which give much internal reflux and drainage (Lemlich, 1968b). Schutz (1946) points out that a sudden decrease (or even an increase) in foam stability while batch foaming may indicate the substantially complete removal of one component and therefore the time to change the receiver to collect the next fraction. Raising the temperature usually decreases adsorption and foam stability, but not always.

Although the minus sign in Eq. (1) implies that substances which lower the surface tension should be adsorbed at the surface, it does not follow that the pool concentration which shows the largest $d\gamma/dC_i$ is necessarily optimal. Other factors are involved. Schutz (1946) recommends foaming at concentrations which correspond to the peaks in foam stability, or in the absence of such peaks, at as low a concentration as possible in order to dilute the liquid carryover and so minimize the surface-inactive impurities in the foam. Selectivity can often be improved by lowering the solubility of the substance, as by adjusting pH. The isoelectric pH may be chosen for proteins (Rubin and Gaden, 1962). Adding a salt may help. Protein separation is discussed in Chapter 9 by Charm.

While air has been used to generate the bubbles in full scale as well as bench scale installations, a prehumidified inert gas such as nitrogen has proven useful in laboratory studies as a means of eliminating spurious evaporative or chemical effects. However, Maas (1969a, b) incorporated certain vapors in his bubbles in order to achieve greater selectivity in separation.

Maas (1969c) also describes "laminae column foaming," a technique for decreasing the liquid carryover in the foam. It involves generating a foam consisting of a single continuous chain of relatively large wall-to-wall bubbles. Such a foam has a large ratio of surface to liquid. It produces a corresponding-ly concentrated foamate which may be concentrated even further through the use of reflux. Of course, the production of overflow is at a slow rate.

With its readily measurable rate of surface production and its relatively highly concentrated foamate, laminae column foaming in the simple mode seems to offer promise as a potentially accurate method for finding the surface excess.

IV. FOAM

A. Drainage and Overflow

Over the past thirty years, the problem of theoretically predicting the rate of interstitial flow (drainage) through foam has occupied the attention of various investigators (Ross, 1943; Brady and Ross, 1944; Miles et al., 1945; Jacobi et al., 1956; Haas and Johnson, 1965, 1967). Generally speaking, these workers employed simplistic models based on flow through vertical circular channels or between essentially parallel planes. The actual cross-sectional geometry of the channels, the mobility of their walls, and the variation in their inclinations were not incorporated in any fundamental quantitative manner. As a result, the final equations in each case include various empirical constants and are accordingly limited in their applic-abilities. A more recent study by Morgan (1969), which treats capillary wall mobility as slip, is also bounded by empiricism.

Leonard and Lemlich (1965a) developed a more realistic model. It is based on polyhedral foams.[5] The bubbles are considered to be regular (congruent pentagonal) dodecahedra with faces bounded by narrow capillaries (Plateau borders) of the cross section shown in Fig. 5(a). The propriety of this cross section is readily evident from a comparison with Fig. 5(b), which shows the filled-in tracing of an electron photomicrograph (Shih, 1969) of the cross section of an actual capillary (strand) in solidified (open-celled polyurethane) foam.

From geometric considerations, the total length of capillaries per unit length of vertical foam column is given by

$$P = 7.81A\lambda/d^2, \tag{11}$$

where λ is the volumetric fraction of gas in the foam.

[5] The low liquid content and interstitial suction of polyhedral foams makes them desirable for foam fractionation.

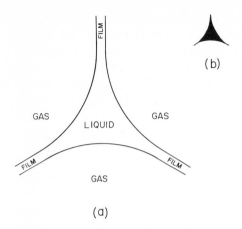

(b)

FIG. 5. (a) Cross section of a liquid capillary (Plateau border) in foam (Leonard and Lemlich, 1965a); (b) filled-in-tracing of an electron photomicrograph showing the cross section of an actual capillary (strand) in solidified foam (Shih, 1969).

(a)

The actual orientation and distribution of the capillaries as a whole is taken as random. From this it can be shown that the probability of a capillary being inclined at angle α to the horizontal is $\cos \alpha \, d\alpha$, and the probability of a capillary inclined at α being intercepted by a hypothetical fixed horizontal plane is $(l \sin \alpha)/H$, where l is the length of the capillary and H is the length (height) of the foam column. The compound probability of the combined event is the product, $(l \sin \alpha \cos \alpha \, d\alpha)/H$. Accordingly, the probable number of all capillaries that are both inclined at α and intercepted by the horizontal plane is $P \sin \alpha \cos \alpha \, d\alpha$. Integrating over all angles from 0 to $\pi/2$ gives $P/2$ as the probable number of capillaries intercepted by a horizontal cross section of the column. Combining $P/2$ with Eq. (11) yields $3.90\lambda/d^2$ as the number of capillaries per unit cross-sectional area, horizontal or otherwise.

To find the interstitial flow, the general differential equation for momentum conservation was solved by digital computation (Leonard and Lemlich, 1965c) for rectilinear laminar Newtonian flow of liquid through a typical capillary, the walls of which are not rigid but flow subject to a surface viscosity μ_s. The resulting local velocities in the capillary were then combined vectorially with the upward velocity (if any) of the bulk foam, and integrated over the horizontal cross section of the vertical column with due regard for the randomness of the capillaries. This is shown in Eq. (12) which gives the net rate of liquid upflow, $U - L$:

$$U - L = P \int_0^{\pi/2} \int_{A_{\text{PBh}}} (u_f - u_z \sin^2 \alpha) \sin \alpha \cos \alpha \, d\alpha \, dA_{\text{PBh}} . \qquad (12)$$

L is the rate of downflow.

Linear velocity u_f is that of the vertical rise of the bulk foam, and u_z is the local relative linear velocity downward within a vertical capillary. Equation

(12) was integrated numerically after first recasting in dimensionless form and simplifying by combination with the expression $A_{\text{PBh}} \sin \alpha = A_{\text{PB}}$, where A_{PBh} is the horizontal cross-sectional area of a capillary, and A_{PB} is its cross-sectional area perpendicular to its longitudinal axis.

The results can be found in the original paper (Leonard and Lemlich, 1965a) in the form of dimensionless curves. To simplify the application of these results to drainage problems, Shih and Lemlich (1967) prepared an information flow diagram to relate the steady drainage rate to the liquid content and other properties of the foam.

The overflow rate for a foam which is of low liquid content can be found from the dimensionless group $\mathscr{L}$ which is $u_Q g d^2 / u_G^2 v$ (or its equivalent, $QAg\rho d^2 / G^2 \mu$), expressed as a function of the dimensionless group $\mathscr{N}$ which is $u_G v^3 / g v_s^2$ (or its equivalent, $\mu^3 G / \mu_s^2 g \rho A$). The theoretical relationship between $\mathscr{L}$ and $\mathscr{N}$ is represented by the curve in Fig. 6 (Fanlo and Lemlich,

FIG. 6. Comparison of theory for the rate of foam overflow with recent experimental results (Shih and Lemlich, 1971).

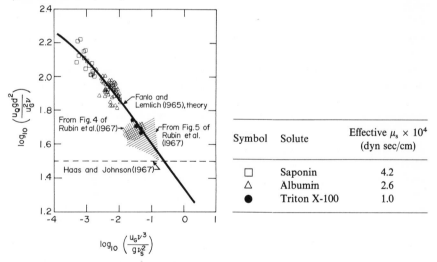

Symbol	Solute	Effective $\mu_s \times 10^4$ (dyn sec/cm)
□	Saponin	4.2
△	Albumin	2.6
●	Triton X-100	1.0

The shaded areas represent the bands of experimental data from the independent work of Rubin *et al.* (1967) with Aresket 300, as discussed by Lemlich (1968e).

1965). Superficial linear velocity u_Q is that of the entire foamate based on the empty column cross section, u_G is that of the gas similarly based, ρ is the liquid density, μ is the liquid viscosity, v is the kinematic liquid viscosity, v_s is the kinematic surface viscosity (μ_s/ρ), and g is the acceleration of gravity.

For foam of moderate liquid content, the theoretical relationships are more complicated (Leonard and Lemlich, 1965a). However, they can be approximated by simply multiplying u_Q obtained from the curve by the factor $(1 + 3u_Q/u_G)$ to give a corrected u_Q. The theory breaks down for foams of very high liquid content, for very unstable foams, or for columns of irregular cross section.

If bubble sizes vary, d^2 in $\mathscr{L}$ and Eq. (11) should be taken as $d_{31}^2 = \Sigma\, n_i d_i^3/ \Sigma\, n_i d_i$. If there is coalescence in the rising foam, bubble sizes should be measured at the top of the vertical portion of the column.

The theory has been tested experimentally with solutions of low μ_s, that is, on the order of 10^{-4} dyn sec/cm, by Fanlo and Lemlich (1965), Leonard and Lemlich (1965b), Hoffer and Rubin (1969), Lemlich (1968e) utilizing the data of Rubin et al. (1967), and Shih and Lemlich (1967) with data from several investigations. For steady drainage through standing foam and for overflow, the agreement between theory and experiment has generally been good. Recent experiments with two aqueous systems of higher μ_s and rounder bubbles also yielded good agreement with theory, provided an effective μ_s considerably lower than that otherwise expected was employed in the theoretical calculations for each system (Shih and Lemlich, 1971). Joly (1964) discusses surface viscosity and points out that the surface viscosity in a foam can be smaller than that measured at the surface of a pool with a surface viscosimeter. Furthermore, a surface viscosimeter typically deals with a comparatively aged surface that is minutes old, while the time of transit from the inlet to the outlet of a foam capillary is typically less than 1 sec. However, as of this writing, it is not clear whether the apparent need for a lower effective μ_s with systems of high μ_s is due to the thixotropic or other non-Newtonian surface behavior, which such systems are known to possess, or due to some idealization in the theory.

Figure 6 shows the results of these recent experiments for foam overflow and compares them with theory. The results also include a few new runs with a previously studied system of low μ_s, namely, the nonionic surfactant Triton X-100 (isooctylphenoxypolyethoxyethanol) in water, as a control. The μ_s of 1.0×10^{-4} dyn sec/cm employed for the aqueous Triton X-100 was measured by Leonard and Lemlich (1965b) with the independent method described by Mysels et al. (1959) which involves timing the rise of thin black spots in the faces of well drained foam bubbles. The shaded regions in Fig. 6 represent the bands of data from the independent work of Rubin et al. (1967) with the anionic surfactant Aresket 300 (monobutylbiphenyl sodium monosulfonate) in water, as discussed by Lemlich (1968e).

For further comparison, Fig. 6 also shows the correlation of Haas and Johnson (1967) put into dimensionless form by the present author for foams of low liquid content. Their correlation appears as a horizontal line because it makes no allowance for differences in surface viscosity.

Figure 7 tests the author's theoretical prediction of overflow rate against

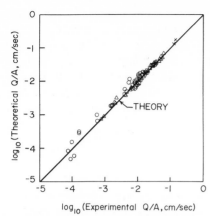

Symbol	Solute	Effective $\mu_s \times 10^4$ (dyn sec/cm)
□	Saponin	4.2
△	Albumin	2.6
○	Triton X-100	1.0
×	Aresket 300	1.8
◇	NaDBS	0.82

FIG. 7. Test of the theoretical prediction of Fanlo and Lemlich (1965) for rate of foam overflow (Shih and Lemlich, 1971). The data for sodium dodecylbenzene sulfonate are from Banfield *et al.* (1965). The other data are from several investigations in the author's laboratory.

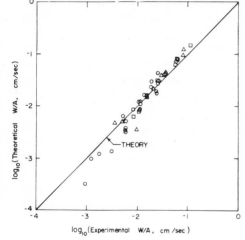

Symbol	Solute	Effective $\mu_s \times 10^4$ (dyn sec/cm)
□	Saponin	4.2
△	Albumin	2.6
◇	Triton X-100	1.0
○	Triton X-100 (earlier studies)	1.0

FIG. 8. Test of the theory of Leonard and Lemlich (1965a) for drainage through a stationary foam at steady state (Shih and Lemlich, 1971). The data are from several investigations in the author's laboratory.

new and old experimental results.[6] Then, using the same values of μ_s as for overflow, Fig. 8 tests the author's theoretical prediction of the rate of steady drainage through stationary foam knowing bulk foam density, bubble sizes, liquid density, and liquid viscosity. Both tests show that the agreement between theory and experiment is generally good, especially in view of the wide range of variables encompassed.

[6]The data of Rubin *et al.* (1967) could not be shown in Fig. 7 because the said paper did not include values for the individual parameters.

The drainage theory also predicts that the liquid content of an *uncoalescing* foam at steady state in an uninterrupted vertical column segment of uniform cross section, whether overflowing or stationary, is uniform along that segment. This has been verified experimentally through electrical conductivity measurements of the foam at various levels within the column (Fanlo and Lemlich, 1965; Shih and Lemlich, 1971). Also, because of back drainage, $Q/G < (1 - \lambda)/\lambda$ in a rising column of foam, even in the absence of coalescence.

B. Coalescence and Collapse

Coalescence in a column of foam has two sources, namely, gas diffusion from smaller bubbles to larger bubbles, and rupture of bubble walls. When coalescence in foam fractionation is severe, the second source is likely to predominate.

Most of the bubble diameters for Figs. 6–8 were measured photographically through the glass wall of the particular foam column. Errors inherent in this type of measurement, as well as the aforementioned subject of coalescence, are dealt with in Chapter 2 by de Vries.

The diameters of bubbles from simple orifices can also be measured stroboscopically in the liquid pool (Leonard and Lemlich, 1965b). Bubble formation at a complex orifice, such as a porous disk, is discussed by Bowonder and Kumar (1970).

The overflowing foam may be collapsed externally by heat (Kishimoto, 1963) if no undesirable chemical changes are so produced, or by chemical antifoaming agents (Ross, 1967) if foamate is not returned as reflux. Foam which is not too stable can be broken by the impact of foamate run back on to it (Brunner and Stephan, 1965). Foam that is more stable can be broken with a high-speed rotating disk (Rubin and Golt, 1970). Foam can also be broken sonically, ultrasonically, and in other ways (Goldberg and Rubin, 1967).

For dephlegmation (deliberate partial collapse) to give reflux, a foam breaker can be installed within the vertical column at its top. If the foam is not too stable, simply widening the column near its top may be sufficient. Of course, the difference between reflux proper, which results from the destruction of bubble surface, and simple drainage, which is always present relative to the rising bubbles, should be borne in mind.

V. CLOSURE

The foregoing has presented some of the principal features of foam fractionation as a method of separation. No attempt has been made in this chapter to detail actual substances which have been separated ; their number is considerable. Such information will be found in subsequent chapters.

For further information, the reader is referred to the reviews of Rubin and Gaden (1962) and Lemlich (1968c). For solved illustrative problems which involve some of the foregoing principles, the reader is referred to other writings by the author (Lemlich, 1968d, 1971b (in press), and especially 1968b). There is also a videotape (WCET, 1970) in which the author presents some simple demonstrations of foam fractionation and bubble fractionation.

SYMBOLS

a' surface divided by liquid volume, cm^{-1}

a_i activity of component i

A horizontal cross-sectional area of empty vertical column, cm^2

A_{PB} transverse cross-sectional area of capillary, cm^2

A_{PBh} horizontal cross-sectional area of capillary, cm^2

C concentration in liquid, gm mole/cm^3

$\overline{C}$ effective concentration in upflow, gm mole/cm^3 on a gas-free basis

$\overline{C}^*$ effective concentration in upflow in equilibrium with C, gm mole/cm^3 on a gas-free basis

d bubble diameter, cm

d_i individual bubble diameter

d_{31} edge average bubble diameter

d_{32} surface average bubble diameter

D volumetric flow rate of collapsed net foam overflow product, cm^3/sec

E relative adsorptive effectiveness of surface surfactant versus micellar surfactant

F volumetric flow rate of feed, cm^3/sec

g acceleration of gravity, cm/sec^2

G volumetric flow rate of gas, cm^3/sec

H length (height) of foam column, cm

HTU   height of a transfer unit, cm

K equilibrium adsorption constant, cm

l length of capillary, cm

$\mathscr{L}$ dimensionless number, $u_Q g d^2/u_G{}^2 v$

n_i number of bubbles of diameter d_i

$\mathscr{N}$ dimensionless number, $u_G v^3/g v_s{}^2$

NTU   number of transfer units

P total length of capillaries per unit height of foam

Q volumetric rate of total foam overflow on a collapsed (gas-free) basis, cm^3/sec

R reflux ratio

$\mathbf{R}$ gas constant, erg/gm mole °K

S bubble surface divided by bubble volume, cm^{-1}

T absolute temperature, °K

u_f upward linear velocity of foam bubbles, cm

u_G superficial linear velocity of gas, G/A, cm/sec

u_Q superficial linear velocity of collapsed total foam, Q/A, cm/sec

u_z local downward linear velocity of liquid flow within a vertical capillary relative to the foam bubbles, cm/sec

U volumetric rate of interstitial liquid upflow, cm^3/sec

W volumetric flow rate of bottoms stream, cm^3/sec

γ surface tension, dyn/cm

Γ surface excess (concentration at surface), gm mole/cm^2

λ volumetric fraction of gas in foam

μ dynamic viscosity of liquid, dyn sec/cm^2

μ_s dynamic viscosity of surface, dyn sec/cm

v kinematic viscosity of liquid, μ/ρ, cm^2/sec

v_s kinematic viscosity of surface, μ_s/ρ, cm³/sec

π 3.14159...

ρ density of liquid, gm/cm³

Subscripts

D   collapsed net foam overflow product

F   feed

Q total foam overflow on a gas-free basis

W bottoms

s   surfactant

sc   critical micelle for surfactant

i   component i

1   in the absence of micelles

2   in the presence of micelles

REFERENCES

Banfield, D. L., Newson, I. H., and Alder, P. J. (1965). *A. I. Ch. E.-I. Chem. E. (London) Symp. Ser. No. 1,* 3.

Bikerman, J. J. (1965). *Ind. Eng. Chem.* **57**(1), 56.

Bowonder, B., and Kumar, R. (1970). *Chem. Eng. Sci.* **25**, 25.

Brady, A. P., and Ross, S. (1944). *J. Amer. Chem. Soc.* **66**, 1348.

Brunner, C. A., and Lemlich, R. (1963). *Ind. Eng. Chem. Fundamentals* **2**, 297.

Brunner, C. A., and Stephan, D. G. (1965). *Ind. Eng. Chem.* **57**(5), 40.

Crits, G. J. (1961). The Crits organic ring test: A simple test for trace organic substances in water, paper presented at Nat. Meeting Amer. Chem. Soc., Div. Water Waste Chem., 140th, Chicago.

Davies, J. T., and Rideal, E. K. (1963). "Interfacial Phenomena," 2nd ed. Academic Press, New York.

Dixon, J. K., Weith, A. J., Argyle, A. A., and Salley, D. J. (1949). *Nature* **163**, 845.

Dorman, D. C., and Lemlich, R. (1965). *Nature* **207**, 145.

Fanlo, S., and Lemlich, R. (1965). *A. I. Ch. E.-I. Chem. E. (London) Symp. Ser. No.* 9, 75. Discussion on p. 85.

Gibbs, J. W. (1928). "Collected Works." Longmans, Green, New York.

Goldberg, M., and Rubin, E. (1967). *Ind. Eng. Chem. Proc. Design Develop.* **6**, 195.

Haas, P. A., and Johnson, H. F. (1965). *A. I. Ch. E. J.* **11**, 319.

Haas, P. A., and Johnson, H. F. (1967). *Ind. Eng. Chem. Fundamentals* **6**, 225.

Hastings, K. (1967). Ph.D. dissertation, Michigan State Univ.

Hoffer, M. S., and Rubin, E. (1969). *Ind. Eng. Chem. Fundamentals* **8**, 483.

Hutchinson, E. (1949). *J. Colloid Sci.* **4**, 600.

Jacobi, W. M., Woodcock, K. E., and Grove, C. S., Jr. (1956). *Ind. Eng. Chem.* **48**, 2046.

Jashnani, I. L. (1970). Ph.D. dissertation in preparation with R. Lemlich, Univ. of Cincinnati.

Joly, M. (1964). *In* "Surface Science" (J. F. Danielli, K. G. A. Pankhurst, and A. C. Riddiford, eds.), Vol. 1, Chapter 1. Academic Press, New York.

Kishimoto, H. (1963). *Kolloid-Z.* **192**, 66.

Lemlich, R. (1965). *J. Chem. Eng. Educ. (Univ. Cincinnati)* **3**(2), 53.

Lemlich, R. (1968a). Tech. Note 1–68. Univ. of Cincinnati Library.

Lemlich, R. (1968b). *In* "Progress in Separation and Purification" (E. S. Perry, ed.), Vol. 1, pp. 1–56. Wiley (Interscience), New York.

Lemlich, R. (1968c). *Ind. Eng. Chem.* **60**(10), 16.

Lemlich, R. (1968d). *Chem. Eng.* **75**(27), 95; errata in (1969), **76**(6), 5.

Lemlich, R. (1968e). *Chem. Eng. Sci.* **23**, 932.

Lemlich, R. (1971a, in press). *J. Coll. Interface Sci.*

Lemlich, R. (1971b, in press). *In* "Chemical Engineers' Handbook" (R. H. Perry and C. H. Chilton, eds.), 5th ed., Sect. 17. McGraw-Hill, New York.

Lemlich, R., and Lavi, E. (1961). *Science* **134**, 191.

Leonard, R. A., and Lemlich, R. (1965a). *A. I. Ch. E. J.* **11**, 18.

Leonard, R. A., and Lemlich, R. (1965b). *A. I. Ch. E. J.* **11**, 25.

Leonard, R. A., and Lemlich, R. (1965c). *Chem. Eng. Sci.* **20**, 790; erratum in (1969), **24**, 1645.

Maas, K. (1969a). *Int. Symp. Separation Methods: Column Chromatography, 5th, Lausanne* Preprint *Chimia* "Schaumchromatographie."

Maas, K. (1969b). *Separ. Sci.* **4**, 457.

Maas, K. (1969c). *Separ. Sci.* **4**, 69.

McBain, J. W., and Swain, R. C. (1936). *Proc. Roy. Soc. (London)* **A154**, 608.

McBain, J. W., Mills, G. F., and Ford, T. F. (1940). *Trans. Faraday Soc.* **36**, 930.

Miles, G. D., Shedlovsky, L., and Ross, J. (1945). *J. Phys. Chem.* **49**, 93.

Morgan, C., Jr. (1969). Ph.D. dissertation, Univ. of Iowa.

Mysels, K. J., Shinoda, K., and Frankel, S. (1959). "Soap Films, Studies of Their Thinning." Pergamon, Oxford.

Osipow, L. I. (1962). "Surface Chemistry, Theory and Industrial Application." Reinhold, New York.

Pethica, B. A. (1954). *Trans. Faraday Soc.* **50**, 413.

Ross, S. (1943). *J. Phys. Chem.* **47**, 266.

Ross, S. (1967). *Chem. Eng. Progr.* **63**(9), 41.

Rubin, E., and Gaden, E. L., Jr. (1962). *In* "New Chemical Engineering Separation Techniques" (H. M. Schoen, ed.), Chapter 5. Wiley (Interscience), New York.

Rubin, E., and Golt, M. (1970). *Ind. Eng. Chem. Process Design Develop.* **9**, 341.

Rubin, E., LaMantia, C. R., and Gaden, E. L., Jr. (1967). *Chem. Eng. Sci.* **22**, 1117.

Schnepf, R. W., Gaden, E. L., Jr., Mirocznik, E. Y., and Schonfeld, E. (1959). *Chem. Eng. Progr.* **55**(5), 42.

Schonfeld, E., and Kibbey, A. H. (1967). *Nucl. Appl.* **3**, 353.

Schutz, F. (1946). *Trans. Faraday Soc.* **42**, 437.

Sebba, F. (1959). *Nature* **184**, 1062.

Sebba, F. (1962). "Ion Flotation." American Elsevier, New York.

Shih, F. S. (1969). Private communication.

Shih, F. S., and Lemlich, R. (1967). *A. I. Ch. E. J.* **13**, 751.

Shih, F. S., and Lemlich, R. (1971). *Ind. Eng. Chem. Fundamentals* **10**, 254.

Smith, B. D., (1963). "Design of Equilibrium Stage Processes." McGraw-Hill, New York.

Snavely, E. S., Schmid, G. M., and Hurd, R. M. (1962). *Nature* **194**, 439.

Wace, P. F., Alder, P. J., and Banfield, D. L. (1969). *Chem. Eng. Progr. Symp. Ser.* **65**(91), 19.

Wace, P. F., and Banfield, D. L. (1966). *Chem. Proc. Eng.* **47**(10), 70.

WCET Television Station (1970). Foam fractionation, 30-minute, half-inch, color, sound, videotape. Referral Service Network Office or College of Engineering, Univ. of Cincinnati.

Weast, R. C., and Selby, S. M., eds. (1967). "Handbook of Chemistry and Physics," 48th ed., p. F148. Chemical Rubber, Cleveland, Ohio.

CHAPTER 4

ION FLOTATION

T. A. Pinfold

Department of Chemistry including Biochemistry
University of the Witwatersrand
Johannesburg, South Africa

I. INTRODUCTION

Ion flotation was first introduced by Sebba (1959), and the basic principles of the technique were outlined in subsequent publications (1960, 1962a, b, 1963, 1964, 1965). The method involves the removal of surface-inactive ions from aqueous solutions by the addition of surfactants, and the subsequent passage of gas through the solutions. As a result of this flotation procedure, a solid, which contains the surfactant as a chemical constituent, appears on the surface of the solution. Usually the surfactant (collector) is an ion of opposite charge to the surface-inactive ion (colligend), and thus cations and anions are floated with anionic and cationic collectors, respectively. It is possible, however, for the collector to be uncharged and to attach itself to the colligend by coordination. For example, Cu^{2+} ions can be floated in this way by octadecylamine molecules.

The collector–colligend product is known as the sublate, and when it ultimately reaches the surface of the solution, it is always present as a solid. Prior to this, however, it may comprise groups of ions held to the surface of the bubble by the surface activity of the collector. Usually the concentrations of collector and colligend are low (about 10^{-5} M), and flotation occurs from a true solution. Raising the concentrations, however, may lead to precipitation of sublate before gas is passed into the solution. Ion flotation carried out under these conditions is a form of precipitate flotation, and its relation to this and other adsorptive bubble separation methods has been explained by Karger et al. (1967), Lemlich (1968), and Pinfold (1970). Ion flotation, therefore, may be considered to occur by one or another of two mechanisms; either the sublate is present in the bulk aqueous solution as a solid, or it only becomes a solid once it leaves the solution and is entrained in the foam.

Besides the investigations of Sebba et al., ten papers on the subject have been published by Grieves and his co-workers. The investigations of these two schools differ primarily in the size of the bubbles involved in the flotation. Sebba has used very small bubbles, slow flow rates, and has produced small foams; Grieves has used coarse bubbles and has often had to contend with copious foams. Ultimately, however, the conclusions reached are similar. Other general publications on the subject are by Dobrescu and Dobrescu (1966, 1968), Charewicz and Niemiec (1966a, b, 1968, 1969), Chernov et al. (1968), and Kuzkin and Semeshkin (1968b). Three of the papers by Rubin et al. (1966, 1967, 1968) and most of the one by Aoki and Sasaki (1966), deal with systems in which no insoluble sublate is formed and hence do not involve ion flotation. Their findings, however, are of interest, as often the effects are very similar to those for known ion flotation systems.

The discussion that follows is an attempt to summarize the present state of knowledge of all aspects of the subject.

II. METHODOLOGY

A. Apparatus

The cells used for ion flotation differ, depending on whether they contain a fine or coarse porosity frit, the average pore diameters being 10 or 50 μ, respectively. As little foam is generated in the former case, the cells are about 2 liters in capacity and normally do not require ducts for the removal of foam. Such a cell, which is also suitable for studies of precipitate flotation, is shown in Fig. 1 of Chapter 5. With coarse-porosity frits, the cells are larger and taller, and provision must be made for removal of large quantities of foam.

An alternative to the use of a frit has been introduced by Grieves and Ettelt (1967); liquid is removed from the cell, subjected to a gas pressure of 40 lb/in.^{2}, and then recycled. The lowering of pressure on return to the cell causes the dissolved gas to escape in a fine stream of bubbles.

The gases normally used are nitrogen or air, and no advantage has been found in replacing these by organic substances. Gas is led to the cell through a buffer vessel which prevents undesirable surges, through a rotameter, and through a bubbler, if prior saturation of the gas is desirable. This last precaution is probably unnecessary. If measurements at a constant temperature are required, the gas is also passed through a heat-exchanging coil, immersed with the cell in a thermostat.

When radioactive solutions are floated, the top of the cell is constricted as a safety measure, and devices for the removal of the active material after flotation are introduced.

The apparatus used for continuous flotation is more complex, and reference should be made to the papers of Davis and Sebba (1966b), Grieves and Schwartz (1966a, b), and Grieves and Ettelt (1967). In each case the flotation cell is the essential feature, with collector and colligend solutions being added through separate flow lines. The processes were followed by samples taken at known intervals, although Davis and Sebba (1966b), who used radioactive solutions, further monitored the effluent stream continuously by using a Geiger–Müller tube connected to a ratemeter and a recorder.

In general, the apparatus required is unsophisticated and not costly.

B. Estimation of the Extent of Flotation

The rate of flotation has been followed in all cases by withdrawal of samples from the cell during flotation, and subsequent analysis to determine the extent to which colligend has been removed. A variety of gravimetric, volumetric, and instrumental methods of analysis have been used, depending on the system under investigation. The method most generally applicable, however, is to add a radioactive tracer of the element being floated and then to follow its removal by the decrease which occurs in the specific activity of the solution. This method was first introduced in ion flotation experiments by Davis and Sebba (1966b) and Charewicz and Niemiec (1966b).

III. PARAMETERS AFFECTING THE PROCESS

A. Collector and Colligend Concentrations

1. Excess of Collector

A leading advantage of ion flotation is the large volume reduction which occurs on collecting the colligend from a dilute solution and concentrating it in a small volume of sublate. The production of a large amount of foam therefore is both unnecessary and undesirable, although a small amount is advantageous in that it supports the sublate and prevents the latter from redispersing into the solution. As the extent to which foam is formed depends on the concentration of collector, the latter must be controlled with some care. In addition, it has been found that the presence of excess of collector inhibits flotation, an effect first noticed by Lusher and Sebba (1965), and subsequently by Davis and Sebba (1966a), Grieves and Bhattacharrya (1968), Grieves et al. (1969), Rose and Sebba (1969), and Spargo and Pinfold (1970). This suppression occurs for a number of reasons. If, for example, the sublate is present in the aqueous solution as a solid, the particles probably become coated with a layer of collector as the concentration of the latter is raised. As this layer will be arranged with the long-chain groups of the collector in contact with the hydrophobic surface of the precipitate and the ionic groups oriented towards the water, the arrangement renders the particles more hydrophilic, and they float less easily. In addition, an increase in the concentration of collector leads to competition between the particles and the collector ions for places on the bubble surfaces, and removal of sublate is impaired.

When the collector and colligend are present in the solution as ions, two reasons have been proposed (Davis and Sebba, 1966a) for the suppression. First, collector–colligend interactions in the bulk of the solution become more likely, and in addition, adsorption of these species on the bubbles becomes more difficult, as the latter are crowded with collector; second, if the concentration of collector exceeds the critical micelle concentration, flotation may be impaired because colligend ions adsorb on the micelles, which are themselves unable to float because of their hydrophilic surfaces. Therefore, in both the above cases, colligend is withheld from the bubbles and flotation is hindered.

2. The Effect of Micelles

The suggestion that micelles are deleterious to ion flotation has been made repeatedly by Sebba but this has been questioned by Rubin et al. (1966) who found that aged collector solutions which contained micelles gave the same removals as others which had been freshly prepared. As micelles have been shown by Kresheck *et al.* (1966) and Jaycock and Ottewill (1967) to disintegrate very rapidly, it may be that their presence is not as important as was first thought, and that the effects on flotation attributed to them are due merely to increased concentrations of collector.

3. The Collector–Colligend Ratio ϕ

Because the sublate formed in ion flotation is a chemical compound of the collector and the colligend, the ratio ϕ of the two required for complete flotation must at least be a stoichiometric one. Therefore, too little collector will have the same overall effect as too much, in that removals of colligend will not be complete. In each system a range of ϕ values exists for which the process is most efficient and, with two exceptions, these ratios have been found to be greater than those required for stoichiometric combination. The excess of collector that is necessary depends on the kind of ion flotation used. For example, if the colligend is precipitated before flotation commences, the amount of collector needed will be close to the stoichiometric requirement. On the other hand, if the mode of collection depends on colligend and collector ions meeting on a bubble surface, the residence time of the bubble must be sufficient to allow this to proceed to completion. If it is not, some collector

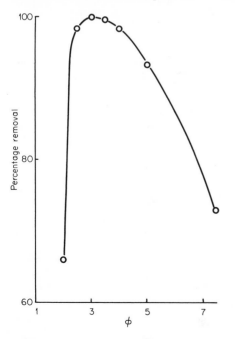

FIG. 1. Dependence of the flotation of Y^{3+} ions on the collector–colligend ratio (ϕ). [From Rose and Sebba (1969).]

floats wastefully and the complete removal of colligend with a stoichiometric amount of collector becomes impossible.

The optimum ϕ ratios that have been reported vary widely. Rose and Sebba (1969), for example, achieved complete removal of Y^{3+} ions with a stoichiometric amount of 2-sulfohexadecanoic acid ($\phi = 3$). The variation of the extent of removal with ϕ is shown in Fig. 1, which is typical for ion flotation systems. Rubin and Lapp (1969) achieved complete removal of lead(II) with an anionic collector at a pH for which the ionic species was $PbOH^+$ and ϕ was unity. The highest optimum ϕ ratio of 44 was recorded by Spargo and Pinfold (1970) but this abnormal value may have been due to the presence of 2-octanol, which was used to collect the sublate by solvent sublation. As the most efficient conditions for ion flotation vary widely, it is clearly necessary to examine each system independently.

4. Pulsed Addition of Collector

The addition of collector can be made in one dose at the commencement of flotation, or by a number of small (pulsed) additions at various intervals thereafter. Rice and Sebba (1965) conclude that there is no difference in the recoveries achieved by the two methods, but Grieves and Wilson (1965) and Razumov et al. (1965) found marked improvements when using pulsed additions. In fact Grieves and Bhattacharrya (1969) showed that when floating complexed cyanides, nearly twice as much of this colligend could be removed by one-fifth less surfactant.

The difference in the above conclusions may arise because of the flow rates used. Pulsed additions of collector imply that, for most of the flotation, colligend is present in excess. This should ensure that the collector does not float wastefully, i.e., without removing a stoichiometric amount of colligend. Although such removals should be more efficient, redispersion may occur to a greater extent than usual if the flow rate is low because little foam is present to support the sublate. The two effects oppose each other and may account for the absence of an overall change. At higher flow rates, however, sufficient collector is removed to form the supporting foam and yet, in addition, flotation is more efficient.

5. Colligend Concentrations

Ion flotation is most efficiently applied when the concentration of colligend is in the range of 10^{-5} to 10^{-3} M. Above these values the quantity of sublate formed is often inconveniently large, while below, the amount of collector present is insufficient to form a supporting foam and flotations are incomplete.

The removal of colligend at a constant value of ϕ passes through a maximum as the concentration of colligend is raised. Typical behavior is shown in Fig. 2 for the flotation of Y^{3+} ions using an anionic collector at a constant value of $\phi = 3$ (Rose and Sebba, 1969). As the concentration of colligend is raised from the lowest values, the collector which floats wastefully is a decreasing

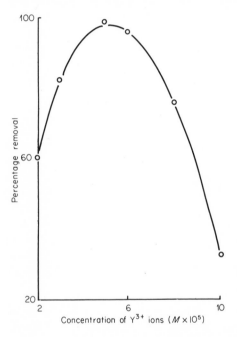

FIG. 2. Dependence of flotation on the concentration of Y^{3+} ions when using a stoichiometric equivalent of collector ($\phi = 3$). [From Rose and Sebba (1969).]

fraction of the whole. In addition, the stability of the supporting foam increases in the presence of more collector, and on both accounts the efficiency rises. It decreases, however, as the concentrations are raised still further because, although ϕ is maintained at a constant value, it is possible that in parts of the system localized collector-to-colligend ratios much in excess of ϕ may occur. Under such conditions, flotation is impaired for the reasons given in Section III,A,1. Therefore, it is clear that the concentration of both collector and colligend should be carefully chosen to ensure the most efficient removals.

B. Flow Rate

Most investigations in ion flotation have been performed using either low flow rates of up to 10 liters/hr and a cell with a fine-porosity frit, or at about 250 liters/hr using a coarse frit. Under the former conditions, recoveries become more rapid on increasing the flow rate, as shown in Fig. 3, which reflects the removal of Sr^{2+} ions using an anionic collector (Davis and Sebba, 1966a). With more rapid rates of flow, however, removals are incomplete because sublate is redispersed into the turbulent solution.

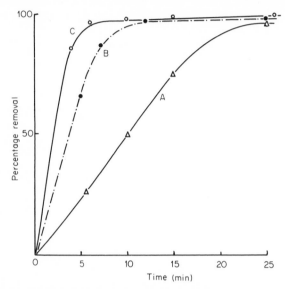

FIG. 3. Variation of the rate of flotation of Sr^{2+} ions with flow rates: (A) 1.25 liters/hr; (B) 2.75 liters/hr, and (C) 6.5 liters/hr. [From Davis and Sebba (1966).]

With coarse frits, the much higher rate of gas flow results in the formation of large amounts of foam which support the sublate more effectively but which entrain appreciable amounts of the aqueous solution. The latter is an unnecessary disadvantage and the use of a fine-porosity frit appears to be the more satisfactory choice.

C. Ionic Strength

It is generally agreed that the presence of neutral salts decreases the efficiency of ion flotation, and that this arises because of competition for collector between the colligend and the added ions.

As expected, ions of a charge opposite to that of the collector have the most effect, but there is difference of opinion on how extensively the charge influences the process. For example, Grieves et al. (1965a, b) and Lusher and Sebba (1965) found that interference when using a cationic collector was more marked in the presence of PO_4^{3-} ions than SO_4^{2-} ions than Cl^- ions. Figure 4 shows this effect and is due to the former authors. Hope (1962), however,

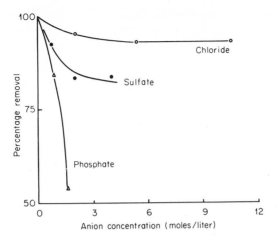

FIG. 4. The effect of anions on the flotation of dichromate. [From Grieves *et al.* (1965a).]

showed that the order was reversed when floating $Fe(CN)_6^{4-}$ ions with a cationic collector, and was able to equate the influence of ion strength μ to the fraction of colligend finally removed, by the equation

$$\log_{10}(\text{fractional recovery}) = A\mu + B \qquad (1)$$

in which B is virtually constant for all determinations and A is a constant for any one anion. Table I shows values of A for a number of ions, introduced into the system as their sodium or potassium salts. The anions of lowest charge clearly have the greatest suppressive influence, which is contrary to the findings of other workers. Hope interprets this in terms of the ionic radii and suggests that the smaller the ion the greater the interfering effects.

TABLE I

THE INFLUENCE OF VARIOUS INERT IONS ON
THE EFFICIENCY OF FLOTATION

Interfering ion	Constant A in Eq. (1)
CN^-	-68.0
NO_3^-	-63.8
Cl^-	-40.2
SO_4^{2-}	-9.3
CO_3^{2-}	-9.5
PO_4^{3-}	-12.4
$P_2O_7^{4-}$	-3.0

The difference may arise because in the latter system the sublate was collected in 2-octanol by solvent sublation. It has been suggested by Spargo and Pinfold (1970) that sublation of $Fe(CN)_6^{4-}$ ions under such conditions is subject to interference by 2-octanol which dissolves in the water during the course of the flotation, and is adsorbed on the bubble surfaces. As flotation of $Fe(CN)_6^{4-}$ requires a group of five ions, four collector and one colligend, for electrical neutrality on the bubble surface, interspersion of 2-octanol molecules keeps the collector ions apart and hinders the formation of the groups. Under these circumstances, the lower the charge on the added ions, the more effectively they neutralize the charge of adsorbed collector ions. With less repulsion between bubbles, the latter are able to leave the solution more readily, and the collector they carry is no longer available for flotation of $Fe(CN)_6^{4-}$ ions. Had the collector ions freedom of movement on the bubble surface, as when 2-octanol was absent, anions with highest charge would have caused the greatest interfering effects.

D. pH

Because changes in pH have marked effects on the nature and charge of both the collector and colligend, ion flotation is particularly susceptible to variations of this parameter. Almost all investigators refer to some aspect of this dependence but the findings will not be considered individually, as all were anticipated by Sebba (1962a) in his original study. The following are the effects to be expected on varying the pH:

1. A change may occur in the charge on the colligend, due to hydrolysis or the formation of other complexes.
2. Changes may occur in the ionization of the collector; acids and amines, for example, may lose their charge at low or high pH values, respectively. They either then cease to be collectors, or their mode of collection changes.
3. The colligend may be precipitated as a hydroxide and then removed by precipitate flotation instead of ion flotation. Variation of the pH therefore can lead to a change in the nature of the process.
4. Flotation may be suppressed by the increased ionic strength which arises on adjusting the pH to extreme values.
5. The stability of the foam supporting the sublate may change, leading to redispersion.

Therefore, for a variety of reasons a knowledge of how pH affects each system is a prerequisite to flotation.

E. Temperature

Studies of how variations in temperature affect ion flotation have been almost completely neglected. Lusher and Sebba (1965) found that recoveries of aluminium increased from 77 to 100% on raising the temperature from 12 to 23°C, respectively. No reason for the effect was suggested. Spargo and Pinfold (1970) found the opposite effect when floating hexacyanoferrate(II) with dodecylpyridinium chloride. Collection was by solvent sublation, and an increase of the temperature from 5 to 35°C caused recoveries to be reduced by half. This considerable decrease cannot be due to increased solubility of sublate, nor to the lower stability of foams. It is attributed to the fact that, as adsorption is an exothermic process, an increase in temperature leads to a decrease in the amount of collector on the bubbles, with a concomitant reduction in recovery. As the effect of temperature appears to be substantial, closer investigation of this parameter is warranted.

F. Presence of Ethanol

Since the inception of ion flotation it has been the practice to add the collector in the form of an ethanolic solution to avoid the formation of micelles. Recent measurements have shown, however, that micelles break down rapidly, and their presence may not, in fact, be detrimental. It is convenient to continue the practice nonetheless, as many of the collectors used are not very soluble and it is necessary to maintain them in solution until they can come in contact with colligend ions. Although adding ethanol has the further advantage that the sizes of the bubbles are smaller because of the lower surface tension of the solution, additions for this purpose are not necessary as adsorption of collector has the same effect, although the process is slower.

Therefore, provided that the collector is soluble, alcohol is not really needed and was, in fact, omitted by Grieves et al. Omission is advisable where possible, as it has been shown by Davis and Sebba (1966b) and Kuzkin and Semeshkin (1968a) that too much alcohol suppresses flotation.

G. Surface Area of Bubbles

The only results available for the surface areas of the bubbles passing during ion flotation are those due to Spargo and Pinfold (1970). The average pore diameter of the frit in the flotation cell was 10 μ. Photographs were taken through a microscope mounted next to the cell, as shown in Fig. 5, and operated in conjunction with a synchronized flash unit. The microscope was fitted with a graticule which was ruled in a pattern of regular squares, an image of which became part of the picture, as shown in Fig. 6. By photographing a wire of known dimensions against the background of the graticule, a scale factor was obtained for converting measurements on a photograph to the actual distances in the cell. Although the smallest bubbles in the system were 40 μ in diameter, the average size was about 100 μ and this value increased with increasing flow rate.

The system investigated was the flotation of $Fe(CN)_6^{4-}$ ions using dodecyl-pyridinium chloride at concentrations of 5×10^{-6} and 2×10^{-5} M, respectively. Removal of the sublate was achieved by solvent sublation into 2-octanol. No precipitate existed in the solution prior to the flow of gas. Table II shows the rate at which new surface area was formed as the bubbles entered the cell at different flow rates, and also gives the average area which was available to each collector ion during the first 5 min of flotation. From these values it is unlikely that the bubble surfaces are saturated with collector at any time during flotation.

If surface areas were measured in this way, and at the same time the rates of removal of both collector and colligend were determined, much could be learned about adsorption of ions at gas–liquid interfaces in contact with a turbulent solution.

IV. CONTINUOUS ION FLOTATION

Most investigations in ion flotation have been on batch operations but the study was extended in two cases to continuous processes. David and Sebba

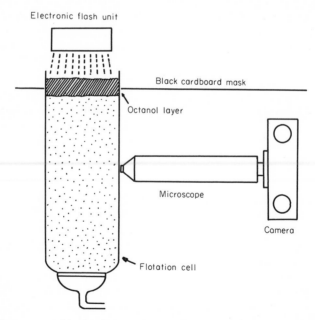

Fig. 5. Apparatus used in the photography of bubbles. [From Spargo and Pinfold (1970).]

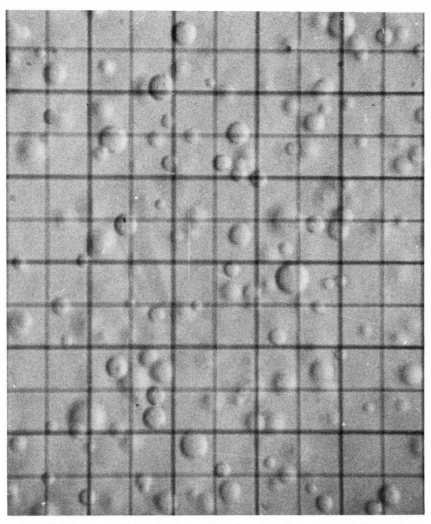

FIG. 6. Bubbles present during flotation at 6 liters/hr, using a fine-porosity frit. [From Spargo and Pinfold (1970).]

TABLE II

THE GAS–LIQUID INTERFACIAL AREAS AVAILABLE AT VARIOUS FLOW RATES

Gas flow rate (liters/hr)	2	4	6	8	10
Total surface area passed into the cell per minute (m^2/min)	1.93	3.57	4.85	6.20	6.72
Average area available to a collector ion ($Å^2$)	400	690	840	970	1000

(1966b) and Grieves and Schwartz (1966a, b) adapted their previous investigations on flotation of Sr^{2+} and $Cr_2O_7^{2-}$ ions, respectively, to continuous processes. Grieves and Ettelt (1967), who also floated $Cr_2O_7^{2-}$ ions, used dissolved instead of dispersed air. They expressed their results in terms of enrichment ratios, defined as the concentration of colligend in the foam divided by the steady state concentration of colligend in the effluent stream.

Strontium was removed by 2-sulfohexadecanoic acid by feeding 5.7×10^{-5} M and 1.8×10^{-2} M solutions of each into a flotation cell of 2-liter capacity at rates of 1250 ml/hr and 10 ml/hr, respectively. (Use of a more concentrated collector solution was necessary, as excessive quantities of ethanol impair flotation.) At a flow rate of 3.5 liters/hr, steady state removals of 98 % were achieved in 6 hr, at an enrichment ratio of 4000.

Typical of the continuous removal of dichromate was the following: With a retention time of 150 min, a feed stream of 50 mg/liter could be split into a foam stream and a residual stream containing 450 and 15 mg/liter, respectively. (Retention time is defined as the volume of solution being floated, divided by the combined flow rate of liquid to the column.) Using such results, the theoretical performance of a multicolumn system was calculated, in which the effluent of one column was the feed stream of the next. With three columns and a feed concentration of 100 mg/liter, the final effluent would contain 8 mg/liter and, in volume, would be 80 % of the initial feed.

In the method of Grieves and Ettelt (1967), colligend and collector solutions were premixed in the ratio of 1:2.7 for 1 hr before flotation, a nonionic polymer was used as a flocculating agent, and dissolved air was introduced by leading off part of the main solution, subjecting it to increased pressure, and returning it to the cell. The efficiency of the process depends on the time the solution remains in the cell, and for a feed solution containing 100 mg/liter, enrichment ratios of 1930, 3810, and 9100 were achieved at detention times of 37, 77, and 131 min, respectively; 95–97% of the dichromate was ultimately removed.

It is generally agreed that ion flotation can be used in continuous processes, although no assessment has been made of how practical it would be economically.

V. ANALYTICAL APPLICATIONS

A. Determination of Surfactants

The first analytical application of ion flotation was made by Tomlinson and Sebba (1962) who removed 9-octadecenoate ions from solutions by the addition of excess of crystal violet dye. The latter was present in a cationic form, and a stoichiometric amount of it was floated from the solution by the 9-octadecanoate ions present. The extent of removal of the latter was then estimated by the decrease in the optical density of solutions before and after flotation, measured at 590 nm where crystal violet absorbs prominently. Dyes have been used before for the determination of charged surfactants but the complex formed between the two had not previously been removed by flotation.

The method was extended by Lovell and Sebba (1966) to the analysis of dodecanoate ions, and also to cationic surfactants using anionic dyes, such as bromophenol blue. Whereas the precision with anionic surfactants was poor (± 5–8%) because of the adsorption of an indeterminate amount of crystal violet on the negatively charged glass surface of the flotation cell, precision when using cationic surfactants, and dyes such as bromophenol blue, was good ($\pm 1\%$).

A flotation procedure has the advantage over methods in which the dye/surfactant complex is removed by extraction in that no emulsification of the organic solvent can occur. The disadvantages, however, are that (1) cationic dyes adsorb on the large surface areas of the glass frit, (2) the dye may have some surface activity and will float independently of the surfactant, and (3) that redispersion of the sublate back into the aqueous phase cannot always be avoided.

B. Determination of Trace Metals

Ion flotation has been used by Mizuike et al. (1969) as a preconcentration technique in the determination of trace elements such as silver, gold, iron, copper, and cobalt. Anionic complex ions of these elements were formed by adding oxalate, cyanide, or thiosulphate ions, and then floated from the solution using a cationic collector. In this way 0.1–1.0 μgm of the trace element could be removed from solutions containing 0.5–3.0 gm of matrix elements such as magnesium, sodium, or zinc which do not form such complexes. After 10 min when flotation was complete, the foam was removed and dissolved in ethanol; the resulting solution was then diluted with water and the flotation process was repeated. The overall yield was greater than 90%, and the concentration factors after one, two, and three separations were 30, 1000, and 30,000, respectively.

The principle of the method was originally envisaged by Sebba (1962a). Although it shows how powerful a concentrating technique ion flotation can be, it suffers from the disadvantages that (1) losses occur on transferring the foam, (2) if removal by flotation is not 100%, each repetition of the process increases the overall error, and (3) the excess of complexing reagent added must be carefully controlled, or competition for the surfactant occurs between the various ionic forms present.

C. Qualitative Analysis of Complex Ions

Sebba (1962a) suggested that ion flotation might be used to identify complex ions in dilute solutions by analysing the sublate removed during flotation. Grieves *et al.* (1968) succeeded in doing this, using a cationic collector and floating $Cr_2O_7^{2-}$ or $HCrO_4^-$ from solutions of sodium dichromate, and $Fe(CN)_6^{4-}$ and $FeFe(CN)_6^{2-}$ from solutions of iron(II) sulfate and sodium cyanide.

When precipitation occurred between the collector and the complex species prior to flotation, as in the above cases, it was shown that the identity of the complexes could be found by determining the molar ratio of surfactant to Cr or to CN in the foam. Where no precipitation occurred prior to flotation, however, for example, with orthophosphates and phenolate ions, the ratios of collector to colligend were indeterminate and markedly in excess of the possible stoichiometric values. This was explained as being due to the simultaneous removal of appreciable quantities of ions other than the complex, and it was concluded that the method could only be used if precipitation occurred before flotation. This is probably not generally true, however, and meaningful results should be obtained with flotations at low flow rates, and in the presence of excess colligend.

VI. THEORETICAL CONSIDERATIONS

A. The Sublation Constant

To explain the selectivity which can be achieved by ion flotation, Sebba (1962a, 1965) derived a function called the sublation constant, which was the product of the activities of the collector and colligend in the aqueous solution when flotation of a sublate first occurred. With collector C^+, for example, and colligend A^{y-}, the sublate formed would be the compound C_yA and the sublation constant F would be equal to $(a_{C^+})^y(a_{A^{y-}})$. No flotation would occur until the activities of C^+ and A^{y-} had been raised to values such that the product given above exceeded F.

It is assumed in the above derivation that sublate either is precipitated prior to flotation or it forms on the bubbles and saturates their surfaces. In both cases the activity of the sublate may be taken as unity, as it is independent of changes of activity in the bulk solution. It is also assumed that the ion with the highest charge has the lowest sublation constant, which is not unreasonable as

$$F = (a_{C^+})^y(a_{A^{y-}}). \qquad (2)$$

As the value of y increases, F decreases. As flotation can always occur from solutions too dilute to form a precipitate in the bulk, only the latter case mentioned above, that in which sublate saturates the bubble surfaces, will be considered in the following discussion.

If two colligends A^{y-} and B^{z-} are present in the solution and collector is added in small doses, the activities of the collector at which the two colligends begin floating simultaneously is given by

$$a_{C^+} = (F_A/a_{A^{y-}})^{1/y} = (F_B/a_{B^{z-}})^{1/z}. \tag{3}$$

It will be assumed that $z \geqslant y$, that $a_{A^{y-}} = a_{B^{z-}} = \rho$, and that the activity of the colligends under the special circumstances given above when the two float together is ρ'. It is easily shown that if $\rho < \rho'$, B^{z-} floats before A^{y-}, and if $\rho > \rho'$, the order is reversed. For example, if dilute, equimolar concentrations of $HCrO_4^-$ and $Cr_2O_4^{2-}$ ions are floated with a cationic collector, CrO_4^{2-} will float preferentially if the activities are below ρ', but $HCrO_4^-$ ions will float first if the solution is more concentrated. Such behavior has never been encountered in ion flotation; instead it has been found that the more highly charged ion floats preferentially from equimolar solutions. (The choice of equal concentrations in the present discussion is a matter of convenience and does not alter the conclusions in any way.)

The sublation constant was originally defined as analogous to the solubility product, the derivation of the two functions being the same. The process of precipitation, however, occurs suddenly on increasing concentrations in a solution, while that of flotation does not. Ions will be floated on a bubble at concentrations well below those required to saturate its surface and hence the concept of a sublation constant has no meaning. The activity of sublate on the bubble is not independent of the activities of the ions in the bulk solution, and cannot be set equal to unity.

It is suggested that the sublation constant be replaced by the "sublation concentration" which is to be defined as the lowest concentration at which flotation of the colligend ion can be detected. A stoichiometric amount of collector would be used, and flotations would occur into an immiscible solvent such as 2-octanol, as the formation of foams will not always be possible at such low concentrations of surfactant. A series of solutions of increasing colligend concentrations would be prepared, stoichiometric amounts of collector added, and each subjected to flotation. The removal of colligend could be detected either by adding a radiotracer and noting the first decrease in the activity of the aqueous solution, or, if the colligend absorbs light of a wavelength to which 2-octanol is transparent, the solvent can be removed, centrifuged, and examined spectrophotometrically. The latter method has, in fact, been used by Pinfold and Smith (1964) and led to a determination of the sublation concentrations of a series of ions of increasing charge, $HCrO_4^-$, CrO_4^{2-}, $Fe(CN)_6^{3-}$, and $Fe(CN)_6^{4-}$. The collector used was dodecylpyridinium chloride and the values found were 4.7×10^{-6}, 2.5×10^{-6}, 1.2×10^{-6}, and $2.3 \times 10^{-7}\,M$, respectively. As expected, the higher the charge on the ion the more readily it is floated because the greater are the number of hydrophobic ions attached to it.

Sublation concentrations will depend on the collector used, the temperature, and also the residence time of bubbles in the solution. From a practical point of view, variations in the latter may not be significant, and sublation concentrations may well become the best way of assessing the floatability of an ion.

B. Kinetics

Only one study has been made of the kinetics of an ion flotation system. Grieves and Bhattacharyya (1969) examined the flotation of $FeFe(CN)_6^{2-}$ ions using a cationic collector in a cell with a coarse-porosity frit. With colligend present in excess at all times, the rate equation obtained was

$$-d\tau/dt = 0.264\tau^{0.93}, \tag{4}$$

where τ is the mass of colligend present which could be floated by the collector available, and t is the time in minutes. The authors refuse to comment on why the process is less than first order, and suggest that speculation is not warranted.

No kinetic studies have been made on ion flotation systems using cells with fine porosity frits, but data is available for foam fractionation performed in such cells. As the processes are the same until the bubbles reach the surface of the solution, provided precipitation does not occur before flotation, the kinetics are comparable. Rubin et al. (1966) found that the flotation of Cu^{2+} ions by sodium dodecylsulfate approximated a reversible first-order reaction and expressed this by

$$\ln[(M - R)/M] = kt/M, \tag{5}$$

where R and M are the fractions of colligend removed at time t and under steady-state conditions, respectively. Sheiham and Pinfold (1970) used this expression and obtained first-order rate constants for the removal of hexadecyltrimethylammonium chloride in solutions of different ionic strength.

Kinetic studies of ion flotation have been sadly neglected, and the omission needs to be righted before further progress is made.

VII. USES

Apart from the investigations described above, a number of publications have appeared in which greater cognizance is taken of practical applications of ion flotation. Two main uses are appearing for the process: the decontamination of water and the extraction of metals. Each will be considered below in detail. One other is the concentration and separation of substances of toxicological and pharmaceutical importance. Hofman (1968), for example, has performed removals of barbiturates, medicinals, and plant materials from dilute solutions.

A. Decontamination of Water

The role of flotation in the treatment of contaminated waters was considered by Argiero and Paggi (1968) and by Grieves et al. (1968). The papers of the latter all involve the removal of dichromate or complex cyanide ions, which occur in the industrial wastes of metal finishing, electroplating, steel, and coke-plant industries. It was shown that ion flotation removes these impurities effectively and that regeneration of the surfactant is feasible. How competitive the process can be made economically is not yet known.

Another aspect of water decontamination is the removal of radioactive elements such as europium, cerium, yttrium, strontium, and cesium from effluents. Cardoza (1963) and Cardoza and De Jonghe (1964), for example, found that Eu^{3+} ions could be readily removed, Sr^{2+} ions less so, and Cs^+ poorly by ion flotation, although Cs^+ ions are easily concentrated by precipitate flotation. Oglaza and Siemaszko (1966) found that whereas cerium ions floated successfully, Sr^{2+} and Cs^+ ions did not. Davis and Sebba (1966a, b), however, have shown that Sr^{2+} ions can be removed successfully both by batch and continuous scale processes. Rose and Sebba (1969) have shown that yttrium too is readily floated.

In view of the above, ion flotation should be considered, along with other adsorptive bubble separation methods, for the removal of radioactive wastes, particularly because of the conveniently small volume into which the active material is concentrated for ultimate storage or disposal.

B. Extraction of Metals

Poltoranina et al. (1963, 1964) separated iron from copper, cobalt, and nickel by precipitation of the former with fatty acids prior to flotation. Of others, Jacobelli-Turi et al. (1963) and Barocas et al. (1964) floated UO_2^{2+} ions from solutions of carbonate; Lusher and Sebba (1965) floated aluminum and separated it from beryllium; Rice and Sebba (1965) concentrated zirconium as the fluorozirconate ion; Usoni (1965) studied ion flotation in connection with the production of yttrium; Razumov et al. (1966) continued the work of Poltoranina et al. on the separation of iron from other metals, while Rubin and Lapp (1969) successfully floated lead.

If to these are added the study by Koizumi (1963) of the flotation of iodine, and the qualitative examinations made by Sebba (1962a), a fair amount of information is available on behavior during flotation.

VIII. CONCLUSION

In the twelve years since its discovery a number of aspects of ion flotation have been studied, although none has been probed deeply. Therefore, the present understanding of the subject is not well developed but most investigations have corroborated the assumptions made initially. As regards practical applications, the process is suited to the extraction of a variety of substances, but its economic feasibility has yet to be established.

REFERENCES

Aoki, N., and Sasaki, T. (1966). *Bull. Chem. Soc. Japan* **39**, 939.
Argiero, L., and Paggi, A. (1968). *Health Phys.* **14**, 581.
Barocas, A., Jacobelli-Turi, C., and Salvetti, F. (1964). *J. Chromatogr.* **14**, 291.
Cardoza, R. L. (1963a). *E.A.E.C. Euratom* 262e.
Cardoza, R. L., and De Jonghe, P. (1963b). *Nature* **199**, 687.
Charewicz, W., and Niemiec, J. (1966a). *Wiad. Chem.* **20**, 693.
Charewicz, W., and Niemiec, J. (1966b). *Krajowe Symp. Zastosow. Isotop. Tech., 3rd, Stettin, Pol.* Sect. 48.
Charewicz, W., and Niemiec, J. (1968). *Neue Huette* **13**, 461.
Charewicz, W., and Niemiec, J. (1969). *Nukleonika* **14.** 17.
Chernov, V. K., Gol'man, A. M., and Klassen, V. I. (1968). *Izv. Akad. Nauk Kirg. SSR.* **6**, 52.
Davis, B. M., and Sebba, F. (1966a). *J. Appl. Chem.* **16**, 293.
Davis, B. M., and Sebba, F. (1966b). *J. Appl. Chem.* **16**, 297.
Dobrescu, L., and Dobrescu, V. (1966). *Rev. Chim.* **17**, 232.
Dobrescu, L., and Dobrescu, V. (1968). *Rev. Minelor* **19**, 231.
Grieves, R. B., and Bhattacharyya, D. (1968). *Separ. Sci.* **3**, 185.
Grieves, R. B., and Bhattacharyya, D. (1969). *J. Appl. Chem.* **19**, 115.
Grieves, R. B., and Ettelt, G. A. (1967). *A. I. Ch. E. J.* **13**, 1167.
Grieves, R. B., and Schwartz, S. M. (1966a). *J. Appl. Chem.* **16**, 14.
Grieves, R. B., and Schwartz, S. M. (1966b). *A. I. Ch. E. J.* **12**, 746.
Grieves, R. B., and Wilson, T. E. (1965). *Nature* **205**, 1066.
Grieves, R. B., Wilson, T. E., and Shih, K. Y. (1965a). *A. I. Ch. E. J.* **11**, 820.
Grieves, R. B., Wilson, T. E., Shih, K. Y., and Schwartz, S. M. (1965b). *Purdue Univ. Eng. Bull. Ext. Ser. No.* **118**, 110.
Grieves, R. B., Jitendra, G., and Bhattacharyya, D. (1968). *J. Amer. Oil Chem. Soc.* **45**, 591.
Grieves, R. B., Bhattacharyya, D., and Conger, W. L. (1969). *Chem. Eng. Progr. Symp. Ser.* **65**, 29.

Hofman, M. (1968). *Mitt. Deut. Pharm. Ges. DDR* **38,** 53.

Hope, C. J. (1962). M.Sc. Thesis, Univ. of the Witwatersrand, Johannesburg, South Africa.

Jacobelli-Turi, C., Barocas, A., and Salvetti, F. (1963). *Gazz. Chim. Ital.* **93,** 1493.

Jaycock, M. J., and Ottewill, R. H. (1967). *Proc. Int. Congr. Chem. Phys. Appl. Surf. Active Subst., 4th* **2,** 545.

Karger, B. L., Grieves, R. B., Lemlich, R., Rubin, A. J., and Sebba, F. (1967). *Separ. Sci.* **2,** 401.

Koizumi, I. (1963). *Kogyo Kagaku Zasshi* **66,** 912.

Kresheck, G. C., Hamori, E., Davenport, G., and Scheraga, H. A. (1966). *J. Amer. Chem. Soc.* **88,** 246.

Kuz'kin, S. F., and Semeshkin, S. S. (1968a). *Izv. Vyssh. Ucheb. Zaved. Tsvet. Met.* **11,** 11.

Kuz'kin, S. F., and Semeshkin, S. S. (1968b). *Fiz. Tekh. Probl. Razrab. Polez. Iskop.* **4,** 131.

Lemlich, R. (1968). *Ind. Eng. Chem.* **60**(10), 16.

Lovell, V. M., and Sebba, F. (1966). *Anal. Chem.* **38,** 1926.

Lusher, J. A., and Sebba, F. (1965). *J. Appl. Chem.* **15,** 577.

Mizuike, A., Fukuda, K., and Susuki, J. (1969). *Bunseki Kagaku* **18,** 519.

Oglaza, J., and Siemaszko, A. (1966). *Nukleonika* **11,** 421.

Pinfold, T. A. (1970). *Separ. Sci.* **5,** 379.

Pinfold, T. A., and Smith, M. P. (1964). Unpublished results.

Poltoranina, T. F. (1963). *Zap. Leningr. Gorn. Inst.* **42,** 78.

Poltoranina, T. F., Illyuvieva, G. V., and Razumov, K. A. (1964). *Obogashch. Rud.* **9,** 11.

Razumov, K. A., Illyuvieva, G. V., and Poltoranina, T. F. (1965). *Obogashch. Rud.* **10,** 14.

Razumov, K. A., Illyuvieva, G. V., and Poltoranina, T. F. (1966). *Obogashch. Rud.* **11,** 16.

Rice, N. W., and Sebba, F. (1965). *J. Appl. Chem.* **15,** 105.

Rose, M. W., and Sebba, F. (1969). *J. Appl. Chem.* **19,** 185.

Rubin, A. J. (1968). *J. Amer. Water Works Assoc.* **60,** 832.

Rubin, A. J., and Johnson, J. D. (1967). *Anal. Chem.* **39,** 298.

Rubin, A. J., and Lapp, W. L. (1969). *Anal. Chem.* **41,** 1133.

Rubin, A. J. Johnson, J. D., and Lamb, J. C. (1966). *Ind. Eng. Chem. Process Design Develop.* **5,** 368.

Sebba, F. (1959). *Nature* **184,** 1062.

Sebba, F. (1960). *Nature* **188,** 736.

Sebba, F. (1962a). "Ion Flotation." American Elsevier, New York.

Sebba, F. (1962b). *Proc. Int. Conf. Coordination Chem., 7th, Sweden* Paper 7J1.

Sebba, F. (1963). Belg. Pat. 623,140.

Sebba, F. (1964). Ger. Pat. 1,163,265.

Sebba, F. (1965). *Proc. A. I. Ch. E. I. Chem. E. Joint Meeting, London, 1965* (1) 14.

Sheiham, I., and Pinfold, T. A. (1970). to be published.

Spargo, P. E., and Pinfold, T. A. (1970). *Separ. Sci.* **5,** 619.

Tomlinson, H. S., and Sebba, F. (1962). *Anal. Chim. Acta* **27,** 596.

Usoni, L. (1965). *Ind. Mineraria* **16,** 23.

CHAPTER **5**   PRECIPITATE FLOTATION

T. A. Pinfold
Department of Chemistry including Biochemistry
University of the Witwatersrand
Johannesburg, South Africa

I. INTRODUCTION

Precipitate flotation includes all processes in which an ionic species is concentrated from an aqueous solution by forming a precipitate which is subsequently removed by flotation. The field includes three such methods defined as follows:

1. Precipitate flotation of the first kind involves the flotation of precipitate particles by a surface-active species; the latter is not a chemical constituent of the precipitate substance and occurs only on the surface of the particles.

2. Precipitate flotation of the second kind uses no surfactant to float the particles but two hydrophilic ions precipitate to form a solid with a hydrophobic surface.

3. A form of ion flotation in which ions are precipitated by surfactants, and the resulting particles are floated. This latter technique has been described in the previous chapter and will not be considered in this discussion.

The first occasion on which particles were floated from the solutions in which they had been formed was in 1938 when Magoffin and Clanton floated colloids with surface-active ions of charges opposite to the ions adsorbed on the particles. The latter became coated with the surfactant, and floated to form a foam in which the coagulated colloid was deposited. It is surprising, therefore, that similar flotations of precipitates were not achieved until 1960 when a number of independent investigators showed that this was possible. Kadota and Matsuno (1960) precipitated magnesium hydroxide from seawater and floated it with sodium dodecylbenzenesulfonate to obtain a solid sample which was 95 % pure. Kanaoka and Iwata (1960) removed insoluble impurities such as calcium carbonate, magnesium oxide, silica, and alumina from a solution of calcium hydroxide by floating the solids with dodecylammoniumacetate and oleic acid. The process required 1 hr and resulted in a calcium hydroxide solution of about 98 % purity. Pushkarev et al. (1960) coprecipitated cesium in the hexacyanoferrates of metals, and floated the latter with gelatin, whereas Skrylev and Mokrushin (1961) precipitated and floated UO_2^{2+} ions in the same way. Voznesenski et al. (1961) adsorbed radioactive isotopes in iron(III) hydroxide and floated the latter with napthalenesulfonic acid. Baarson and Jonaitis, acting on a private communication from Sebba (1962), floated copper(II) hydroxide with an anionic collector, while Sebba et al. (1962) described flotation of the hydroxides of iron(III), copper, chromium(III), and zinc with 2-sulfododecanoic acid. Further investigation of the coprecipitation of cesium, this time in iron hexacyanoferrates, was made by Skrylev and Pushkarev (1962).

Other early publications in the field were made by Svoronos (1963) who floated aluminum hydroxide, Baarson and Ray (1963) who investigated separations based on the flotation of hydroxides, Cardoza (1963) and Cardoza and De Jonghe (1963) who co-precipitated cesium in copper hexacyanoferrate (II) and floated the solid with sodium dodecylsulfate, Pushkarev et al. (1964a, b) who adsorbed strontium, yttrium, or niobium ions in iron(III) or aluminum hydroxides, or barium sulfate, and floated the precipitates with suitable collectors, and finally Sebba et al. (1965) who once more described separations based on selective precipitation of hydroxides with subsequent flotation.

Since these publications, Lusher and Sebba (1966) investigated the separation of aluminum from beryllium by forming an insoluble fluorocomplex of the former element and floating it with quaternary ammonium collectors. In addition, a series of papers by Koyanaka (1965, 1966a, b, c, d, 1967, 1969a, b) and Mukai et al. (1968, 1969) have appeared on the concentration of radioactive material by precipitate flotation, while further examinations of the removal of cesium by coprecipitation in copper hexacyanoferrate(II) have been made by Oglaza and Siemaszko (1969) and Davis and Sebba (1966), the latter authors having examined continuous removal by precipitate flotation. Finally, Fukuda and Mizuike (1968), and Wainai and Susuki (1968) precipitated silver with α-dimethylaminobenzylidenerhodanine and floated it with sodium dodecylbenzenesulfonate, while Grieves *et al.* (1969) made a comparison of precipitate, colloid, and ion flotation by examining systems in which each occurred.

The investigations described above were known collectively as precipitate flotation, and in each case the removal necessitated the use of a surfactant. It was then shown by Mahne and Pinfold (1966) that the flotation of precipitates could be achieved without the use of surfactants. Two hydrophilic ions, for example, could precipitate to form particles which had hydrophobic surfaces and which floated readily. That some solids are naturally floatable has been known for a long time, but the formation of such substances by precipitation of hydrophilic species to effect concentration from very dilute solutions has not been described before. Clearly this technique was also precipitate flotation. It was called precipitate flotation of the second kind to distinguish it from the older method which used surfactants, which was named precipitate flotation of the first kind. Mahne and Pinfold (1966, 1968a, b, 1969a) examined precipitate flotation of the second kind in some detail and showed that it has marked advantages over the older method.

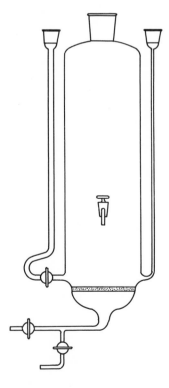

FIG. 1. The flotation cell.

The only searching investigations of the parameters affecting precipitate flotation of the first kind were made by Rubin *et al.* (1966, 1967, 1968) and by Sheiham and Pinfold (1968). The relationship which precipitate flotation bears to other adsorptive bubble separation techniques has been discussed by Karger *et al.* (1967), Lemlich (1968), and Pinfold (1970).

The apparatus required is the same as that described for ion flotation (Section II,A, Chapter 4), both for batch and continuous processes. The type of flotation cell used for batch flotations is shown in Fig. 1; two side arms are included, one for the precipitant and the other for the collector solutions. A tap in the side of the cell allows the removal of samples, the analysis of which indicates the extent of removal during flotation. The average diameter of the pores in the frit at the base of the cell was about 10 μ.

The discussion which follows is an attempt to summarize the present state of knowledge of all aspects of the subject.

II. PRECIPITATE FLOTATION OF THE FIRST KIND

It is generally agreed that the mechanism of precipitate flotation of the first kind involves adsorption from the solution onto the particles of one or other of the ions of which the precipitate is constituted. In this way the surface becomes charged, and may be rendered hydrophobic and hence suitable for flotation by the addition of surfactant ions of opposite charge to those on the surface. Coulombic attraction between surfactant ions and the particle causes the surface to become covered with surfactant, and the precipitate is carried upwards by a fine stream of bubbles.

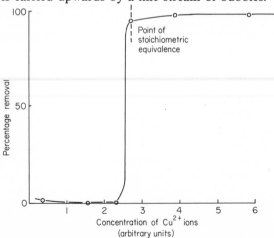

FIG. 2. The effect of surface charge on flotation of a $Cu_2[Fe(CN)_6]$ precipitate. [From Davis and Sebba (1967).]

The importance of the surface charge is clearly shown in Fig. 2, from Davis and Sebba (1966), in which copper hexacyanoferrate(II) is floated with the anionic collector, 2-sulfohexadecanoic acid. When the concentration of Cu^{2+} ions is below the point of stoichiometric equivalence, an excess of $Fe(CN)_6^{4-}$ ions is present, the surface of the particles are negatively charged, and no flotation with the anionic collector is possible. However, once Cu^{2+} ions are in excess, the surface is positively charged and attachment of the collector occurs readily.

The parameters which are of most importance to the process will be discussed.

A. Collector and Colligend Concentrations

The ion to be precipitated and eventually concentrated by flotation will be known as the colligend, a term originally introduced by Sebba (1962a). Similarly, the surfactant used for flotation will be termed the collector, and the ratio of collector to colligend existing at the commencement of flotation will have the symbol ϕ.

One of the most striking features of precipitate flotation of the first kind is the low values of ϕ required. This arises because each macroscopic particle contains a large number of colligend ions but only requires a monolayer of collector on its surface for efficient flotation. If this were the only consideration, however, values of ϕ below 0.0001 would be commonplace. Instead they are usually between 0.005 and 0.1 because it is necessary to form a foam to support the precipitate on the surface of the solution and prevent its redispersion into the bulk. Such low, substoichiometric values of ϕ are a marked advantage over those required for foam fractionation and ion flotation, which are appreciably higher and in excess of stoichiometric requirements.

The lower limit of the colligend concentrations for which flotations remain efficient is set by several factors, the most important being the solubility of the precipitate. The colligend concentration, for example, must be sufficiently high to ensure that virtually complete precipitation occurs. A large excess of precipitant, added to ensure complete precipitation, is unacceptable as increases in ionic strength are believed to affect the process adversely. Second, it is inevitable that a small amount of precipitate will avoid flotation, and if this is an appreciable part of the total solid present, severe errors could occur. Third, unless adverse effects due to an excess of collector, that is, too large a value of ϕ, are to be avoided, collector concentrations must be lowered correspondingly as colligend concentrations decrease. The absence of a supporting foam, and the consequent redispersion of the precipitate, then results in poor recoveries.

Redispersion appears to be unavoidable at low concentrations. On the one hand, it is a penalty paid for the use of substoichiometric amounts of collector, but on the other, it is unreasonable to suppose that the concentration of the collector can be reduced to very low values without this occurring. In an attempt to overcome redispersion, nonionic surfactants, which are uncharged and hence should not interfere, were added to the solutions. It was hoped that the foams they formed would support the floated precipitate when the collector was too sparse to do so. All attempts failed, however, because these surfactants had a greater tendency to float than the coated precipitate. When the latter eventually did collect at the surface, it was forced back into the solution by the mass of foam above it, and redispersion was even more marked.

Removal of the precipitate by suction was also unsuccessful as the lifting of a finely divided, often colorless, precipitate from the surface seldom could be achieved efficiently and always resulted in appreciable quantities of the bulk solution being withdrawn at the same time. A special flotation cell was also tried (Sheiham and Pinfold, 1968) in which the gas was forced through a small orifice in an inverted hollow cone in the hope that the precipitate would be deposited on the upper surfaces of the cone and not return to the solution. This method was no more successful than the others and it is evident that a certain amount of supporting foam is an essential.

As the collector concentration is increased beyond the optimum values for efficient removal, the rate of flotation decreases and therefore passes through a maximum with increasing ϕ. Rubin et al. (1966), when floating copper hydroxide with sodium dodecylsulfate, concluded that neither the rate nor the total removal was significantly affected at high values of ϕ. Davis and Sebba (1967), however, floated copper hexacyanoferrate(II) and inferred that values of ϕ greater than unity were undesirable as the rate of flotation was progressively retarded. Aoki and Sasaki (1966) floated iron with sodium dodecylsulfate at ϕ values of 40–100 and obtained reasonable recoveries at low pH. On changing the acidity and precipitating iron hydroxide, however, flotation was completely inhibited, probably due to the large excess of collector present. Sheiham and Pinfold (1968) have shown (Fig. 3) that the flotation of strontium carbonate with hexadecyltrimethylammonium chloride is markedly impaired by too much collector, the values declining above $\phi = 0.1$. Two reasons are suggested for this behavior: (1) the precipitate particles are increasingly excluded from the bubble surface by the excess collector, and (2) once a monolayer of collector ions has adsorbed on a particle, a second layer may adsorb in which the orientation of the collector is reversed. The charged ends in this second layer then point outwards to the water and the particle becomes hydrophilic once more. Of these two suggestions, doubt has been cast on the validity of the second (Sheiham and Pinfold,

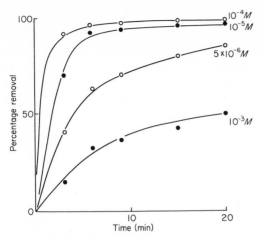

FIG. 3. The influence of collector concentration on the flotation of strontium carbonate precipitates. [From Sheiham and Pinfold (1968).]

1970) but the first appears to be correct. It is evident, therefore, that the concentration of both collector and colligend should be carefully chosen to ensure the most efficient removals.

B. Temperature and Induction Time

If a precipitate is not floated immediately after its formation, but a time of induction is allowed between the two events, subsequent recovery by

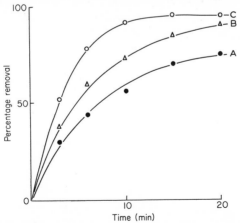

FIG. 4. The influence of temperature and induction time on the flotation of strontium carbonate precipitates: (A) 15°C, 10 min induction time, (B) 15°C, 20 min induction time, and (C) 25°C, 20 min induction time. [From I. Sheiham (1970), Ph.D. thesis, Univ. of the Witwatersrand, Johannesburg, South Africa.]

flotation is more efficient. An increase in temperature in the range 15–30°C also leads to more rapid recovery, and it has been suggested by Sheiham and Pinfold (1968) that this arises because increases in either induction time or temperature lead to the formation of larger particles. Provided the latter are not too large, collection by flotation must be more efficient because of the collection by coagulation which occurs beforehand. The effects are shown in Fig. 4 for the flotation of strontium carbonate by a cationic collector; an increase in induction time at constant temperature and an increase in temperature at constant induction time both lead to more efficient flotation.

C. Effect of pH

The influences of pH on precipitate flotation of the first kind are similar to those for ion flotation and may be summarized as follows:

1. The pH of the solution must lie within a range of values ·suitable for complete precipitation of the ion to be removed.

2. The charge on the surfaces of the particles may vary with pH and will then determine the nature of the collector to be used. For example, at a pH of 6.5, copper hydroxide particles are positively charged by virtue of adsorption of Cu^{2+} ions and require an anionic collector; at higher values, however, the particle surfaces are covered by adsorbed OH^- ions and a cationic collector must be used. The effect is clearly shown in Fig. 5 for the flotation of $Cu_2[Fe(CN)_6]$; up to a pH of 6.5, Cu^{2+} ions are adsorbed on the exterior of the particles which can then be floated with 2-sulfohexadecanoic acid. At higher pH values, however, a secondary precipitation reaction occurs in

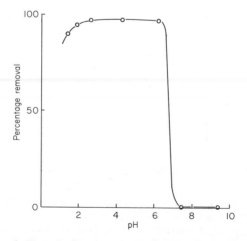

FIG. 5. The effect of pH on the flotation of a $Cu_2[Fe(CN)_6]$ precipitate. [From Davis and Sebba (1967).]

which $Cu(OH)_2$ forms on the particles. Hydroxyl ions are then adsorbed on this layer, giving it a negative charge and causing a sharp decline in flotation.

3. The pH should be such that the collector is in the desired state of ionization. For example, amines and weak acid collectors will not be ionized in solutions of high or low pH, respectively.

4. Flotation may be suppressed by the increased ionic strength which arises on adjusting the pH to extreme values.

5. The stability of the foam supporting the precipitate may change with pH, leading to redispersion.

As the pH of the solution can influence the process in a number of ways, its control is of considerable importance.

D. Influence of Ionic Strength

Rubin *et al.* (1966) have concluded that the flotation of copper hydroxide with sodium dodecylsulfate is not affected by increases in the ionic strength of the solution. This is a surprising result as the attachment of collector to the particles occurs by Coulombic attraction and should be reduced by the presence of added salts. Flotation is then impaired because of a decrease in the hydrophobic nature of the surface.

Davis and Sebba (1967) also experienced no difficulty in floating copper hexacyanoferrate(II) in the presence of added quantities of divalent cations, but as the concentrations of the latter did not exceed 4×10^{-4} M, the effects probably went unnoticed.

Sheiham and Pinfold (1968) found the flotation of strontium carbonate to be markedly impaired by increases in ionic strength. The reductions in recoveries on adding 4×10^{-2} M solutions of univalent cations are shown in Table I. A period of 20 min was allowed between the addition of precipitant and that of the collector, as it was shown that precipitation was appreciably slower in the presence of the added salt and was not complete in under 20 min.

TABLE I

EFFECTS OF INCREASED IONIC STRENGTH
ON PERCENTAGE RECOVERIES OF Sr^{2+} IONS

Cation	Concentration of cations (M)	
	4×10^{-3}	4×10^{-2}
Li^+	97%	86%
Na^+	98	84
K^+	99	86
Cs^+	99	90

Reductions in recoveries probably arise for the following reasons:

1. The attachment of collector to the precipitate particles is less secure. A zeta potential exists because of the CO_3^{2-} ions adsorbed on the precipitate, and is responsible for attracting and holding the surfactant ions. When its value is reduced by increased ionic strength, however, attachment is less secure.

2. Flotation of the collector is more rapid. A probable reason for this is described by Sheiham and Pinfold (1970) and involves a reduction in repulsions between bubbles which enables them to leave the solution more easily.

3. There is an increase in drainage from the foam. The stability of the foam is partially due to repulsion of positive surfactant ions adsorbed on the inside and outside surfaces of the bubble film. The increased concentration of anions in the film reduces this repulsion and allows them to become thinner; the bubbles are thus more susceptible to rupture, and drainage from them is more rapid. In consequence, redispersion occurs more freely and sometimes prevents complete recovery.

The effects of increased ionic strength should be examined more thoroughly with a variety of precipitates and collectors, as such repressions of flotation may impose limitations on the use of the technique.

E. Kinetics

Only Rubin (1968) has attempted a kinetic study of precipitate flotation of the first kind, and concludes that the process does not follow any of the usual rate laws, not even as an approximation. He has used the equation

$$\log(M - R) = \log B - m \log t, \tag{1}$$

where m and B are constants, mB is the rate constant, and M and R are the fractions removed finally and after time t, respectively. Data for the flotation of copper hydroxide using sodium dodecylsulfate satisfy the above expression quite well, but being a logarithmic relationship and relatively insensitive to changes, this may not be significant.

As removals by precipitate flotation are rapid and virtually complete within a few minutes, kinetic studies are difficult and may require special techniques.

F. Uses

Precipitate flotation of the first kind has been used in two ways, either directly by precipitating the ion to be removed, or indirectly by coprecipitating it with larger amounts of other ions and then floating the resulting precipitates. Examples of the former case nearly all involve hydroxides; separations are effected by controlling the pH and causing some elements to precipitate while others do not. However, more use has been made of the second alternative, mainly for the purpose of removing radioactive ions from reactor effluents. Besides the investigations of Pushkarev et al., described in Section I under the historical development of the subject, similar studies have been made by Koyanaka. In 1965 and 1966, isotopes such as ^{95}Zr, ^{95}Nb, ^{106}Ru, ^{106}Rh, and ^{141}Ce were removed by precipitation in iron(III) and cobalt(II) hydroxides, which were subsequently floated using sodium oleate. Removals were in excess of 98% when a single operation was used; with four successive flotations on the same solution, recoveries greater than 90%, and decontamination and volume reduction factors of 10^3–10^5 and 100, respectively, were achieved.

In 1967 and 1969 Koyanaka developed a preferential flotation method in which some elements in a mixture were coprecipitated and floated with certain reagents, the remainder being removed similarly in subsequent stages. Mixtures such as ^{90}Sr, ^{95}Zr, ^{106}Ru, ^{137}Cs, and ^{144}Ce could be partially separated by coprecipitating ^{137}Cs with $Cu_2[Fe(CN)_6]$ and collecting with octadecylamine acetate, then removing ^{95}Zr, ^{106}Ru, and ^{144}Ce with $Fe(OH)_3$ and sodium oleate at pH 6.0–8.0, and finally ^{90}Sr with $Fe(OH)_3$ and sodium oleate at pH 9.5–10.5. Calcium phosphate was also used as a carrier in similar separations.

The technique of coprecipitation and flotation appears to be quite promising as a means of removing trace elements. Sheiham and Pinfold (1968) have removed 90% of ^{89}Sr from solutions of 10^{-8} M by precipitating in calcium carbonate and floating with cationic collectors. They have suggested the name coflotation for such separations.

III. PRECIPITATE FLOTATION OF THE SECOND KIND

The only publications in this field are those of Mahne and Pinfold (1966, 1968a, b, 1969). Their investigations allow a reasonable assessment of the process to be made and show that it has many advantages over precipitate flotation of the first kind. No surface-active reagents are required, but the solid formed when colligend and collector (precipitant) ions react must have a hydrophobic surface. It is believed that this occurs because the precipitant ion has both polar and nonpolar parts. When present in aqueous solution, the influence of the former predominates and the ion is hydrophilic and surface inactive. On precipitation with colligend, however, the charges are neutralized, leaving the nonpolar groups to play the dominant role and to give the newly formed particle a hydrophobic exterior. The precipitants that have been used, and the elements that have been floated by them, are as follows: benzoinoxime, Cu; benzoylacetone, U; α-furildioxime, Ni; 8-hydro-xyquinoline, Cu, Zn; nioxime, Ni, Pd; α-nitroso-β-naphthol, Ag, Co, Pd, U; phenyl-α-pyridylketoxime, Au.

A. Influence of Various Parameters

That the collector is not surface active, and therefore does not adsorb on the bubbles, is the most significant feature of the process. No competition arises between precipitate and precipitant for positions on the bubbles, and flotation is not impaired by large excesses of collector. In fact it is often advantageous to raise the concentration of collector to the point where it itself precipitates. Particles of collector and of collector–colligend product then float together, the former supporting the latter on the surface and reducing redispersion.

A further consequence of the absence of adsorbed ions on the bubbles is that the latter do not repel each other, and hence they coalesce fairly readily. This results in the formation of larger bubbles which have little difficulty in leaving the solution, carrying with them their adsorbed precipitate. The process is consequently more rapid than either ion flotation or precipitate flotation of the first kind, and requires only 3 min to float the ions from 10^{-5} M solutions under optimum conditions.

Another significant feature is that the process is independent of ionic strength. Removals of 98% of Ni^{2+} ions from 1.5×10^{-5} M solutions, for example, are possible even in the presence of a 10,000-fold excess of strontium chloride. This is possible because the hydrophobic nature of the particle surface is independent of the presence of a surfactant and is therefore not influenced by the intervention of ions.

As with precipitate flotation of the first kind, recoveries improve on (1) raising the temperature in the range 20–40°C, (2) introducing an induction time between precipitation and flotation, and (3) introducing the collector in an ethanol solution. In addition, an optimum pH range exists for each system, values outside of which cause undesirable changes in the nature of the precipitant. Excellent separations can be achieved by the use of a collector which precipitates one ion from a solution but not others. No flotation of the latter ions can occur because hydrophobic properties only manifest themselves from the time the precipitate is formed.

Two main disadvantages can be cited. First, the collectors are probably impracticable for use in industry although they are usually available in the laboratory. Second, the absence of a surfactant ensures that no supporting foam is present on the surface of a solution. Redispersion then often goes unchecked, particularly if the precipitate is dense and finely divided. It may sometimes be overcome even in such cases by using an excess of collector as explained above or by utilizing the technique of extractive solvent sublation described in Section III,C.

B. Bubble Measurements

The average diameters and total surface areas of the bubbles passed during flotations were determined by taking photographs through a microscope mounted next to the cell. The technique differed from that used for ion flotation studies (Section III,G, Chapter 4) in that the flash illumination came from behind the cell rather than from above it. The bubbles therefore were silhouetted against the light and their profiles were consequently sharper than in the original method. Fewer of them were visible on each photograph, however, and a greater number of exposures had to be made.

The results show that, unlike flotations performed in the presence of a surfactant, the distribution of bubble diameters changes with increasing height above the frit, favoring larger bubbles. As has been explained in Section III,A, coalescence is responsible for this increase, and it arises because no appreciable zeta potential exists at the surfaces of the bubbles in the absence of surface-active ions, and mutual repulsions are not appreciable. The surface areas passed are large nevertheless, and do not differ greatly from comparable systems that contain surfactants.

C. Extractive Solvent Sublation

One method examined as a possible means of avoiding redispersion was to spread a thin layer of an immiscible organic solvent, such as 2-octanol or benzene, on the surface of the water. Sebba (1962a) had introduced this method to collect material removed by ion flotation, and had called it solvent sublation. The floated material either dissolved in the organic layer or remained in it as a suspension. The solids removed by precipitate flotation of the second kind, however, are not sufficiently soluble to dissolve and not hydrophobic enough to remain suspended in such a solvent. Most of the particles gather at the interface and are then even more susceptible to redispersion than before.

It was found that if a very small quantity of surfactant was dissolved in the organic layer, solubilization of the precipitate occurred and the suspension became more stable. The effect is shown in Fig. 6 for the flotation of silver by α-nitroso-β-naphthol into benzene. After 4 min the removal of silver had only reached 73% and began declining because of redispersion. A few milliliters of benzene containing a cationic collector were then added to the organic layer and within a short time removals increased to satisfactory values. The surfactant used must be of a charge opposite to that of the ions adsorbed on the precipitate. As an excess of α-nitroso-β-naphthol was present, the particles were negatively charged and needed cationic surfactants for their solubilization.

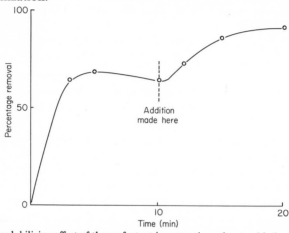

Fig. 6. The solubilizing effect of the surfactant in extractive solvent sublation. [From Pinfold and Mahne (1969).]

It has been found that in some cases solubilization is also achieved by dissolving either a complexing agent or the precipitant in the benzene layer. In the above case, for example, salicylaldoxime, which forms a complex with Ag^+ ions, was as effective as the surfactant. Similarly, nioxime dissolved in benzene was found to solubilize nickel nioximate precipitates. The mechanisms by which these occur is not clear but then little is understood generally about solubilizations in nonaqueous media.

Extractive solvent sublation overcomes redispersion in a number of cases but not in others, and it is clear that the problem has not yet been satisfactorily solved.

IV. CONCLUSION

Precipitate flotation has many attractive aspects and warrants a more thorough investigation than it has received to date. The mechanisms of the process are probably relatively simple and the descriptions of them may well be correct, but little has actually been proved. With the large number of precipitants that are available, the field should be extensive and may well afford some interesting study.

REFERENCES

Aoki, N., and Sasaki, T. (1966). *Bull. Chem. Soc. Japan* **39**, 939.
Baarson, R. E., and Ray, C. L. (1963). *Met. Soc. Conf. (Proc.)* **24**, 656.
Cardoza, R. L. (1963). *European At. Energy Comm. EURATOM* 262e.
Cardoza, R. L., and De Jonghe, P. (1963). *Nature* **199**, 687.
Davis, B. M., and Sebba, F. (1966). *J. Appl. Chem.* **7**, 40.
Davis, B. M., and Sebba, F. (1967). *J. Appl. Chem.* **17**, 41.
Fukuda, K., and Mizuike, A. (1968). *Bunseki Kagaku* **17**, 319.
Grieves, R. B., Bhattacharyya, D., and Conger, W. L. (1969). *Chem. Eng. Progr. Symp. Ser.* **65**, 29.
Kanaoka, K., and Iwata, S. (1960). Japan. Pat. 3657 (1963).
Kadota, M., and Matsuno, T. (1960). Japan. Pat. 13454 (1962).
Karger, B. L., Grieves, R. B., Lemlich, R., Rubin, A. J., and Sebba, F. (1967). *Separ. Sci.* **2**, 401.
Koyanaka, Y. (1965). *Nippon Genshiryoku Gakkaishi* **7**, 621.
Koyanaka, Y. (1966a). U.S. At. Energy Comm. KURRI-TR-21.
Koyanaka, Y. (1966b). *Radioisotopes* **15**, 77.
Koyanaka, Y. (1966c). U.S. At. Energy Comm. KURRI-TR-19.
Koyanaka, Y. (1966d). U.S. At. Energy Comm. KURRI-TR-24.
Koyanaka, Y. (1967). Rep. Japan At. Energy Res. Inst. NSJ-TR-90.
Koyanaka, Y. (1969a). *Nippon Genshiryoku Gakkaishi* **11**, 198.
Koyanaka, Y. (1969b). *J. Nucl. Sci. Tech. (Tokyo)* **6**, 607.
Lemlich, R. (1968). *Ind. Eng. Chem.* **60**(10), 16.
Lusher, J. A., and Sebba, F. (1966). *J. Appl. Chem.* **16**, 129.
Mahne, E. J., and Pinfold, T. A. (1966). *Chem. Ind. (London)* **30**, 1299.
Mahne, E. J., and Pinfold, T. A. (1968a). *J. Appl. Chem.* **18**, 52.
Mahne, E. J., and Pinfold, T. A. (1968b). *J. Appl. Chem.* **18**, 140.
Mahne, E. J., and Pinfold, T. A. (1969). *J. Appl. Chem.* **19**, 57.
Mukai, S., and Koyanaka, Y. (1968). *Suiyokai-Shi* **16**, 530; *Chem. Abstr.* **70**, 63195 (1969).
Mukai, S., Tsutsui, T., and Koyanaka, Y. (1969). *Mem. Fac. Eng. Kyoto Univ.* **31**, 115.
Oglaza, J. and Siemaszko, A. (1969). *Nukleonika* **14**, 275.
Pinfold, T. A. (1970). *Separ. Sci.* **5**, 379.
Pinfold, T. A., and Mahne, E. J. (1969). *J. Appl. Chem.* **19**, 188.
Pushkarev, V. V., Skrylev, L. D., and Bagretsov, V. F. (1960). *Zh. Prikl. Khim.* **33**, 81.
Pushkarev, V. V., Bagretsov, V. F., and Puzako, V. C. (1964a). *Radiokhimiya* **6**, 120.
Pushkarev, V. V., Egorov, Y. V., Tkachenko, E. V., and Zoltavin, V. L. (1964b). *At. Energ.* **16**, 48.
Rubin, A. J. (1968). *J. Amer. Water Works Assoc.* **60**, 832.
Rubin, A. J., and Johnson, J. D. (1967). *Anal. Chem.* **39**, 298.
Rubin, A. J., Johnson, J. D., and Lamb, J. C. (1966). *Ind. Eng. Chem. Process Des. Develop.* **5**, 368.
Sebba, F. (1962). "Ion Flotation". American Elsevier, New York.
Sebba, F., Jonaitis, C. W., Ray, C. I., and Baarson, R. E. (1962). Ger. Pat. 1,175,622.
Sebba, F., Jonaitis, C. W., Ray, C. L., and Baarson, R. E. (1965). U.S. Pat. 3,476,553.
Sheiham, I., and Pinfold, T. A. (1968). *J. Appl. Chem.* **18**, 217.
Sheiham, I., and Pinfold, T. A. (1970). Unpublished manuscript.
Skrylev, L. D., and Mokrushin, S. G. (1961). *Zh. Prikl. Khim. (Leningrad)* **34**, 2403.
Skrylev, L. D., and Pushkarev, V. V. (1962). *Kolloid. Zh.* **24**, 738.

Svoronos, J. J. (1963). Brit. Pat. 975,860.
Voznesenskii, S. A., Serada, G. A., Baskov, L. I., Tkachenko, E. V., and Bagretsov, V. F. (1961). *Kernenergie* **4**, 316.
Wainai, T., and Susuki, Y. (1968). *Bunseki Kagaku,* **17**, 319.

CHAPTER 6   PRINCIPLES OF MINERAL FLOTATION

D. W. Fuerstenau and T. W. Healy*
Department of Materials Science and Engineering
University of California
Berkeley, California

*Present Address: Department of Physical Chemistry, University of Melbourne, Parkville, Victoria, Australia.

I. INTRODUCTION

The froth flotation process for the recovery of valuable minerals from ores represents the oldest and the largest practical adsorptive bubble separation process. Froth flotation is that unit operation which is used to separate one kind of particulate solid from another through their selective attachment to air bubbles in an aqueous suspension. The volume to commemorate the 50th anniversary of froth flotation (Fuerstenau, 1962) shows very clearly how the vast national mineral development of the United States, Canada, Australia, Africa, and the U.S.S.R. began with the introduction of froth flotation. Each day in the United States, for example, the unit operation of froth flotation is responsible for the treatment of nearly one million tons of materials (Fuerstenau, 1962). The source of most of the common base metals are the flotation concentrates of metal sulfides, PbS, $CuFeS_2$, ZnS, etc., and continuous flotation produces large tonnages of these materials economically by fairly standard techniques. In recent years however, there has been an increasing need by modern technology for the so-called nonmetallic minerals, including the ceramic oxides, silicates, and clays, the phosphate fertilizers, iron ores, the high-temperature refractories, nuclear minerals, and others. Consequently, there is an increasing need to provide these important raw materials by such methods as selective flotation separation. In the United States in 1960, for example, phosphate materials made up one-third of the total 21 million tons of all flotation concentrates produced (Fuerstenau, 1962). This was almost twice the tonnage of any other single material produced by flotation treatment.

The present and the rapidly expanding use of nonmetallic flotation in mineral processing has stimulated research into the fundamental surface chemistry of selective flotation separation. Since, for example, the valuable constituent and the gangue in certain nonmetallic flotation separations may both be silicates or both oxides, detailed knowledge of the interactions at the mineral–water interface is required in order to effect selective separation.

Flotation science and technology has two main areas of concern : (1) finding proper physical chemical conditions for achieving appropriate selectivity between mineral species, and (2) determining conditions for optimum process kinetics. In this chapter, the principles upon which the selective separation of particulate solids can be achieved by froth flotation will be developed primarily in terms of the results of recent advances in the surface chemistry of nonmetallic mineral systems. Particular emphasis will be placed on the principles involved in making solid particles selectively hydrophobic. Also, factors that control the kinetics of bubble–particle attachment will be discussed.

II. MEASUREMENT OF FLOTATION BEHAVIOR

First of all, in the literature of froth flotation there has developed a standard terminology to describe components and units in the process. The common flotation technology terms are the following:

1. A *collector* is the surface active agent (e.g., fatty acid, long-chain sulfate, sulfonate, amine) that is added to the flotation pulp where it is selectively adsorbed on the surface of the desired mineral to render it hydrophobic. An air bubble preferentially adheres to such a surface and concentrates the mineral into the froth layer at the top of the cell.

2. A *frother* is again a surface active agent (e.g., short-chain alcohol, aromatic alcohol, ethylene oxide surfactant), added to the flotation pulp primarily to stabilize the froth in which the hydrophobic minerals are trapped. When long-chain soaps or amines are used as collectors, a frother may not be necessary.

3. An *activator* is a material added to promote selective adsorption of the collector on a particular mineral. Metal ions, for example, can activate the anionic surface of the mineral quartz (Fuerstenau et al., 1963) so that anionic collectors (e.g., alkyl sulfonates) can adsorb and induce flotation.

4. A *depressant* is a material added to prevent the collector from adsorbing on a particular mineral in the mixture in the flotation pulp. Metal ions, for example, may act as depressants by preventing adsorption of a cationic surfactant on an otherwise anionic surface of a particular mineral. Sodium silicate can depress minerals that tend to respond to anionic collectors.

In the actual flotation of complex ores in practice, the operator must optimize the effect of a large number of interacting variables. In order to understand the basic chemistry of the unit operation of flotation, and in particular the surface chemistry that is so critical in obtaining selective separation, a laboratory flotation device is needed in which chemical and mechanical variables can be closely controlled. Such a device, known as the Hallimond tube was originally designed by Hallimond (1944, 1945), revised first by Ewers (Sutherland and Wark, 1955) and by Fuerstenau et al. (1957) until it has evolved to its present design shown in Fig. 1. The mineral of interest is first conditioned in the absence of air with the reagents to be studied and the solution and mineral is poured into the tube so that the mineral settles onto the sintered glass disk at the base of the tube. A small magnetic stirrer is used to insure uniform mixing of the particles with the incoming gas bubbles. A controlled volume of nitrogen (or other gas) is passed at a controlled rate through the sintered glass disk and into the agitated bed of mineral. The

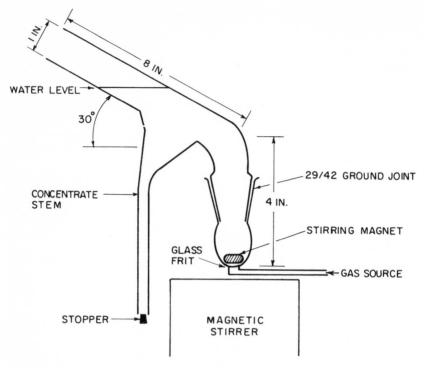

FIG. 1. Schematic drawing of the modified Hallimond tube which can be used for controlled testing of flotation behavior. [After Fuerstenau *et al.* (1957).]

bubbles rise with their load of particles and since no frother is used the bubbles burst at the water surface. The "concentrate" so formed then drops back into the side arm and can be recovered at the end of an experiment, weighed, and compared with the weight of unfloated mineral. By keeping the gas flow and stirring rates constant and varying the amount of collector, for example, the response of the mineral to the collector can be evaluated.

More recently, Fuerstenau (1964) devised a small-scale froth flotation cell to which frother is added to maintain a stable froth layer on the liquid. Because of the presence of the froth, flotation conditions more nearly approximate practical operation. As a physical chemical research tool, however, this method may be somewhat less satisfactory than the modified Hallimond tube because the frother molecules may adsorb cooperatively with the collector.

The "bubble-pickup" technique has also been of value, again with all conditions held constant except that one variable under consideration (Cooke, 1949). Essentially it consists of pressing a captive bubble against a bed of particles and then counting or weighing the load of particles attached to the bubble.

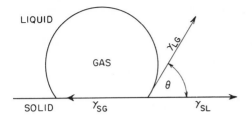

FIG. 2. Schematic representation of the equilibrium contact between an air bubble and a solid immersed in a liquid. The contact angle θ is the angle between the liquid–gas and liquid–solid interfaces, measured through the liquid.

In vacuum flotation studies (Schuhmann and Prakash, 1950) the air or gas saturated mineral pulp is subjected to partial vacuum to form bubbles in site. It is in essence a measure of "incipient flotation" and must be interpreted as such (Somasundaran and Fuerstenau, 1968). This technique is used to evaluate the limits for flotation.

Measurement of contact angles has been a standard research technique in flotation studies over many years and a great deal of information has been obtained by the technique (Sutherland and Wark, 1955). A captive or sessile bubble is placed against the polished surface of the solid and the contact angle (see Fig. 2) is measured with a goniometer microscope or from a photographic image. In the past very little analysis or use has been made of the difficult question of hysteresis of the contact angle, namely whether one is considering the advancing (θ_A) or receding (θ_R) contact angle, and which of these approximates the "equilibrium" contact angle. Some recent studies by Gaudin *et al.*

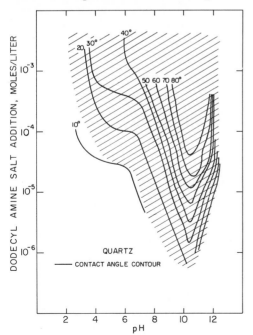

FIG. 3. An isogonic map for the contact angle of quartz as a function of pH for the addition of dodecyl amine salt (acetate or chloride salts). The shaded portion shows the region in which flotation is possible; the minimum contact angle for flotation is slightly greater than 10°. [After Smith (1963).]

(1964) do attempt to use the hysteresis value ($\theta_R - \theta_A$) to characterize air–water–solid and oil–water–solid contact conditions. A fairly standard representation of contact angle results is in the form of a contact/no contact domain diagram (Sutherland and Wark, 1955). Other workers have attempted to add more detail by including an intermediate region variously labeled as partial contact or "cling" (Gaudin and Morrow, 1957; Sagheer, 1966). Smith (1963) has published a contact angle contour diagram showing the effect of collector concentration and pH on the contact angle of quartz. On this diagram (Fig. 3) the results of a vacuum flotation study on the same material are also included. In the shaded area flotation is possible, which indicates that a minimum contact angle of approximately 10° is required for flotation in this system, and also shows solution conditions required to attain this condition. Actual flotation recovery contour diagrams are also useful to assess the complex practical behavior of a given system in detail (Iwasaki, 1965).

III. EQUILIBRIUM CONSIDERATIONS IN FLOTATION SYSTEMS

The froth flotation process is controlled by two distinct sets of variables:

1. Thermodynamic or equilibrium considerations involving solid, liquid, and gas phases with solute components such as the surfactants (both collector and frother), activators, depressants, and "contaminant" species from the ore.

2. Kinetic considerations involving bubble–particle collision in the turbulent pulp, thinning of the interfacial film to allow bubble–particle contact, shear forces acting at the line of bubble–particle contact, and so on.

Generally, the kinetics of reagent adsorption is not significant in flotation technology and adsorption kinetics will not be discussed.

A. Thermodynamics of Wetting

The present section will be concerned with the thermodynamics or equilibrium which establish the energy considerations that make bubble–particle contact possible.

The general thermodynamic condition for three-phase contact is defined by Young's equation for the system depicted schematically in Fig. 2:

$$\gamma_{SG} = \gamma_{SL} + \gamma_{LG} \cos \theta, \tag{1}$$

where the γ terms are the interfacial tensions of the solid–gas, solid–liquid, and liquid–gas interfaces, respectively, and θ is the contact angle. The free energy change, on a unit area basis, for the attachment of a bubble to a mineral particle in an aqueous suspension is given by

$$\Delta G = \gamma_{SG} - (\gamma_{SL} + \gamma_{LG}) \qquad (2a)$$

or applying Young's equation,

$$\Delta G = \gamma_{LG}(\cos \theta - 1). \qquad (2b)$$

Thus, for any finite value of the contact angle, there will be a free energy decrease upon attachment of a mineral particle to an air bubble.

Because of the polar nature of most mineral surfaces, clean solids are wet by water and therefore the contact angle is nil. Only a few solids are naturally hydrophobic and therefore respond to flotation without adding a collector. These solids include such materials as graphite, sulfur, talc, and pyrophyllite (which are silicate minerals), molybdenite and stibnite (which are sulfide minerals), and the organic polymers.

The wettability of solids can be controlled through the adsorption of collectors, which can change the interfacial tensions so that the contact angle becomes finite. The change in the interfacial tension at any interface is given by the Gibbs equation, at constant T and P:

$$d\gamma = - \sum_{\text{all } i} \Gamma_i \, d\mu_i, \qquad (3a)$$

where Γ_i is the surface excess (adsorption density) of species i in moles per square centimeter, and μ_i is the chemical potential of species i. This can also be written as

$$d\gamma = - RT \sum_{\text{all } i} \Gamma_i \, d \ln a_i, \qquad (3b)$$

where a_i is the activity of species i, and R is the gas constant.

The control of wettability in flotation systems can be interpreted in terms of Eqs. (1) and (3). To illustrate the kind of result that comes from Eqs. (1) and (3), let X be a surfactant species in water which is present at all three interfaces (de Bruyn *et al.*, 1954; Aplan and de Bruyn, 1963; Smolders, 1961). The solvent H_2O is considered to be uniform right up to the surface such that $\Gamma_{H_2O} = 0$ (Gibbs' convention) and the adsorption of X is referred to this standard state, that is, for each interface

$$d\gamma = - RT\Gamma_X \, d \ln a_X \qquad (4)$$

and together with Eq. (1),

$$\Gamma_X^{SG} = \Gamma_X^{SL} + \cos \theta \, \Gamma_X^{LG} - (\gamma_{LG}/RT)(d \cos \theta / d \ln a_X). \qquad (5)$$

Several general observations are pertinent to this relationship (Smolders, 1961). For example, it is observed that with increasing surfactant concentration, $\cos \theta$ decreases and Γ_X^{SL} increases. Now for most flotation systems γ_{LG} is unaffected by the low concentrations of surfactant used, and hence

$$\Gamma_X^{SG} \simeq \Gamma_X^{SL} + (\text{constant})(d \cos \theta / d \log C_X), \qquad (6)$$

where activity can be replaced by concentration since C_X is usually much less than $10^{-3} M$ or $10^{-2} M$. The second term on the right-hand side of Eq. (6) is negative to yield the general condition for stable bubble contact in a solid–liquid–gas system, namely that $\Gamma_X^{SG} > \Gamma_X^{SL}$ for finite contact (Smolders, 1961). This is illustrated in the results of Smolders (1961) for the system mercury–hydrogen gas–aqueous solutions of sodium dodecylsulfate (SDS). The saturation adsorption density at the Hg–gas interface was 4.4×10^{-10} mole/cm^2 while the adsorption density at the Hg–aqueous interface was 1.6×10^{-10} mole/cm^2.

This general result, namely, that for finite contact Γ_X^{SG} is greater than Γ_X^{SL}, is of importance in several areas of flotation science. It emphasizes the fact that adsorption at all interfaces must be considered (de Bruyn et al., 1954; Aplan and de Bruyn, 1963; Somasundaran, 1968a, b; Digre and Sandvik, 1968), and that the contact angle will depend on the adsorption densities of all surface-active species, e.g., both the collector and the frother. In fact, one author (Wada et al., 1969) has shown that it is advantageous to add the collector and frother to the dry solid as an aerosol in order to promote solid–gas interfacial saturation before the mineral is immersed in water.

In the following sections, attention is focused on the role of the structure of the surface-active agent in adsorption and flotation. Since nearly all flotation reagents are added to the aqueous phase, then adsorbed by the mineral before air bubbles are introduced, it is necessary to consider the detailed structure of the interface and then the mechanism of surfactant adsorption at that interface.

B. The Electrical Double Layer at Mineral-Water Interfaces

The process of immersing a solid into an aqueous solution is known to produce a region of electrical inhomogeneity at the solid–solution interface. An excess $(+$ or $-)$ charge apparently fixed at the solid surface is exactly balanced by a diffuse region of equal but opposite charge on the liquid side. This region, the surface charge and its counterions, is called the electrical double layer.

Since adsorption of collectors at mineral–water interfaces are controlled in many cases by the electrical double layer, we must be concerned with factors responsible for the surface charge on the solid and with the behavior of ions that adsorb as counter ions to maintain electroneutrality. Figure 4 is a schematic representation of a simple model of the electrical double layer at a

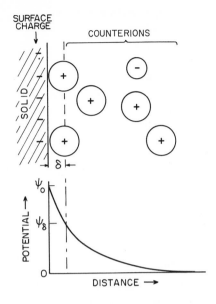

SURFACE
CHARGE

COUNTERIONS

SOLID

ψ_0

ψ_δ

POTENTIAL →

DISTANCE →

FIG. 4. Schematic representation of the electrical double layer and the potential drop through the double layer at a mineral–water interface.

mineral–water interface (Kruyt, 1952; Parsons, 1954), showing the charge on the solid surface and the diffuse layer of counterions extending out into the aqueous phase. This figure also shows the drop in potential across the double layer. The closest distance of approach of counterions to the surface δ is called the Stern plane. The double-layer potential or the surface potential is ψ_0, and that at the Stern plane is ψ_δ; from the Stern plane out into the bulk of the solution, the potential drops exponentially to zero. As will be discussed in more detail later, the electrokinetic or zeta potential is taken as an approximation of ψ_δ.

In the case of the ionic solids such as AgI, Ag_2S, $BaSO_4$, CaF_2, $CaCO_3$, and others, the surface charge can arise when there is an excess of one of the lattice ions at the solid surface. Equilibrium is attained when the electrochemical potential of these ions is constant throughout the system. Those particular ions which are free to pass between both phases and therefore are able to establish the electrical double layer are called *potential-determining ions*. In the case of AgI, the potential-determining ions are Ag^+ and I^-. For a solid such as calcite, $CaCO_3$, the potential-determining ions are Ca^{2+} and CO_3^{2-}, but also H^+, OH^-, and HCO_3^- because of the equilibria between these latter ions and CO_3^{2-}.

For oxides, hydrogen and hydroxyl ions have long been considered to be potential determining (Fuerstenau, 1962; Kruyt, 1952) although the mechanism as to how pH controls the surface charge on oxides is still speculative. It is well known that oxides possess a hydroxylated surface in the presence of water. Adsorption–dissociation of H^+ from the surface hydroxyls can account for the surface charge on the oxide by the following mechanism:

$$MOH_{(surf)} \rightleftharpoons MO^-_{(surf)} + H^+_{(aq)}$$

$$MOH_{(surf)} + H^+_{(aq)} \rightleftharpoons MOH_2^+{}_{(surf)}$$

Thus, in the case of an oxide in water, changing the pH of the solution will markedly affect the magnitude and even the sign of the surface charge.

The single most important parameter that describes the electrical double layer of a mineral in water is the *point of zero charge*, pzc. The pzc is expressed as the particular value of the activity of the potential-determining ions, $(a_{M^+})_{pzc}$ or $(a_{A^-})_{pzc}$, which cause the surface charge to be zero. Assuming that potential differences due to dipoles, etc., remain constant, the total double-layer potential or the surface potential, ψ_0, is considered to be zero at the pzc (Kruyt, 1952). The value of the surface potential at any activity of potential-determining electrolyte is given by

$$\psi_0 = (RT/zF) \ln[(a_{M^+})/(a_{M^+})_{pzc}], \tag{7}$$

where F is the Faraday constant, T is the temperature in $°K$, and z is the valence of ion under consideration. For silver iodide, M^+ and A^- are simply Ag^+ and I^-; for an oxide M^+ and A^- are H^+ and OH^-. Table I presents the pzc's of a number of ionic (salt-type) materials that have been investigated. Calcite, fluorite, and barite are positively charged in their saturated solution at neutral pH, whereas the others listed are negative, except for hydroxyapatite which is uncharged at this pH.

TABLE I

THE POINT OF ZERO CHARGE OF SOME IONIC SOLIDS

Material	pzc	Ref.
Barite, $BaSO_4$	pBa 6.7	Buchanan and Hayman, 1948
Calcite, $CaCO_3$	pH 9.5[a]	Somasundaran and Agar, 1967
Fluorapatite, $Ca_5(PO_4)_3(F,OH)$	pH 6[a]	Somasundaran, 1968c
Fluorite, CaF_2	pCa 3	Parks, 1967
Hydroxyapatite, $Ca_5(PO_4)_3(OH)$	pH 7[a]	Parks, 1967
Scheelite, $CaWO_4$	pCa 4.8	O'Connor, 1958
Silver chloride, $AgCl$	pAg 4	Kruyt, 1952
Silver iodide, AgI	pAg 5.6	Kruyt, 1952
Silver sulfide, Ag_2S	pAg 10.2	Freyberger and de Bruyn, 1957

[a]From the hydrolysis equilibria and solubility data, the activities of the other potential-determining ions can be calculated.

A detailed evaluation of pzc values of oxides and silicates has been given by Parks (1965, 1967) and some representative examples for oxides are given in Table II. The pzc values for oxides depend markedly on the history of the sample and its method of pretreatment. For a given pretreatment condition it has been shown (Healy and Fuerstenau, 1965; Healy *et al.*, 1966) for several pure inorganic oxides that the pH of the pzc increases as the crystal field of the lattice increases; qualitatively pzc increases from about pH 1-2 for amorphous solids to 6-7 for the covalent transition metal oxides and to 12 for the more ionic oxides. Thus, we shall see that pH should be a very important variable in effecting flotation separations between various oxides.

TABLE II
A SELECTION OF pzc VALUES OF
SOME INORGANIC OXIDES[a]

Material	pzc
SiO_2, silica gel	pH 1-2
SiO_2, α-quartz	pH 2-3
SnO_2, cassiterite	pH 4.5
ZrO_2, zirconia	pH 4
TiO_2, rutile	pH 5.8-6.7
Fe_2O_3, hematite (natural)	pH 4.8-6.7
Fe_2O_3, hematite (synthetic)	pH 8.6
FeOOH, goethite	pH 6.8
Al_2O_3, corundum	pH 9.1
AlOOH, boehmite	pH 7.8-8.8
MgO, magnesia	pH 12

[a]From Parks (1965, 1967).

In considering the double layer on complex aluminosilicates and clays it is important to point out that such minerals are frequently unstable in water in that Al(III) and to a lesser extent Si(IV) and other species leach out from the crystal into the bulk solution (Smolik *et al.*, 1966; Buchanan and Oppenheim, 1968). Fortunately however, the resultant surface is essentially an oxide and pH represents a convenient parameter once again with which to define the pzc. Thus, kaolinite and bentonite have pzc values, in this sense, at about pH 3. In the case of the layer silicates, that is the clays and the micas, because of the substitution of Al(III) for Si(IV) in the silica tetrahedra or Mg(II) for Al(III) in the octahedral layer of the crystal lattice, the surfaces of these crystal faces may carry a residual negative charge that is independent of solution conditions. This phenomenon is utilized in the flotation separation of mica from quartz (Fuerstenau, 1962).

The importance of the pzc is that the sign of the surface charge has a major effect on the adsorption of all other ions and particularly those ions charged oppositely to the surface because these ions function as the counterions to maintain electroneutrality. In contrast to the situation in which the potential-determining ions are special for each system, any ions present in the solution can function as the counter ions. If the counter ions are adsorbed only by electrostatic attraction, they are called *indifferent electrolytes*. As has been well established (Kruyt, 1952), the counterions occur in a diffuse layer that extends from the surface out into the bulk solution. The "thickness" of the diffuse double layer is $1/\kappa$, where κ is given by

$$\kappa = (8\pi F^2 z^2 C/\varepsilon RT)^{1/2}, \tag{8}$$

where ε is the dielectric constant of the liquid and z is the valence of the ions in the double layer. For a 1–1 valent electrolyte, $1/\kappa$ is 1000 Å in 10^{-5} M, 100 Å in 10^{-3} M, and 10 Å in 10^{-1} M solutions, for example.

The charge in the diffuse double layer σ_d given by the Gouy–Chapman relation (Kruyt, 1952) as modified by Stern (for a symmetrical electrolyte) is

$$\sigma_d = -\sigma_s = -[(2\varepsilon RT/\pi)C]^{1/2}\sinh zF\psi_\delta/2RT. \tag{9}$$

Further, if ψ_δ does not change appreciably, this equation shows that the adsorption density of counter ions should vary as the square root of the concentration of added electrolyte, as de Bruyn (1955) has found for the adsorption of dodecyl ammonium acetate on quartz at low concentrations.

On the other hand, some ions exhibit surface activity in addition to electrostatic attraction and adsorb strongly in the Stern plane because of such phenomena as covalent bond formation, hydrophobic bonding, hydrogen bonding, solvation effects, etc. Because of their surface activity, the charge of *such surface-active counterions* adsorbed in the Stern plane can exceed the surface charge. In such cases,

$$\sigma_s = -(\sigma_\delta + \sigma_d), \tag{10}$$

where σ_δ is the charge due to adsorption in the Stern plane δ. Flotation collectors generally function as surface-active counterions.

C. Experimental Measurement of Electrical Double-Layer Properties

1. Direct Measurement of Surface Charge

First of all, the total double-layer potential ψ_0 can be measured in principle directly by constructing an electrode of the mineral. However, except for mercury and a few simple conducting or semiconducting minerals (AgI, ZnO, Ag$_2$S), this technique is prohibited by the high electrical resistance of most minerals, especially the insulating oxides and silicates.

Alternatively it is possible to determine adsorption isotherms of the potential-determining ions, for example, by means of a potentiometric titration technique (Kruyt, 1952; Freyberger and de Bruyn, 1957; Parks and de Bruyn, 1962). The unique role of pH in determining, often indirectly, the double layer potential means that a glass electrode can be used to monitor the activity of H^+ and OH^- ions in solution during such a titration. Thus, since the surface charge σ_s is given by the *difference* in adsorption of positive and negative potential-determining ions, we can write

$$\sigma_s = zF(\Gamma_{M^+} - \Gamma_{A^-}). \tag{11}$$

This definition of surface charge requires that all M^+ and A^- be adsorbed onto the surface and not occur in the diffuse layer. To insure this, titration experiments must be carried out at high ionic strength (Kruyt, 1952). At the pzc the adsorption densities of potential-determining ions (M^+ and A^-) are equal. Figure 5 presents the results of such titrations of synthetic ferric

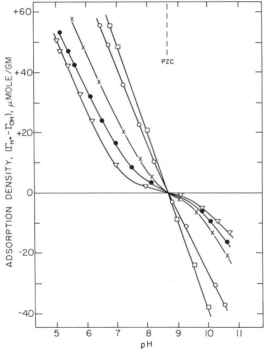

FIG. 5. The adsorption density of potential-determining ions on ferric oxide as a function of pH and ionic strength using KNO_3 as the indifferent electrolyte concentrations: ($\triangledown$) 10^{-4} M, ($\bullet$) 10^{-3} M, ($\times$) 10^{-2} M, ($\bigcirc$) 10^{-1} M, ($\square$) 10 M. [From G. A. Parks and P. L. de Bruyn, *J. Phys. Chem.* **66**, 967 (1962).]

oxide (hematite) with hydrogen and hydroxyl ions in the presence of KNO_3 as the supporting electrolyte (Parks and de Bruyn, 1962). This figure clearly shows that the surface charge on ferric oxide reverses its sign at pH 8.6 and that it increases in absolute magnitude with increasing ionic strength and increasing concentration of potential determining ion. Thus, at pH values below 8.6, the counter ions on this material will be anions, and above pH 8.6 they will be cations.

2. Electrokinetic Methods

The electrokinetic or zeta potential (ζ) is the potential at the plane of shear in the double layer when the solid is moved relative to the liquid. Care must be taken in interpreting the meaning of this double-layer potential, but it is useful to approximate ζ with ψ_δ. The major importance of electrokinetic methods is that they can be performed rapidly, and if indifferent electrolyte only is present, the pzc is that activity of potential-determining ions at which the electrokinetic potential is zero (Fuerstenau, 1962). Collectors generally are specifically adsorbed, and because of this strong adsorption, they can reverse the sign of ζ.

The two common methods (Kruyt, 1952) for evaluation of electrokinetic potentials are electrophoresis (which involves measurement of the rate at which fine particles migrate in an electric field) and streaming potential (which involves measurement of the potential set up when a liquid is forced through a bed of particles). Such methods have been widely used for research in flotation chemistry.

For understanding the flotation system in detail, it is fruitful to use a number of experimental methods in the same system. Figure 6 presents a summary of the results of several studies for the quartz–dodecylammonium acetate system, including measurement of flotation recovery with a Hallimond tube, contact angles, adsorption of collector with radiotracers, and zeta potentials with a streaming potential apparatus. This figure illustrates how measurements that reflect conditions at the solid–liquid interface (adsorption and zeta potential) can be correlated directly with surface phenomena that reflect complex conditions at the solid–liquid gas interfaces (contact angle and flotation). The significance of the shape of these curves will be discussed in detail later.

IV. ADSORPTION OF ORGANIC AND INORGANIC IONS AT MINERAL-WATER SURFACES

A. Properties of Surfactants in Solution

Collectors for nonmetallic minerals are long-chain electrolytes, either anionic or cationic. Both the hydrocarbon chain and the ionic head on the chain control the chemical and physical properties of the collector. The ionic head determines whether the collectors are strong electrolytes (e.g., sulfon-

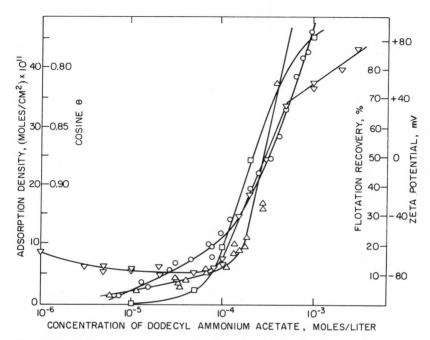

FIG. 6. Correlation among contact angle, adsorption density, flotation response, and zeta potential for quartz as a function of dodecylammonium acetate concentration at pH 6–7: (○) contact angle, (△) adsorption density, (▽) zeta potential, (□) flotation recovery. [After Fuerstenau *et al.* (1964).]

ates) which ionize completely in solution, or weak electrolytes which ionize only slightly or hydrolyze to form neutral molecules.

In the case of weak electrolytes, the dissociation constant is important because it determines the proportion of neutral molecules in the system. Soaps hydrolyze to yield carboxylic acids

$$RCOO^- + H^+ \rightleftharpoons RCOOH; \quad pK = 4.7 \text{ at } 25°C.$$

Primary amine salts hydrolyze in alkaline solutions to yield amines in solution according to the reaction

$$RHN_3^+ \rightleftharpoons RNH_2 + H^+; \quad pK = 10.7 \text{ at } 25°C.$$

Similarly, it will be shown that the hydrolysis of inorganic cations may have an appreciable effect in flotation systems.

Because of the tendency of hydrocarbon chains to be expelled from water, surfactant ions in aqueous solution tend to associate into micelles, namely, into clusters of ions with the charged heads oriented towards the water and the association chains thereby removed from the water. The balance between the free energy decrease upon removal of the chain from water and the free energy increase due to the electrostatic repulsion between the charged heads controls micelle size, and other factors. The length of the hydrocarbon chain is the dominant factor and controls the concentration in solution at which micelles begin to form. This concentration is called the critical micelle concentration, or cmc. For the various numbers of carbon atoms in the chains of aminium chlorides (Fuerstenau, 1962), the values of the cmc are as follows: 10C, 3.2×10^{-2} M; 12C, 1.3×10^{-2} M; 14C, 4.1×10^{-3} M; 16C, 8.3×10^{-4} M; and 18C, 4×10^{-4} M.

B. Adsorption of Inorganic Ions at Mineral-Water Interfaces

The excess charge σ_d in the diffuse layer can be calculated from Eq. (9); and from this same theory, it is possible to derive an equation (Kruyt, 1952) for the adsorption density Γ_{d+} of positively charged counterions (or an analogous equation for anions):

$$\Gamma_{d+} = \sigma_{d+}/zF = [(\varepsilon RT/2\pi)C]^{1/2}[\exp(zF\psi_\delta/RT) - 1]. \tag{12}$$

Under conditions where ψ_δ does not change appreciably, the adsorption density of indifferent counterions should be proportional to the square root of their bulk concentration.

In flotation, it is those ions adsorbed in the Stern plane which are of primary interest since they are the ions anchored to the solid. The adsorption density of counter ions in the Stern plane can be evaluated by the following relationship (Kruyt, 1952):

$$\Gamma_\delta = 2rC\exp(- \Delta G^\circ_{ads}/RT), \tag{13}$$

where r is the radius of the adsorbed counter ion, C is the number of moles of ions per cubic centimeter in bulk. Here Γ_δ is the adsorption density in moles per square centimeter in the Stern plane, and ΔG°_{ads} is the standard free energy of adsorption. If ions are adsorbed only because of electrostatic interaction, then the standard free energy of adsorption will be given by

$$\Delta G^\circ_{ads} = zF\psi_\delta. \tag{14}$$

On oxides, alkali cations appear to be surface inactive, that is, only electrostatic interactions appear to be operative. As for anions, nitrate ions have been shown to be surface inactive on SnO_2, TiO_2, Al_2O_3, and Fe_2O_3. Chloride ions are not surface active on Al_2O_3 but appear to be specifically adsorbed on Fe_2O_3.

When an ion exhibits surface activity, the standard free energy of adsorption has additional terms

$$\Delta G^\circ_{ads} = zF\psi_\delta + \Delta G^\circ_{spec}, \tag{15}$$

where ΔG°_{spec} represents the specific interaction terms. In terms of the model of the electrical double layer, it is possible to estimate ΔG°_{spec} at the reversal of the electrokinetic potential (i.e., when $\psi_\delta = 0$). A number of inorganic electrolytes are surface-active on oxides, for example, sulfate ions reverse the sign of ζ when the surface of alumina is positively charged but they lose their surface activity when the surface is negatively charged (Fuerstenau, 1962). On the other hand, barium ions are specifically adsorbed on negatively charged alumina but not on the positively charged solid (Fuerstenau, 1962). Other examples of the reversal of ζ by alkaline earth cations are barium ions on quartz at pH 7 (Gaudin and Fuerstenau, 1955b) and calcium ions on rutile at pH's ranging from 7 to 11 (Gaudin and Fuerstenau, 1955b). In the case of quartz, ζ reverses at 2×10^{-2} M Ba(II) at pH 6.5, whereas upon adding Al(III) to water initially at pH 6.5, ζ reverses at a concentration of 7×10^{-7} M. In the latter case, Al(III) was beginning to hydrolyze. During the last ten years there has been an extensive amount of research directed towards understanding the role of hydrolyzed metal ions in coagulation processes (Matijevic, 1967), and this work has clearly demonstrated that hydrolyzed cations exhibit exceptionally strong surface affinity.

At this stage, the mechanism by which inorganic ions acquire their surface activity is not understood. Some have suggested that chemisorption of such ions as Ba^{2+} on silica results from the formation of a barium silicate bond (Gaudin and Fuerstenau, 1955b). However, in the general sense, any multi valent inorganic ion can reverse the zeta potential of an oxide when the surface is charged oppositely to the cation; perhaps this merely results from local electric forces between the multivalent ion and the single site on the charged surface. In the case of hydrolyzed cations, surface activity must result from a combination of Coulombic and specific adsorption effects. The very great surface activity of hydrolyzed metal ions has been suggested to result from hydrogen bonding with surface hydroxyl and surface oxygen sites.

C. Ion Exchange in the Double Layer

The structure of the diffuse part of the double layer depends upon the composition of the bulk solution and any change in the composition of the bulk solution can cause ion exchange within the double layer (Kruyt, 1952). If two different kinds of counter ions, a and b, are present, then

$$\Gamma_a/\Gamma_b = K(C_a/C_b), \tag{16}$$

where Γ_a and Γ_b are the adsorption densities in the double layer, C_a and C_b are their respective bulk concentrations, and K is a constant. For indifferent counter ions such as K^+ and Na^+, K has a value close to unity (Kruyt, 1952). However, ions having a strong affinity for the surface will tend to concentrate in the double layer and thus specifically adsorbed ions should predominate in the double layer over simple indifferent electrolytes. Where appreciable adsorption occurs in the Stern layer, the competition for sites is given by the following relation:

$$\frac{\Gamma_a}{\Gamma_b} = \frac{C_a}{C_b} \exp\left[\frac{(\Delta G^\circ_{ads})_b - (\Delta G^\circ_{ads})_a}{RT}\right].$$ (17)

If a is an ion that is adsorbed only electrostatically (e.g., Na^+ on quartz), and b is an ion that has specific affinity for the surface, then the exponential term would simply be $\exp[(\Delta G^\circ_{spec})_b/RT]$.

D. Collector Adsorption at Mineral-Water Interfaces

The amphipathic nature of organic ions has a marked effect on their adsorption. The nature of the polar head on the molecule or ion controls whether there is any chemical interaction with the mineral, whereas the structure of the hydrocarbon chain determines the extent of its interaction with the aqueous media.

The standard free energy of adsorption of an organic ion in the Stern plane at a mineral–water interface is given by

$$\Delta G^\circ_{ads} = \Delta G_{elec} + \Delta G^\circ_{CH_2} + \Delta G^\circ_{chem} + \cdots,$$ (15a)

where $\Delta G_{elec} = zF\psi_\delta$, $\Delta G^\circ_{CH_2}$ represents the interaction due to association of hydrocarbon chains of adsorbed ions at the surface, ΔG°_{chem} represents the free energy due to the formation of covalent bonds with the surface, etc.

At this point, it may be of value to discuss terminology concerned with the mechanism of adsorption of organic ions at solid–water interfaces. If the ions are adsorbed only with such forces as electrostatic attraction and hydrophobic bonding (van der Waals interaction between the hydrocarbon chains), the process should be termed *physical adsorption*. If the surfactant forms covalent bonds with metal atoms in the surface, then the process should be termed *chemisorption*. Much of the detailed research that has been carried out on surfactant adsorption at mineral–water interfaces has been concerned with the two physical adsorption systems alkylammonium–quartz and alkyl sulfonate–alumina, in which the adsorption isotherms (de Bruyn, 1955) and electrokinetic (Gaudin and Fuerstenau, 1955a) behavior has strongly suggested that at low concentrations the surfactant ions are adsorbed as individual counter ions, but that at higher concentrations they associate through interaction of the hydrocarbon chains of those ions adsorbed in the Stern layer. These associated, adsorbed species (two-dimensional aggregates) at the surface have been termed *hemi-micelles* (Gaudin and Fuerstenau, 1955a). The sharp break upwards in the adsorption isotherm shown in Fig. 6 reflects hemi-micelle formation, which, because of the higher concentration at the interface, occurs at a bulk concentration of about 10^{-4} M dodecyl ammonium acetate, i.e., at about 1 % of the bulk critical micelle concentration for this material. Figure 6 also clearly shows that a sharp change and reversal in the zeta potential of quartz occurs upon formation of hemi-micelles, indicating strong adsorption in the Stern plane. Likewise there is a marked change in wettability, as indicated by the contact angle and flotation behavior.

Recently, adsorption in the alumina–alkyl sulfonate system has been studied in considerable detail (Somasundaran and Fuerstenau, 1966; Wakamatsu and Fuerstenau, 1968). Figure 7 presents the effect of the con-

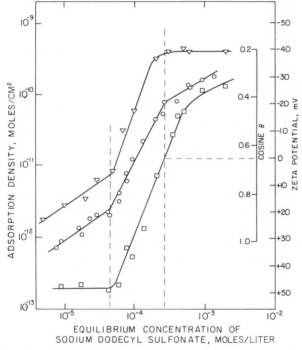

FIG. 7. The adsorption density of sodium dodecyl sulfonate on alumina, the zeta potential of alumina, and the contact angle on alumina as a function of the equilibrium concentration of sodium dodecyl sulfonate at pH 7.2 and ionic strength of 2×10^{-3} N at $24 \pm 1°$ C controlled with NaCl: ($\triangledown$) contact angle, ($\bigcirc$) adsorption density, ($\square$) electrophorytic mobility.

centration of dodecyl sulfonate on the electrophoretic mobility (zeta potential), adsorption density, and contact angle on alumina at pH 7.2 and at an ionic strength of 2×10^{-3} N controlled by NaCl (Wakamatsu and Fuerstenau, unpublished data). From this figure, it can be seen that these adsorption phenomena can be divided into three distinct regions. At low concentrations, adsorption of sulfonate ions occurs by exchange with chloride ions in the double layer; during the exchange, the zeta potential remains constant. In this region only the electrostatic adsorption potential is active. In the second region, the adsorbed ions begin to associate, with adsorption increasing markedly due to the enhanced adsorption potential because $\Delta G_{CH_2}^{\circ}$ is active. The third region is reached when the zeta potential reverses. At concentrations higher than this, the electrostatic interaction opposes the specific adsorption effects, with the resulting decrease in the slope of the adsorption isotherm.

Somasundaran and Fuerstenau (1966) have clearly demonstrated that the adsorption of anionic dodecyl sulfonate ions occurs only when the alumina is positively charged and decreases sharply as the pH of alumina is increased towards the pzc. Iwasaki et al. (1962) showed the adsorption of dodecyl sulfate ions to occur significantly only when magnetite (Fe_3O_4) is positively charged, and conversely dodecyl ammonium ions adsorb when magnetite is negatively charged.

If the specific adsorption of an amine salt on quartz or a sulfonate on alumina arises from association of the hydrocarbon chains of adsorbed ions, hemi-micelle formation should depend directly on chain length. In a series of experiments involving alkyl sulfonates with 8, 10, 12, 14, and 16 carbon atoms, Wakamatsu and Fuerstenau (1968) found that the three regions of the adsorption isotherms on alumina shifted to more dilute surfactant concentrations as the chain length increased. For C8 detergent, only region 1 exists, that is, hemi-micelles do not form with this short-chain compound. By means of the Stern–Grahame model of the double layer, the contribution of the cohesive energy per mole of CH_2 groups to the adsorption potential can be evaluated. If the standard free energy for removing one mole of CH_2 groups from water through association is ϕ, then the total contribution is $N\phi$ if N is the number of CH_2 groups in the chain. Thus, the contribution of this hydrophobic bond to the adsorption process is

$$\Delta G_{CH_2}^{\circ} = N\phi. \tag{18}$$

The adsorption density Γ_{δ} of surfactant ions in the Stern plane in the absence of chemisorption will be given by

$$\Gamma_{\delta} = 2rC \exp(-\Delta G_{ads}^{\circ}/RT) = 2rC \exp[(-zF\psi_{\delta} - N\phi)/RT]. \tag{19}$$

From Wakamatsu's results, ϕ has been evaluated to be about $-RT$ (about 0.6 kcal/mole of CH_2 groups) in agreement with values obtained from solubility data and micelle formation.

Perhaps the most widely studied system involving only physical bond formation is the adsorption of alkyl sulfonates on alumina. Infrared spectroscopy shows no evidence of chemical bond formation between the sulfonate and the alumina surface. Furthermore, adsorption only occurs at pH values less than 9, under which conditions the surface of the solid is positively charged. However, in a number of oxide mineral–surfactant systems, chemisorption occurs. Chemisorption is often arbitrarily referred to as an adsorption process in which the adsorbate attaches to the surface of the adsorbent with a molar free energy of approximately 10 kcal or greater. In terms of Eq. (15a), chemisorption occurs when $\Delta G^{\circ}_{\text{chem}}$ has some finite value. In flotation systems chemisorption is of primary interest since selectivity may be obtained if there is a specific collector–mineral adsorption reaction rendering a single mineral or group of minerals air avid.

In nonmetallic flotation systems several examples of chemisorption may be cited (Fuerstenau, 1962). Hexanethiol chemisorbs on the surface of zincite (ZnO) and willemite (Zn_2SiO_4), forming a strong zinc–mercaptan bond. Oleic acid will chemisorb on fluorite (CaF_2), forming strongly adsorbed films which are difficult to remove. Infrared spectroscopy applied to the study of adsorbed monolayers has been particularly useful in characterizing several adsorption processes since vibrational modes of the adsorbate *in situ* may be obtained by this technique. Using such methods, Peck *et al.* (1966) recently demonstrated the chemisorption of oleate on hematite by means of infrared spectroscopy; they showed that ferric oleate bonds are formed by the displacement of surface hydroxyls on hematite.

V. SOME PHYSICAL CHEMICAL VARIABLES IN FLOTATION

In the first parts of this section, the discussion will be concerned with flotation in systems involving physical adsorption of the collector and thus flotation response will be closely related to double-layer phenomena since collector ions function as counter ions. Particularly, we will be concerned with factors that affect adsorption in the Stern layer since the flotation rate clearly correlates with the adsorption of collector ions in the Stern layer, that is, with hemi-micelle formation in that region.

A. The Effect of the Point of Zero Charge and Surface Charge

As first pointed out by Fuerstenau and Modi (1956), determining the sign of the charge on oxide mineral surfaces is necessary in order to ascertain the response of minerals to flotation with anionic or cationic collectors. An example of the dependence of flotation on the sign and magnitude of the surface charge is presented in Fig. 8, which is based on the experiments of Iwasaki, *et al.* (1960b). In their research, they conducted an investigation of the electrokinetic and flotation behavior of goethite (FeOOH) in the presence

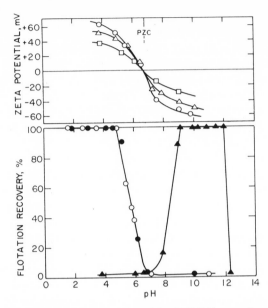

FIG. 8. The dependence of the flotation of goethite (FeOOH) on surface charge. The upper curves show the zeta potential as a function of pH at different concentrations of NaCl, indicating the pzc to be at pH 6.7 : (○) 10^{-4} M, (△) 10^{-3} M, (□) 10^{-2} M. The lower curves are the flotation recoveries in 10^{-3} M solutions of dodecylammonium chloride (○), sodium dodecyl sulfate (●), or sodium dodecyl sulfonate (▲). [After Iwasaki et al. (1960b).]

of a cationic collector and two anionic collectors. Since H^+ and OH^- are potential-determining for oxides, pH is the most important variable in this system. The electrokinetic experiments indicate that the pzc of goethite occurs at pH 6.7, and thus in acidic solutions the surface of goethite is positively charged and in basic solutions it is negatively charged. With 10^{-3} M solutions of dodecyl ammonium chloride, sodium dodecyl sulfate, and sodium dodecyl sulfonate as collectors, Fig. 8 shows that the collector must be anionic when the surface of the solid is positively charged and cationic when the surface is negatively charged. This figure also indicates that the flotation response at a fixed collector concentration is greater, for example, at pH 5 than at pH 6. (Because an experiment is run at fixed flotation time in the modified Hallimond tube, the increase in flotation rate below pH 5 would not be reflected in these particular data.) Increased flotation response at lower pH's with anionic collectors results from the increase in σ_s, and thus larger amounts of collector are adsorbed as counterions in accordance with the results given in Fig. 5. Flotation ceases above pH 12.2 because of hydrolysis of the collector to the amine molecule. Flotation falls off for two reasons: The concentration of RNH_3^+ is diminished as the pH is increased and at the same time there is ionic competition from the increasing Na^+ concentration. (Under these conditions of solubility equilibria (Fuerstenau, 1962), the concentration of RNH_2 is $2 \times 10^{-5} M$, RNH_3^+ is but $4 \times 10^{-7} M$, and Na^+ is $1.6 \times 10^{-2} M$.)

The behavior of numerous mineral–collector systems similar to those shown in Fig. 8 have been published, including the flotation of alumina (Fuerstenau and Modi, 1956; Modi and Fuerstenau, 1960), hematite and magnetite (Iwasaki et al., 1962), ilmenite (Choi et al., 1967), aluminosilicates (Smolik et al., 1966) and rutile (Choi et al., 1967). All of the above studies were conducted with collectors with 12 carbon atoms. Iwasaki et al. (1960a) showed that increasing the collector chain length to 18 carbon atoms caused hematite flotation to occur at pH values above the pzc; this particular chain length effect is not fully understood at the present time. From a technical point of view, increasing the length of the hydrocarbon chain thus decreases the selectivity of the collector in making flotation separations.

As for other types of minerals, an early study of collection of minerals with amine salts by Taggart and Arbiter (1946) showed that the contact angle on barite with dodecyl ammonium chloride increased with increasing SO_4^{2-} concentration and decreased with increasing Ba^{2+} concentration in solution. These results can be interpreted in terms of the double layer model since Ba^{2+} and SO_4^{2-} are potential determining in this system.

In summary, separation of minerals can be achieved by finding conditions where minerals may be oppositely charged so that a cationic or anionic collector adsorbs only on the desired mineral. For example, in considering the flotation separation of a mixture of quartz and rutile, the two minerals are oppositely charged between approximately pH 3 and 6. Within this pH range quartz can be floated from rutile with an alkylammonium collector or, on the other hand, rutile can be floated from quartz with a sulfonate collector. Above pH 6, however, both minerals are negatively charged, and a separation can not be achieved with these collectors.

B. Effect of the Hydrocarbon Chain on the Collector

Because the length of the hydrocarbon chain controls the interaction of the collector with water molecules, the chain length has a pronounced effect on the adsorption of a surfactant at solid–water and air–water interfaces, as has already been discussed. Of particular importance is the effect of the alkyl chain on association phenomena at the surface. Figure 6 shows that rapid flotation begins when the hydrocarbon chains of adsorbed collector ions associate into hemi-micelles. Figure 9 presents the results of an investigation of the flotation response of quartz at neutral pH using alkylammonium acetates as the collector (Fuerstenau et al., 1964). This figure shows that the

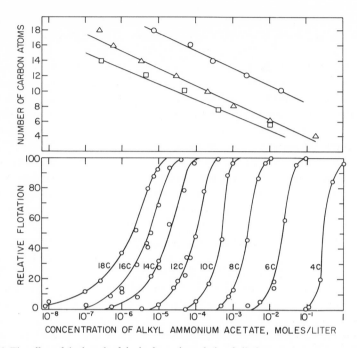

FIG. 9. The effect of the length of the hydrocarbon chain of alkylammonium acetate collectors on the surface and flotation properties of quartz at pH 6–7. The upper curves show similar dependence of incipient flotation (□), the onset of rapid flotation (△) (hemi-micelle formation), and the reversal of the zeta potential on the chain length (○). [After Somasundaran, *et al.*, (1964); Somasundaran and Fuerstenau (1968); Fuerstenau *et al.* (1964).]

concentration of collector required for incipient flotation ranges from about 10^{-8} M for C18 to about 10^{-1} M for the C4 collector. By defining the onset of hemi-micelle formation by extrapolation of the sharp vertical rise in the flotation curve to zero flotation (Fuerstenau *et al.*, 1964), one can plot C_{HM} (the concentration at which hemi-micelles form versus the number of carbon atoms in the alkyl chain). Also in this figure are plotted the collector concentrations required for incipient flotation versus the number of carbon atoms, using vacuum flotation test data (Somasundaran and Fuerstenau, 1968). Similarly, the collector concentration required to bring the zeta potential of quartz to zero (Somasundaran *et al.*, 1964) is plotted versus the alkyl chain length in the same figure. These three curves are displaced towards higher concentrations in the order of the concentration required for incipient flotation, the onset of rapid flotation (hemi-micelle concentration), and for the reversal of the zeta potential (Somasundaran *et al.*, 1964). Interpretation of each of these results can be made in terms of some suitable modification (Somasundaran and Fuerstenau, 1968; Somasundaran *et al.*, 1964) of Eq. (19).

From the slopes of the straight lines given in Fig. 9, the value of the standard free energy ϕ for the removal of hydrocarbon chains from water is approximately $-RT$ or -0.6 kcal/mole of CH_2 groups for the zeta potential and Hallimond tube flotation results. From the incipient flotation data, ϕ is about -0.7 kcal/mole.

Considerable attention has been given to the role of double bonds in the collector chain. Several investigations have been concerned with the sequence: stearic, oleic, linoleic, and linolenic acids. The effectiveness of these collectors on hematite was found to decrease in this same order by Iwasaki *et al.* (1960a). The greater the number of double bonds, the greater is the polar character of the chain; the end result is that the collector has a reduced tendency to adsorb at a mineral water interface. Purcell and Sun (1963b) observed similar effects of these collectors on rutile, as will be discussed later.

C. Effect of the Polar Head Group of the Collector

Several properties of the head on the collector molecule are important in flotation: (1) its electrical charge, (2) its size, and (3) its chemical properties. The role of electrical charge has already been discussed. Ionic size determines how strongly an ion is adsorbed in the Stern layer. The chemical nature of the ionic head most importantly determines the extent of hydrolysis, and also any chemical interaction.

In a detailed investigation of the flotation of alumina with various anionic and cationic collectors, Modi and Fuerstenau (1960) showed that those collectors which are weak electrolytes and hydrolyze are effective at somewhat lower bulk collector concentrations. For example, xanthates and carboxylates are more effective than sulfonates and sulfates; similarly dodecyl ammonium chloride is more effective than trimethyl dodecyl ammonium chloride. As will be seen in the following section, coadsorption of neutral molecules increases the hydrophobicity of the mineral.

Recently Schubert and Schneider (1969) conducted a detailed investigation of the flotation of quartz at pH 8 with a series of cationic nitrogen surfactants. These collectors were dodecyl ammonium chloride, $R-NH_3Cl$ (N1); methyl dodecylammonium chloride, $R-NH_2CH_3Cl$ (N2); dimethyl dodecyl ammonium chloride, $R-NH(CH_3)_2Cl$ (N3); trimethyl dodecyl ammonium chloride, $R-N(CH_3)_3Cl$ (N4); and dodecyl pyridinium chloride,

$$R-N\overset{\frown}{\underset{\smile}{}}Cl \quad (N5);$$

and dodecyl morpholine chloride,

$$R-N\begin{array}{c} CH_2-CH_2 \\ \\ CH_2-CH_2 \end{array}OCl \quad (N6).$$

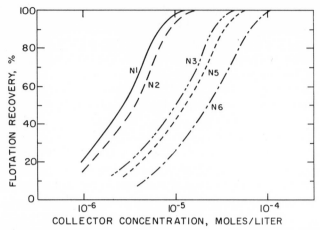

FIG. 10. The effect of the polar group of cationic surfactants on the flotation of quartz at pH 8 : Nl, dodecylammonium chloride; N2, methyldodecylammonium chloride; N3, dimethyl-dodecylammonium chloride; N5, dodecylpyridinium bromide; N6, dodecylmorpholine chloride. [After Schubert and Schneider (1969).]

Figure 10 shows the flotation response of quartz with these collectors, and the order of their effectiveness is N1 > N2 > N3 > N4 > N5 > N6. They also studied flotation response with these compounds having 18 carbon atoms in the alkyl chain, and found the order of effectiveness to be N1 > N2 > N3 > N5 > N4 > N6. N4 and N5 are strong bases whereas the pK for amine hydrolysis is 10.7 for N1, N2, and N3. Alkylmorpholines are somewhat weaker bases with their pK being 8.4. Although collector adsorption in these various systems is somewhat complicated by collector hydrolysis, these results indicate that the onset of marked adsorption leading to rapid flotation (i.e., hemi-micelle formation) generally occurs at higher bulk concentrations with increase in head group size. In addition to steric effects, an explanation of this phenomenon can be made in terms of the double-layer model. The surface charge (Kruyt, 1952) is proportional to the potential gradient across the Stern layer, $\sigma_s \sim - d\psi/dx$. Since σ_s will be about the same in all cases at a given collector concentration and pH, then $d\psi/dx$ is similarly about the same. Thus, for systems with the larger collector ion, the Stern plane will lie farther from the surface and ψ_δ must be less. Since ψ_δ is less, a higher bulk concentration will be required to achieve the same Γ_δ according to Eq. (19).

D. Effect of Neutral Molecules

The contact angle measurements shown in Fig. 3 shows that quartz becomes more hydrophobic when the dodecyl ammonium ions hydrolyze as the pH becomes alkaline. The maximum contact angle is about 40° in this region below the pH of neutral molecule formation. As the pH is increased for the same total concentration of DDA, for example, the contact angle increases to a maximum value at pH 10 to 10.5 almost independently of total amine concentration (Smith, 1963). At pH 6 the contact angle increases from zero at 10^{-6} M DDA to a constant value of about 40° at 10^{-3} M and above. At pH 10–10.5, however, the contact angle is about 30° at 10^{-6} M and rises to 85° at 10^{-4} M DDA and above. Above pH 12 the contact angle drops to zero as a result of amine molecule precipitation and because the neutral amine alone cannot adsorb.

The important role of the neutral amine molecule is to increase the contact angle to a maximum at about the pK_a of pH 10.7 where the aminium ion/amine molecule concentrations are equal. Coverage at pH values around 6 corresponds to a contact angle of 30–40° and are submonolayer (Shergold et al., 1968), whereas monolayer coverage at pH 10–11 corresponds to a contact angle 80–85°. The role of the neutral molecule in increasing the contact angle and surface coverage is emphasized by some results of Smith on contact angles on quartz with dodecyl amine–dodecyl alcohol mixtures (Smith, 1963). With 1.4×10^{-5} M dodecyl alcohol at pH 6 and 4×10^{-4} M dodecylammonium acetate, the contact angle is increased from 30 to around 70°.

An interesting example of the same neutral molecule–ion coadsorption is provided by the recent study (Wottgen, 1969) of alkylphosphonic acids adsorbing at the surface of cassiterite (SnO_2) which was found to have a pzc at pH 5.6. In this system, there is marked adsorption as the pH is lowered towards the pK_a of phosphonic acid, approximately 2.8. The maximum adsorption density occurs at the pK_a where neutral molecule coadsorption most efficiently neutralizes any adsorbed anion–anion repulsion at the positive SnO_2 surface.

Several studies have highlighted kinetic considerations (e.g., time dependence of contact angle) and a detailed understanding of alkylammonium–amine molecule systems must include these rate processes (Smith and Lai, 1966). At pH 6–7 where essentially there are no neutral molecules present the establishment of the condition $\Gamma_{SG} > \Gamma_{SL}$ upon formation of contact involves rearrangement of charged groups at the various interfaces. Alternatively, around the pK_a of the amine, the $\Gamma_{SG} > \Gamma_{SL}$ condition upon contact can be obtained by rearrangement of neutral molecules, energetically a much easier process.

Neutral molecules in flotation systems are often provided by the added frother and several authors (Leja and Schulman, 1954) have considered neutral molecule effects in terms of "frother–collector" interaction. If conditions exist for hemi-micelle formation of adsorbed collector ions, then the presence of uncharged long-chain frother molecules should enhance hemi-micelle formation. This is especially so since uncharged polar heads on an alcohol frother will lower the electrostatic energy of repulsion between adjacent charged polar groups of the collector in the two-dimensional aggregate. This effect is clearly evident in the results of a study of the flotation response of alumina as a function of dodecyl sulfate concentration for various additions of decyl alcohol (Fuerstenau and Yamada, 1962), as shown in Fig. 11. As the alcohol concentration increases, hemi-micelle formation, and hence

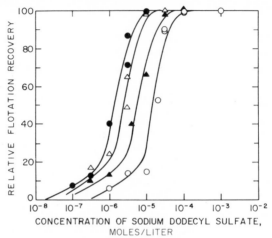

CONCENTRATION OF SODIUM DODECYL SULFATE, MOLES/LITER

FIG. 11. The effect of neutral organic molecules (decyl alcohol) on the flotation of alumina with sodium dodecyl sulfate as collector at pH 6. Decyl alcohol concentration: ($\bigcirc$) O, ($\blacktriangle$) 4.4 × 10^{-5} M, ($\triangle$) 1.1 × 10^{-4} M, ($\bullet$) 2.2 × 10^{-4} M. [After Fuerstenau and Yamada (1962).]

the characteristic rapid rise in flotation response, occurs at successively lower alkyl sulfate concentrations. The coadsorption of ionic and neutral surfactants has additional importance in real flotation systems. Commercial surfactants, particularly the alkyl sulfate and sulfonate types, invariably contain amounts of alkyl alcohol as an impurity. As discussed above, the alcohol molecules will be carried into the mineral–water interface by the adsorbing collector ions. This mechanism effectively lowers collector requirements but, on the other hand, the same mechanism may inhibit selectivity. Even minor adsorption of the collector ion onto the gangue mineral may carry in sufficient alkyl alcohol to cause a decrease in grade of the concentrate.

E. Flotation Depression with Inorganic Ions

Because collector ions function as counter ions in the double layer, their adsorption density will depend on competition with any other counter ions in solution. Thus, the presence of excessive amounts of dissolved salts can inhibit flotation. In the flotation of goethite with quaternary amine salts at pH 11, by adding about 0.03 M NaCl flotation is reduced to about nil (Iwasaki et al., 1960b) and thus the reduction in flotation in this system at very high pH's results from ionic competition from the Na$^+$ added by the NaOH. A detailed study (Onoda and Fuerstenau, 1964) of the depression of quartz flotation with low concentrations of dodecyl ammonium acetate as the collector showed that Ba^{2+} and Na$^+$ both inhibit flotation, the effect being considerably greater than the divalent salt. The depressant effect of inorganic ions on the flotation of quartz with this collector depends on whether the collector ions are adsorbed as individual counterions in the double layer or whether they are associated into hemi-micelles in the Stern plane. At low collector concentrations inorganic ions depress flotation through ionic competition for sites in the double layer in accordance with Eq. (17); but at high collector concentrations where the specific adsorption potential of the collector is high due to association at the surface, inorganic ions have little effect on flotation.

For multivalent ions to depress flotation, they must be charged similarly to the collector. Figure 12 illustrates the depressant effect of Cl$^-$ and SO$_4^{2-}$ on the flotation of alumina at pH 6 with sodium dodecyl sulfate as collector (Modi and Fuerstenau, 1960). This figure shows that both Cl$^-$ and SO$_4^{2-}$ inhibit the flotation of positively charged alumina, but the effect of SO$_4^{2-}$ is nearly 500 times that of Cl$^-$. This results from the specific adsorption

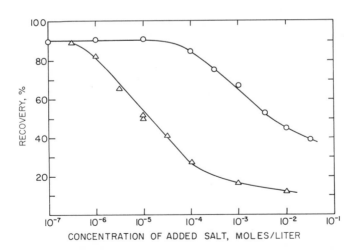

FIG. 12. The effect of inorganic electrolytes [sodium chloride (○) and sodium sulfate (△)] on the flotation of alumina with sodium dodecyl sulfate (4×10^{-5} M) as collector at pH 6. [After Modi and Fuerstenau (1960).]

potential of SO_4^{2-} on alumina ; and since the collector ions have not formed hemi-micelles, their specific adsorption potential is nil and they can be readily displaced in accordance with Eq. (17). Such phenomena become technologically important in cationic flotation if seawater is used in the processing plant.

F. Flotation Activation with Inorganic Ions

In the previous section, we have discussed the effects of sodium sulfate in inhibiting the flotation of positively charged alumina with an anionic collector. At pH 6, alumina does not respond to flotation with dodecyl ammonium chloride as collector since the organic cations do not adsorb on the positively charged solids. As already stated, by adding sufficient SO_4^{2-} to the system, the charge in the Stern plane on alumina can be reversed and the solid can now adsorb organic cations. In the presence of $5 \times 10^{-3} \ M$ Na_2SO_4 at pH 6, alumina readily responds to flotation with a cationic collector (Modi and Fuerstenau, 1960). In flotation terminology, sulfate ions are said to function as an activator ; they serve as a link between a surface and a collector which are charged similarly.

Shergold et al. (1968) recently discovered that at very low pH's hematite can be activated for flotation with dodecyl amine salts. If perchloric acid were used for pH regulation, no flotation would occur. However, at pH's between 0.8 and 2.0, sulfuric acid, hydrochloric acid, or hydrofluoric acid each gave excellent flotation response. Their interpretation was that a complex such as $Fe(Cl)_5OH^{-3}$ might serve as the activator.

In a very recent investigation of the processing of ilmenite sands, Nakatsuka et al. (1970) investigated the surface chemistry and the flotation of hematite, ilmenite, magnetite, quartz, and feldspar using HCl and also H_2SO_4 as the pH regulator. Figure 13, which is based on their results, shows the flotation behavior of ilmenite ($FeTiO_3$) with dodecyl ammonium acetate as collector when NaOH and HCl or H_2SO_4 are used as pH regulators. In the presence of either HCl or H_2SO_4, as the pH is lowered towards that of the pzc, namely pH 5.2, the flotation of ilmenite decreases in the expected manner. However, at pH's below the pzc in the presence of H_2SO_4, flotation rises sharply again because SO_4^{2-} activates ilmenite for cationic flotation. Almost identical behavior was observed with hematite. Their more detailed work showed that both sulfate and phosphate ions in acid pH's activate ilmenite and hematite but do not activate quartz, feldspar, hypersthene, and, interestingly, magnetite.

In nonmetallic mineral flotation, the most common examples of activation include the use of alkaline earth cations to promote the flotation of oxides

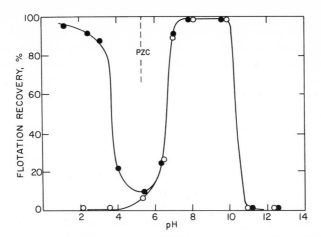

FIG. 13. The effect of hydrochloric acid and sulfuric acid on the flotation of ilmenite with dodecylammonium acetate as collector, illustrating activation with sulfate ions : (○) HCl/NaOH, (●) H_2SO_4/NaOH. [After Nakatsuka *et al.* (1970).]

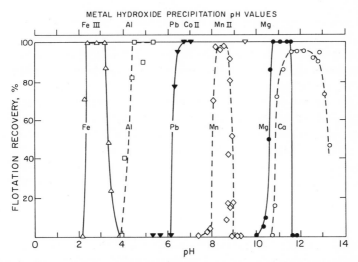

FIG. 14. The effect of cation hydrolysis in the activation (2×10^{-4} M) of quartz for flotation with a sulfonate (6×10^{-5} M) as collector. [After Fuerstenau *et al.* (1963).]

with soaps as the collector. In these processes, the role of cation hydrolysis has been exceedingly important. Although the significance of hydrolysis in the activation of minerals was pointed out over twenty years ago (Gutzeit, 1946), only more recently has the concept been pursued with vigor in flotation research (Fuerstenau *et al.*, 1963). Figure 14 presents the flotation response of quartz (Fuerstenau *et al.*, 1963) as a function of pH in the presence of a number of activators (about 2×10^{-4} M activator) with a sodium alkylaryl sulfonate as collector (about 6×10^{-5} M collector). Under these conditions, the role of hydrolysis of the activator is clearly evident. The activator begins to function effectively as the metal ion begins to hydrolyze, and flotation ceases when the metal hydroxide begins to precipitate. In Figure 14, the decreasing side of the flotation recovery curve for aluminum ions and lead ions are omitted for reasons of clarity. Flotation ceases (Fuerstenau *et al.*, 1963) with aluminum activation at about pH 7, and with lead activation at pH 12 due to the formation of plumbite ion.

Interpretation of the flotation behavior of the complex silicate minerals may become complicated because these minerals may respond to flotation by a process of autoactivation, that is, the silicate may slightly dissolve, with the released cations subsequently hydrolyzing and readsorbing as a flotation activator.

G. Systems Involving Chemisorption

In the foregoing sections, we have discussed collector–mineral systems involving physical adsorption phenomena. In many instances, the collector adsorbs through the formation of chemical bonds with ions at the crystal surface. Often chemisorption is desirable, since this gives the possibility of achieving greater selectivity.

In the section on adsorption phenomena, examples of chemisorption systems were presented and briefly discussed, including the infrared studies of Peck *et al.* (1966) on the hematite–oleate system. Iwasaki *et al.* (1967) investigated the flotation of hematite and goethite with the series of 18-carbon aliphatic acids: oleic acid (1 double bond), linoleic (2 double bonds), and linolenic (3 double bonds). Interest in these unsaturated acids is due to their ready availability as natural products and because of the increased solubility effected by the double bonds. Stearic acid, the C_{18} saturated acid, is essentially insoluble.

Figure 15 presents the flotation response of goethite (Iwasaki *et al.*, 1960a) as a function of pH for the three unsaturated fatty acids. The pzc of goethite occurs at pH 6.7 but rapid flotation with these collectors occurs up to about pH 9 with linolenic, pH 10 with linoleic, and pH 11 with oleic acid. For this kind of behavior, strong chemisorption (in agreement with the infrared studies) must be occurring, otherwise adsorption could not take place under conditions where the surface potential is so negative. The linolenic, linoleic, oleic sequence observed for goethite and hematite was also found by Purcell and Sun (1963b) for the flotation of rutile and by Choi *et al.* (1967) for the

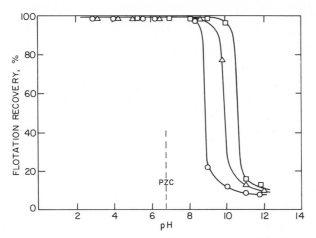

FIG. 15. The effect of pH on the flotation of goethite with oleic (□), linoleic (△), and linolenic (○, 10^{-4} M) acids as collector. [After Iwasaki *et al.* (1960a).]

flotation of ilmenite. The explanation for the sequence is that the increased hydrophilic content of the collector chain introduced per double bond reduces the extent of adsorption and subsequent flotation.

Recent research has shown that potassium octyl hydroxamate, a chelating agent (R—COON—K) strongly chemisorbs on hematite (Fuerstenau *et al.*, 1970). The effectiveness of this collector increases by conditioning the pulp at

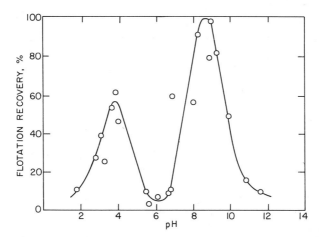

FIG. 16. The effect of pH on the flotation of pyrolusite (MnO_2) with potassium oleate (10^{-4} M) as collector, showing a physical adsorption region and a chemisorption region. [After Fuerstenau and Rice (1968).]

higher temperatures, in common with other chemisorption systems in aqueous media.

An interesting example of a collector adsorbing physically under some conditions and chemically under others is the oleate–pyrolusite (MnO_2) system (Fuerstenau and Rice, 1968). With dodecyl amine, pyrolusite flotation occurs essentially only in alkaline solutions whereas with a sulfonate it occurs essentially only in acid solutions. However, behavior with oleate as the collector is far more complex (Fig. 16). This complex curve indicates physical adsorption of oleate ions plus oleic acid molecules below pH 6; but above pH 6 chemisorption of the collector must be occurring. Because sulfonate showed no such increase in flotation at these pH's, this phenomenon is not the result of autoactivation. The authors (Fuerstenau and Rice, 1968) interpret the chemisorption process in terms of a mechanism involving hydrolysis of the cations constituting the lattice. They infer that chemisorption will be a maximum at the pH where the mineral cations hydrolyze. On the other hand, Peck *et al.* (1966) considers that maximum chemisorption occurs when the mineral is at its pzc.

In general, such salt-type minerals as apatite, barite, calcite, fluorite, and rhodochrosite, are concentrated by flotation with long-chain fatty acids or soaps which chemisorb at these mineral surfaces. Chemisorption of this collector on the surface of these minerals has been confirmed by infrared studies, for example, with the formation of calcium oleate on fluorite (Fuerstenau, 1962) or calcium laurate on calcite (Fuerstenau and Miller, 1967).

Separation between calcium salt-type minerals is difficult because the selectivity of chemisorption of fatty acids or soaps must first be achieved. Tannins, particularly quebracho, have been used to depress calcite during soap flotation of fluorite (Sutherland and Wark, 1955). Recently, separation between such minerals as apatite and calcite or apatite and gypsum has become important. Selectivity between apatite and calcite has been achieved using shorter-chain fatty acids or certain aromatic fatty acids which are too soluble to chemisorb onto calcite (Fuerstenau, 1962).

VI. SOME EXAMPLES OF TECHNOLOGICAL FLOTATION SEPARATIONS

Minerals often occur in nature in very complex mixtures, and flotation selectivity can only be achieved through the use of depressants that inhibit the flotation of undesired minerals.

Hydrofluoric acid and sodium fluoride are used in the flotation of silicate minerals, particularly in the separation of quartz and feldspar. Sodium silicate is widely used in nonmetallic mineral flotation as a depressant and as a dispersant for slimes. Its role must be rather complex, although a part of its effectiveness most likely results from its ability to form polysilicate colloidal aggregates in solution. These polymers may then coat mineral particles because of the chemical similarity between a siliceous mineral surface and the polymers. The hydrophilic organic colloidal materials such as gelatin, guar, gum arabic, proteins, and starch are also used as depressants in flotation. These large macromolecules adsorb on certain minerals, making them hydrophilic, and thus assist the flotation engineer in making the desired mineral separations.

Flotation of minerals that respond to flotation only through activation can be controlled by preventing adsorption of activator ions by reducing their bulk concentration through precipitation or by complexing them with a sequestering agent. Complexing agents are particularly useful in nonmetallic flotation where often only subtle differences in the response of minerals are available for their separation. The prime function of these reagents is to sequester those ions in solution that might otherwise adsorb on the surface of an undesirable mineral and cause it to adsorb collector. Examples of such reagents include acid, phosphates, sodium carbonate, tannic acid, quebracho, and EDTA.

To illustrate the kinds of separations that can be made by flotation, a few examples will be given (Fuerstenau, 1962). In the case of iron ores, the usual problem is to separate hematite from quartz. Six different procedures have been developed, the first four of which have been put into industrial practice :

1. Flotation of hematite using a sulfonate as the collector at pH 2–4.
2. Flotation of hematite with a fatty acid as the collector at pH 6–8.
3. Flotation of quartz with an amine as collector at pH 6–7.
4. Flotation of quartz activated with calcium ions at pH 11–12, using a soap as the collector together with starch to depress the hematite.
5. Flotation of hematite with an amine as collector at pH 1.5 in the presence of hydrochloric acid or sulfuric acid (Shergold et al., 1968).
6. Flotation of hematite with a hydroxamate as collector at pH 8.5 and methylisobutylcarbinol as a frother (Fuerstenau et al., 1970).

Briefly, the flotation of hematite with a sulfonate results from adsorption of the anionic collector on positively charged hematite at pH 2–4 (quartz is negatively charged at these pH's). In the second process, the fatty acid chemisorbs on the hematite and does not adsorb on quartz. In the third process, the cationic amine adsorbs on the highly negatively charged quartz but not on the essentially uncharged hematite. In the fourth industrial process, the flotation of hematite is prevented with the hydrophilic starch macromolecules that chemisorb onto the hematite through their carboxyl groups. Without the starch, hematite would also float under these conditions.

Ores containing mica, feldspar, and quartz can be separated by flotation by a method utilizing the crystal chemistry of mica. The first step involves floating the mica with an amine at pH 3, under which conditions quartz and feldspar are depressed. Since mica is a layer silicate with a fixed negatively charged surface, the cationic amine adsorbs on mica. Afterwards, in the presence of HF as an activator at pH 2.5 using an amine as collector, the feldspar can be floated from the quartz.

Phosphate ores generally are mixtures of apatite and quartz. The usual flotation procedure involves flotation of the apatite with a fatty acid as collector. In this separation step, the fatty acid chemisorbs onto the apatite and does not adsorb onto the negatively charged quartz. In order to make a final upgrading of the phosphate, the concentrate is blunged with acid to remove the collector. Then, using an amine as collector, the concentrate is subjected to flotation again for removal of contaminating silica. In this latter step, which is carried out at neutral pH, the amine adsorbs on the negatively charged quartz and not on the uncharged apatite.

The concentration of potash (KCl) from halite (NaCl) is a unique concentration process in that two water-soluble minerals are separated by flotation in a saturated brine, using an amine as collector. Amines selectively adsorb on the surface of KCl and not on NaCl, so the separation is not a difficult one.

VII. FLOTATION KINETICS

Flotation kinetics is the study of the variation in the amount of material overflowing in the froth product with flotation time, and the quantification of rate-controlling variables. Sutherland and Wark (1955) have tabulated the dozens of variables that affect the flotation process, dividing the list into three general groups: (1) properties of the ore and its constituent minerals, (2) the reagent treatment given the pulp, and (3) the characteristics of the flotation machines and their operation. Because of this very large number of variables, experimental analysis of flotation kinetics has been difficult.

The separation of materials by the flotation process involves two phases within the flotation cell, a pulp phase and a froth phase, with the contents of the flotation cell divided between them. Industrially, flotation is carried out in continuous systems with a number of cells arranged in series (the volume of each cell ranging up to as much as 200 ft^3). Thus, kinetic analysis of an industrial flotation operation must involve flow into and out of each cell and the interchange between the pulp and the froth column within each cell.

Even though the process is more complex, as indicated above, the most common approach to flotation kinetics has involved consideration of the flotation cell as a perfectly mixed single phase in which the solid particulates are separated into two products. By analogy with chemical kinetics, the majority of authors have utilized the following equation for expressing concentration changes within the cell as a function of time:

$$dC(t)/dt = - k[C(t)]^n, \tag{20}$$

where $C(t)$ is the concentration of floatable material, k is the rate constant, and n is the "order" of the rate equation. As reviewed by a number of authors (Arbiter and Harris, 1962; Mika and Fuerstenau, 1969), the general consensus is that the process is basically first order (although some authors have taken n to be 2 while others have considered it to be variable). Actually, any rate equation based on the assumption that the flotation rate depends directly on the frequency of particle–bubble collision must be first order.

Because not all mineral particles behave the same in a flotation cell, the concept of a "terminal concentration" was introduced to account for the apparently floatable material that may remain in the flotation cell after long times (Morris, 1952; Bushell, 1962). For such situations, the rate equation has been written as

$$dC/dt = - k[C - C_\infty]^n, \tag{21}$$

where C_∞ represents the terminal concentration of mineral being floated.

Another variable that has received considerable attention is the effect of particle size on flotation rates, beginning with the investigation of Schlechten (Gaudin et al., 1942). Using a cell somewhat related to the concept of the Hallimond tube in which a froth column is absent, Tomlinson and Fleming (1965) investigated a number of the variables associated with flotation rates, including particle size. Figure 17 shows a summary of their results on the effect of particle size on the rate of flotation of hematite and quartz with 1.5×10^{-5} M potassium oleate as the collector at pH 8, with all physical operating variables (except for the feed size and mass) being maintained constant. A number of phenomena encountered in flotation practice can be explained in terms of results given in this figure. First of all, the recovery of mineral particles in flotation plants is generally nearly complete in the size range of 20–100 μ, decreasing for both coarser and finer particles. Particles larger than about 0.5–1.0 mm generally cease to be recovered in flotation because they are too heavy to be levitated by air bubbles in the cell. Figure 17 shows that with increasing particle size, the rate of flotation increases, attaining a maximum value, and then decreasing. Because the contact angle on hematite is considerably larger than that on quartz [the quartz in this case must have been activated by the impurity cations], the maximum size floated is necessarily large. At a particle size of around 100 μ (the most

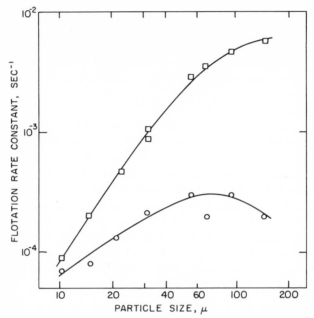

FIG. 17. The effect of particle size on the first order flotation rate constants of quartz (○) and hematite (□) using oleate as collector at pH 8. [After Tomlinson and Fleming (1965).]

abundant size usually encountered in flotation practice), the rate constant of hematite exceeds that of quartz by a factor of 10–20. To achieve the required purity of the final product, the froth product may have to be refloated two or three times. Also, in flotation practice, selectivity is often difficult to achieve with particles of fine size. (Figure 17 shows that the rate constants of the two minerals approach each other as particle size is decreased.) Because of this, in industrial iron ore flotation the ore is generally deslimed and material finer than about 15 μ is discarded (Fuerstenau, 1962).

Recent approaches to flotation kinetics include the formal partitioning of the flotation cell contents into the pulp and the froth phase (Arbiter and Harris, 1962). In formulating this model, Arbiter and Harris (1962) assumed first-order kinetics for transfer into and return from the froth phase, and thus the cell to be operating as a perfect mixer both with respect to the pulp and froth phases. Harris and Rimmer (1966) established that frequently the apparent nonlinearities associated with using the simple model [Eq. (20)] for analysis of experimental (batch) flotation data can be eliminated by the two-phase viewpoint.

More recently, in analyzing the kinetics of an actual flotation feed consisting of material with a distribution of particle sizes, particle shapes, and surface characteristics, the important concept of a distributed set of rate constants was introduced (Imaizumi and Inoue, 1965; Woodburn and Loveday, 1966). It was proposed that the first-order rate constant is not single valued, but distributed in a continuous fashion over a range of values. This treatment in which the mineral particles contain a range of floatabilities seems much more meaningful than analyses involving two discrete values of the rate constant, corresponding to the extremities of floatable and non-floatable particles.

Finally, in order to better study the dynamics of the flotation process and its time dependent behavior, a microscopic viewpoint has been proposed to relate kinetic parameters with the physical and chemical variables of the system (Mika and Fuerstenau, 1969). The treatment entails modeling the flotation cell as an idealized turbulent regime in which the interaction between particles and bubbles reflects the influence of specific material (primarily surface-chemical) variables upon close approach of the particle to the bubble. This complex model should provide a framework within which to investigate details of phenomena that affect flotation process dynamics.

VIII. SUMMARY

By making particulate solids selectively hydrophobic through the selective adsorption of surfactants, it is possible to separate them by flotation. Efficient separations can generally be achieved by determining factors that control the surface chemistry of the various minerals in aqueous media. The kinetics of bubble–particle attachment have also been briefly considered. For more detailed information on collector–mineral interactions, separation techniques, and flotation technology, the reader is referred to a number of textbooks that have been written on flotation (Fuerstenau, 1962; Sutherland and Wark, 1955; Gaudin, 1957; Klassen and Mokrousov, 1963).

REFERENCES

Aplan, F. F., and de Bruyn, P. L. (1963). *Trans. A.I.M.E.* **226**, 235.
Arbiter, N., and Harris, C. C. (1962). *In* "Froth Flotation—50th Anniversary Volume," (D. W. Fuerstenau, ed.), p. 215. A.I.M.E., New York.
Buchanan, A. S., and Heyman, E. (1948). *Proc. Roy. Soc. (London)* **A195**, 150.
Buchanan, A. S., and Oppenheim, R. C. (1968). *Aust. J. Chem.* **21**, 2367.
Bushell, C. H. G., (1962). *Trans. A.I.M.E.* **223**, 266.
Choi, H. S., and Kim, Y. S., and Paik, Y. H. (1967). *Can. Mining Met. Bull.* **60**, 217.
Cooke, S. R. B. (1949). *Trans. A.I.M.E.* **184**, 306.
de Bruyn, P. L. (1955). *Trans. A.I.M.E.* **202**, 291.
de Bruyn, P. L., Overbeek, J. Th., and Schuhmann, R., Jr. (1954). *Trans. A.I.M.E.* **199**, 519.

Digre, M., and Sandvik, K. (1968). *Trans. I.M.M.* **77**, C61.

Freyberger, W. L., and de Bruyn, P. L. (1957). *J. Phys. Chem.* **63**, 1475.

Fuerstenau, D. W. ed. (1962). "Froth Flotation—50th Anniv. Volume," Amer. Inst. Mining and Met. Eng., New York.

Fuerstenau, D. W., and Modi, H. J. (1956). Adaptation of New Research Techniques to Mineral Engineering Problems, NYO 7178, MITS 31, 30 April, 1956; NYO 7179, MITS 32, 31 July, 1956, Massachusetts Inst. of Technology.

Fuerstenau, D. W., and Yamada, B. J. (1962). *Trans. A.I.M.E.* **223**, 50.

Fuerstenau, D. W., Metzger, P. H. and Seele, G. D. (1957). *Eng. Mining J.* **158**, 93.

Fuerstenau, D. W., Healy, T. W., and Somasundaran, P. (1964). *Trans. A.I.M.E.* **229**, 321.

Fuerstenau, M. C. (1964). *Eng. Mining J.* **165**, 108.

Fuerstenau, M. C., and Miller, J. D. (1967). *Trans. A.I.M.E.* **238**, 153.

Fuerstenau, M. C., and Rice, D. A. (1968). *Trans. A.I.M.E.* **241**, 453.

Fuerstenau, M. C., Martin, C. C., and Bhappu, R. B. (1963). *Trans. A.I.M.E.,* **226**, 449.

Fuerstenau, M. C., Harper, R. W., and Miller, J. D. (1970). *Trans. A.I.M.E.* **247**, 69.

Gaudin, A. M. (1957). "Flotation." McGraw Hill, New York.

Gaudin, A. M., and Fuerstenau, D. W. (1955a). *Trans. A.I.M.E.* **202**, 958.

Gaudin, A. M., and Fuerstenau, D. W. (1955b). *Trans. A.I.M.E.* **202**, 66.

Gaudin, A. M., and Morrow, J. G. (1957). *Trans. A.I.M.E.* **208**, 1365.

Gaudin, A. M., Schuhmann, R., Jr., and Schlechten, A. W. (1942). *J. Phys. Chem.* **46**, 902.

Gaudin, A. M., Witt, A. F., and Biswas, A. K. (1964). *Trans. A.I.M.E.* **229**, 1.

Gutzeit, G. (1946). *Trans. A.I.M.E.* **169**, 276.

Hallimond, A. F. (1944). *Mining Mag.* **70**, 87.

Hallimond, A. F. (1945). *Mining Mag.* **72**, 201.

Harris, C. C., and Rimmer, H. W. (1966). *Trans. I.M.M.* **75**, C153.

Healy, T. W., and Fuerstenau, D. W. (1965). *J. Colloid Sci.* **20**, 376.

Healy, T. W., Herring, A. P., and Fuerstenau, D. W. (1966). *J. Colloid Sci.* **21**, 435.

Imaizumi, T., and Inoue, T. (1965). *Proc. Int. Mineral Process. Congr., 6th, Cannes, 1963,* p. 581.

Iwasaki, I. (1965). *Trans. A.I.M.E.* **232**, 383.

Iwasaki, I., Cooke, S. R. B., and Choi, H. S. (1960a). *Trans. A.I.M.E.* **217**, 237.

Iwasaki, I., Cooke, S. R. B., and Colombo, A. F. (1960b). Rep. Investigation No. 5593, U.S. Bureau of Mines.

Iwasaki, I., Cooke, S. R. B., and Kim, Y. S. (1962). *Trans. A.I.M.E.* **223**, 113.

Klassen, V. I., and Mokrousov, V. A. (1963). "An Introduction to the Theory of Flotation" (Transl. J. Leja and G. W. Poling). Butterworth, London and Washington, D.C.

Kruyt, H. R. (1952). "Colloid Science," Vol. I. Elsevier, Amsterdam.

Leja, J., and Schulman, J. H. (1954). *Trans. A.I.M.E.* **199**, 221.

Matijevic, E. (1967). "Principles and Applications of Water Chemistry" (S. D. Faust and J. V. Hunter, eds.), p. 328. Wiley, New York.

Mika, T. S., and Fuerstenau, D. W. (1969). *Proc. Int. Mineral Process. Congr. 8th Leningrad, 1968,* Vol. 2, p. 246. Mechanobr Institute, Leningrad.

Modi, H. J., and Fuerstenau, D. W. (1960). *Trans. A.I.M.E.* **217**, 381.

Morris, T. M. (1952). *Trans. A.I.M.E.* **193**, 794.

Nakatsuka, K., Matsuoka, I., and Shimoiizaka, J. (1970). *Proc. Int. Process. Congr., 9th, Prague,* p. 251.

O'Connor, D. J. (1958). *Proc. Int. Congr. Surf. Activity, 2nd* **1**, p. 311.

Onoda, G. Y., and Fuerstenau, D. W. *Proc. Int. Mineral Process. Congr., 7th, New York, 1964,* p. 301.

Parks, G. A. (1965). *Chem. Rev.* **65**, 177.

Parks, G. A. (1967). *A.C.S. Advan. Chem. Ser.* **67**, 121.

Parks, G. A., and de Bruyn, P. L. (1962). *J. Phys. Chem.* **66**, 967.

Parsons, R. (1954). "Modern Aspects of Electrochemistry" (B. E. Conway and J. O'M Bockris, eds.), Vol. 1, p. 103. Butterworths, London and Washington, D.C.

Peck, A. S., Raby, L. H., and Wadsworth, M. E. (1966). *Trans. A.I.M.E.* **235**, 301.

Purcell, G., and Sun, S. C. (1963a). *Trans. A.I.M.E.* **226**, 6.

Purcell, G., and Sun, S. C. (1963b). *Trans. A.I.M.E.* **226**, 13.

Sagheer, M. (1966). *Trans. A.I.M.E.* **235**, 60.

Schubert, H., and Schneider, W. (1969). *Proc. Mineral Process. Congr., 8th, Leningrad, 1968* **2**, p. 315.

Schuhmann, R., Jr., and Prakash, B. (1950). *Trans. A.I.M.E.* **187**, 591.

Shergold, H. L., Prosser, A. P., and Mellgren, O. (1968). *Trans. I.M.M.* **77**, C166.

Smith, R. W. (1963). *Trans. A.I.M.E.* **226**, 427.

Smith, R. W., and Lai, R. M. W. (1966). *Trans. A.I.M.E.* **235**. 413.

Smolders, C. A. (1961). *Rec. Trav. Chem.* **80**, 651, 699.

Smolik, J. S., Harman, and Fuerstenau, D. W. (1966). *Trans. A.I.M.E.* **235**, 367.

Somasundaran, P. (1968a). *Trans. A.I.M.E.* **241**, 105.

Somasundaran, P. (1968b). *Trans. A.I.M.E.* **241**, 341.

Somasundaran, P. (1968c). *J. Colloid Interface Sci.* **27**, 659.

Somasundaran, P., and Agar, G. (1967). *J. Colloid Interface Sci.* **24**, 233.

Somasundaran, P., and Fuerstenau, D. W. (1966). *J. Phys. Chem.* **70**, 90.

Somasundaran, P., and Fuerstenau, D. W. (1968). *Trans. A.I.M.E.* **241**, 102.

Somasundaran, P., Healy, T. W., and Fuerstenau, D. W. (1964). *J. Phys. Chem.* **68**, 3562.

Sutherland, K. L., and Wark, I. W. (1955). "Principles of Flotation," Australian Inst. Mining Met., Melbourne.

Taggart, A. F., and Arbiter, N. (1946). *Trans. A.I.M.E.* **169**, 266.

Tomlinson, H. S., and Fleming, M. G. (1965). *Proc. Int. Mineral Process. Congr., 6th, Cannes, 1963* p. 563.

Wakamatsu, T., and Fuerstenau, D. W. (1968). *A.C.S. Advan. Chem. Ser.* **79**, 161.

Wada, M., Kono, S., Nadatani, A., and Suzuki, H. (1969). *Proc. Int. Mineral Proc. Congr., 8th, Leningrad, 1968* **1**, p. 452.

Woodburn, E. T., and Loveday, B. K. (1966). *J. South Africa Inst. Mining Metall.* **66**, 649.

Wottgen, E. (1969). *Trans. I.M.M.* **78**, C91.

Robert Lemlich
Department of Chemical and Nuclear Engineering
University of Cincinnati
Cincinnati, Ohio

I. INTRODUCTION

Bubble fractionation is the name given by Lemlich (1964) to the partial separation of components *within* a solution, which results from the selective adsorption of such components at the surfaces of rising bubbles. It is not necessary for the solution to foam (Dorman and Lemlich, 1965).

Figure 1(a) shows a nonfoaming bubble fractionation column in batchwise operation. Gas enters at the bottom and bubbles up through the solution. The adsorbed solute is carried up and, if nonvolatile, is then deposited at the top of the liquid as the gas exits. There its concentration builds up to a steady-state value which is limited by the downward transport of axial dispersion. The overall result is enrichment in the upper portion of the column and stripping in the lower portion.

Figure 1(b) shows how a nonfoaming bubble fractionation column can be operated continuously. The feed should enter at the level of matching concentration so as to minimize its disturbing effect. As with batch operation, the taller the column, the greater the degree of separation.

Unlike foam fractionation which requires a certain minimum concentration of surfactant in order to foam, bubble fractionation theoretically has no lower

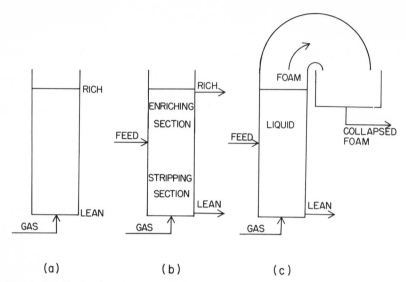

FIG. 1. Bubble fractionation: (a) batchwise operation; (b) continuous flow operation; (c) continuous booster operation for foam fractionation.

concentration limit for operability. It can operate with mere traces of a solute which is naturally surface active, or which can be made effectively surface active by means of an appropriate additive.

As shown in Fig. 1(c), bubble fractionation can also be employed in tandem with foam fractionation (Harper and Lemlich, 1965), perhaps to raise the concentration up to the foaming threshold. Such a combination multiplies the separation ratios of the two techniques. However, when combined with foam fractionation which is truly operating in the simple mode (no internal coalescence), the *enriching* action of the bubble fractionation (as distinct from its stripping action) may be lost because the adsorbed solute is entirely carried off by the foam with no opportunity for redeposition at the top of the liquid. A similar loss can theoretically occur when the foam column is a very tall efficient stripper, enricher, or both (Lemlich, 1968).

Bubble fractionation is perhaps the simplest of the various adsubble techniques. It is, of course, not limited to true solutions, but can also be applied to colloidal and macroparticle suspensions. As a technique of separation, it suggests itself as a method for concentrating or removing substances present at low levels of concentration without the necessity of foaming.

Bubble fractionation can also occur "incidentally," that is, not by design. This could happen in many liquid mixtures or solutions under the appropriate conditions. The gas need not even be injected; the bubbles could evolve by the liberation of gas previously under pressure, or by electrochemical or chemical reaction.

Along with Anderson and Quinn (1970), the present author cannot help but wonder how many studies in bubbled liquid beds involved significant effects of bubble fractionation that went unnoticed.

II. THEORY

A. Batchwise Operation

1. Lumped Parametric Approach for the Bubble

Lemlich (1966) considered two sets of cases: one where the aforementioned downward transport is expressed in terms of gross circulation currents, and the other where it is expressed in terms of a diffusion coefficient. The latter set is outlined below. For the former, the reader is referred to the original paper.

A material balance around either end of the column in Fig. 1(a) yields for steady state

$$af\Gamma = DA \, dC/dh. \tag{1}$$

The surface excess Γ can be viewed as being essentially the concentration of adsorbed solute at the bubble surface. The boundary condition is $C = C_B$ at $h = 0$.

Essentially, D is the eddy diffusivity because the molecular diffusivity plus the free mass convection are likely to be relatively small, as shown by the persistence of the vertical concentration gradient after shutdown.

In Eq. (1) D is based on the full A. Alternatively, it could have been based on only the liquid cross section by simply multiplying A by the quantity $(1 - \lambda)$. In any case, λ is generally comparatively small, so the difference between the two versions of D is also small.

Three subcases are considered for the adsorption at the bubble surface: a constant Γ, a linear isotherm such as Eq. (2), and a reversible first order adsorption as expressed by Eq. (3):

$$\Gamma = KC, \tag{2}$$

$$d\Gamma/dh = (k/v)(C - \Gamma/K). \tag{3}$$

At low concentrations, Eq. (2) or Eq. (3) appears more realistic that the assumption of a constant Γ.

Solving Eq. (1) with the given boundary condition yields for the subcase of constant Γ

$$C - C_B = a\Gamma f h/DA. \tag{4}$$

Utilizing Eq. (2), instead of constant Γ, gives

$$C/C_B = \exp(Kaf h/DA). \tag{5}$$

Utilizing Eq. (3) instead of constant Γ, together with an additional boundary condition of $\Gamma = 0$ at $h = 0$, yields

$$\frac{C}{C_B} = \frac{E + N}{2E} \exp\left(\frac{E - N}{2}h\right) + \frac{E - N}{2E} \exp\left(\frac{-E - N}{2}h\right), \tag{6}$$

where $N = k/vk$, $M = kaf/DAv$, and $E = (N^2 + 4M)^{1/2}$.

Equations (4)–(6) represent possible theoretical vertical profiles of concentration along a batch column at steady state.

Neglecting the small holdup of solute on the rising bubbles at any instant, a simple material balance against the initial (charge) concentration C_i gives

$$\int_O^H C \, dh = C_i H. \tag{7}$$

Then, by combining Eq. (7) with either Eqs. (4), (5) or (6), C_B can be eliminated, and the profile expressed in terms of C_i. For example, the result with Eq. (5) is

$$\frac{C}{C_i} = \frac{JH \exp(Jh)}{\exp(JH) - 1}, \tag{8}$$

where $J = Kaf/DA$.

2. Distributed Parametric Approach for the Bubble

Cannon and Lemlich (1971) used the hydrodynamics of a moving bubble to describe its local surface coverage with solute from a dilute solution. The bubble was viewed as being surrounded by two regions of interest : a laminar upstream region and a turbulent discharging wake. Penetration theory was used to describe the adsorption from the upstream region, with the molecules so adsorbed being swept into the wake. A random surface renewal model was employed for the adsorption at the bubble surface in contact with the wake. Material balances around the rising bubble and wake, and around either end of the column, then yielded the concentration profile summarized by

$$\frac{C}{C_B} = \frac{\beta_2 \exp(\beta_1 h) - \beta_1 \exp(\beta_2 h)}{\beta_2 - \beta_1}. \tag{9}$$

The parameters β_1 and β_2 are peculiar to the particular system and operating conditions involved. The evaluation of these parameters in terms of more fundamental quantities is described in the original report (Cannon, 1970).

B. Continuous Flow Operation

The lumped parametric approach was readily extended by Cannon and Lemlich (1971) to continuous flow operation such as shown in Fig. 1(b), with feed entering at the match point. The feed, entering at rate F, splits within the column into T which rises and B which descends. B was then incorporated into the material balance around the bottom of the column, and T was incorporated into the balance around the top. Equations for the concentration profile were then developed for continuous flow situations corresponding to Lemlich's batchwise subcases previously mentioned.

For example, for the stripping section; that is, the portion of the column below the feed level, Eq. (6) is replaced by Eq. (10) in which m_1, $m_2 = -\frac{1}{2}\{(B/DA) + N \pm [E^2 - (2BN/DA) + (B/DA)^2]^{1/2}\}$:

$$\frac{C}{C_B} = \left[\frac{Kaf}{Kaf - B}\right]\left[\frac{m_2 \exp(m_1 h)}{m_2 - m_1} + \frac{m_1 \exp(m_2 h)}{m_1 - m_2}\right] + \frac{B}{B - Kaf}. \quad (10)$$

In defining M and N for Eq. (10), v is taken for the stripping section.

Similarly, when $\Gamma = KC$, Eq. (5) is replaced by Eq. (11) for the stripping section, and by Eq. (12) for the enriching section:

$$\frac{C}{C_B} = \frac{B}{B - Kaf} + \left(1 - \frac{B}{B - Kaf}\right)\exp\left(\frac{Kaf - B}{DA}\right)h, \quad (11)$$

$$\frac{C}{C_T} = \frac{T}{T + Kaf} + \left(\frac{C_F}{C_T} - \frac{T}{T + Kaf}\right)\exp\left[\left(\frac{T + Kaf}{DA}\right)(h - h_F)\right]. \quad (12)$$

Equating C to C_F in Eq. (11) and solving for h gives the optimum feed location h_F. Additional results can be found in Cannon's report.

Lee (1969) also studied continuous flow operation, but with feed entering the top of the column rather than at the optimum location. His analysis was somewhat similar to that which led to Eq. (10). It involved mass transfer to an equilibrium surface (which is essentially Eq. (3) with k as the mass transfer coefficient) together with axial dispersion. However, his bubbles originated in a bottom recycle stream which passed through an external bubble generator. Accordingly, at $h = 0$, he employed surface equilibrium; that is, $\Gamma = KC_B$, as one boundary condition, and $d^2C/dh^2 = 0$ as the other. The results of his analysis are similar in overall form to Eq. (10), but with different parameters that result from his particular set of boundary conditions.

III. EXPERIMENTAL RESULTS

A. Batchwise Operation

Dorman and Lemlich (1965) demonstrated the feasibility of bubble fractionation by partially separating aqueous solutions of several ppm of technical monobutyldiphenyl sodium monosulfonate in batchwise operation. A single stream of prehumidified nitrogen bubbles was employed in glass columns 1.3 cm to 8.9 cm in diameter, and about 50 cm to 130 cm in height. The separation ratio C_T/C_B achieved at steady state was found to increase with column height, to decrease with column diameter, and to be substantially independent of bubble frequency over the experimental range of 120–250 bubbles/min with bubble diameters in the neighborhood of 0.37 cm. The separation was also found to increase with charge concentration, but only modestly. Separation ratios of up to 4.5 were obtained.

Harper and Lemlich (1966) also studied batch operation, but with crystal violet chloride as a convenient solute in water. The concentration gradient

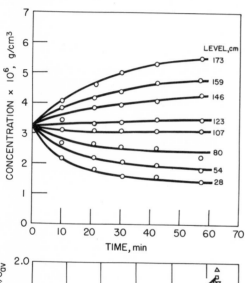

Fig. 2. Typical transient approach to steady state (Shah and Lemlich, 1970): Aqueous solution of crystal violet in a 2.22-cm. i.d. column with a fritted glass bubbler at a gas flow rate of 2.52 cm³/sec.

Symbol	$C_i \times 10^6$ (gm/cm³)	G (cm³/sec)
▽	2.00	6.63
△	2.00	23.88
○	2.00	33.83
□	0.96	15.32
◇	2.00	15.32
⊘	4.00	15.32

Fig. 3. The profile of concentration ratio at steady state for an aqueous solution of crystal violet chloride in a 4.76-cm. i.d. column with a fritted glass bubbler (Shah and Lemlich, 1970).

was thus readily visible to the eye, and samples were withdrawn for colorimetric analysis. Prehumidified nitrogen entered the column through a sintered glass dispersion tube. Separation ratios greater than 10 were obtained at steady state under appropriate conditions.

The dye technique was improved upon by Shah and Lemlich (1970) in a more extensive study. They devised an external, vertically traveling, colorimeter to measure local concentration directly within the glass column without the need to sample or probe its contents. As before, the bubbles were generated by injecting prehumidified nitrogen through a fritted glass sparger at the bottom of the column.

Figure 2 illustrates their results for the transient approach to steady state at various levels within a column. Figure 3 shows the profile at steady state in another column. The insensitivity of the profile with G accords with the largely cancelling effects of G on f and D in Eqs. (4)–(6), and on f and circulation in the remainder of their theory. The insensitivity of the profile with C_i agrees with the form of their equations with respect to concentration. (The final average concentration C_{av} was used instead of C_i to normalize the profile because some dye was lost due to chemical instability. Were it not for this, C_{av} would equal C_i.)

As indicated by the theory, increasing the total length of the column increases the overall separation ratio. This effect is clearly evident in Fig. 4.

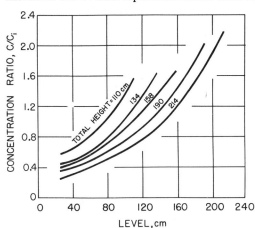

FIG. 4. The effect of total column height on the profile of concentration ratio at steady state in a 2.22-cm. i.d. column with a gas flow rate of 2.53 cm^3/sec and a charge of 4.8×10^{-6} gm/cm^3 crystal violet chloride in water (Shah and Lemlich, 1970). Data points are omitted for clarity; they are shown in Fig. 5.

The separate profiles in Fig. 4 can be effectively merged into a single curve by plotting against the relative level defined by Eq. (13), rather than against the absolute level h:

$$h_r = h - h_0. \tag{13}$$

The level h_0 is that at which $C = C_i$, so that $C/C_i = 1$ at $h_r = 0$. Combining

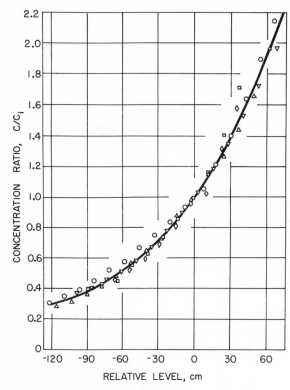

FIG. 5. The correlating effect of relative level on concentration ratio (Shah and Lemlich, 1970). The data are from Fig. 4.

Total column height, cm :	110	134	158	190	214
Symbol :	◇	□	▽	△	○

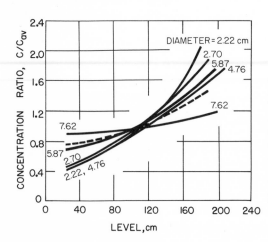

FIG. 6. The effect of column diameter on the profile of concentration ratio at steady state for aqueous solutions of crystal violet chloride (Shah and Lemlich, 1970). Gas flow rate ranges from 2.5 to 32 cm^3/sec, and charge concentration ranges from 0.4×10^{-6} to 5×10^{-6} gm/cm^3. Data points are omitted for clarity. (The dashed curve is for aqueous patent blue in the 2.22-cm. i.d. column.)

Eqs. (5), (7), and (13) is one way of supporting from theory the applicability of relative level (Shah and Lemlich, 1970). The success of relative level in merging the profiles is evident from Fig. 5.

Figure 6 shows that the degree of separation decreases with increasing column diameter. This is attributed to the increase in bubble driven circulation and/or axial eddy diffusion which disperses the concentration gradient.

B. Various Improvements and Continuous Flow Operation

Shah and Lemlich (1970) were unable to reduce this undesirable circulation or eddy diffusion in a worthwhile manner through the use of column packing. The packing promoted bubble coalescence with subsequent loss of adsorptive surface.

Chemical additives yielded better results. Replacing distilled water with tap water increased the separation ratio of crystal violet fivefold! Deliberately adding certain salts was even more effective. Figure 7 shows how the addition

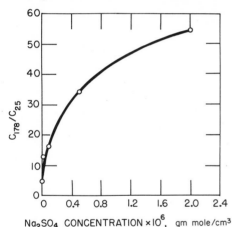

FIG. 7. The effect of sodium sulfate on the separation of crystal violet chloride at a charge concentration of 3.2×10^{-6} gm/cm^3 in water, in a 2.22-cm. i.d. column with gas flowing at the rate of 2.53 cm^3/sec through a fritted glass bubbler (Shah and Lemlich, 1970). The ordinate is the concentration 178 cm up the column divided by the concentration 25 cm up the column.

of Na_2SO_4 can increase the separation ratio of the dye to more than 50. This is a considerable separation for such a simple technique as bubble fractionation.

Lee (1969) employed high shear in his external vortex bubble generator in an effort to form smaller bubbles. This was followed by some aging of the bubbles in order to inhibit subsequent coalescence in the column, plus swirling motion at the bottom of the column to stabilize the flow locally. Using an aqueous solution of sodium dodecylbenzene sulfonate in various columns fed continuously at the top, he reported good agreement between measured profiles and his aforementioned theoretical predictions. For the latter, he utilized mass transfer rates estimated from an independent correlation, and dispersion coefficients determined from separate experiments with ink.[1] He found a separation ratio as high as 61.2 in a column of 1-in. diameter and 180 cm in height. This is a considerably greater separation than he obtained in an earlier study (Lee, 1965).

Kown (1971) also conducted experiments with continuous bubble fractionation, and similarly found higher separation ratios in narrow columns. His principal system was crystal violet chloride in water. A few experiments with detergent and with lignin were carried out as well (Kown, 1970).

In an interesting footnote, Maas (1969) describes his novel "booster bubble fractionation" of nonfoaming dye solutions. This procedure involves the addition of certain volatile organic compounds to the gas in order to improve the selectivity of the separation.

Solvent sublation (Sebba, 1962), which involves the use of an immiscible liquid atop the main liquid to catch the solute, can also be viewed in a sense as being an improvement on bubble fractionation which is suitable for certain purposes. Karger discusses solvent sublation in detail in Chapter 8.

SYMBOLS[2]

a surface of a bubble, cm^2

A column cross section, cm^2

B rate of bottoms withdrawal, cm^3/sec

C concentration of solute in liquid, gm/cm^3

D effective axial diffusivity, cm^2/sec

E parameter, $(N^2 + 4M)^{1/2}$, cm^{-1}

f bubble frequency, sec^{-1}

F rate of feed, cm^3/sec

G volumetric flow rate of gas, cm^3/sec

h level (height) along column, cm

H total height of column, cm

J parameter, Kaf/DA, cm^{-1}

k kinetic rate constant or mass transfer coefficient, cm/sec

K constant, which *may* be for equilibrium, cm

m_1 first parameter for Eq. (10), cm^{-1}

m_2 second parameter for Eq. (10), cm^{-1}

M parameter, Kaf/DAv, cm^{-2}

N parameter, k/vK, cm^{-1}

T rate of overflow, cm^3/sec

v bubble velocity, $dh/d\tau$, cm/sec

β_1 first parameter for Eq. (9), cm^{-1}

β_2 second parameter for Eq. (9), cm^{-1}

Γ surface excess (concentration of adsorbed solute at bubble surface), gm/cm^2

λ volumetric fraction of gas

τ time, sec

Subscripts

av   average

B   bottom

F   feed

i   initial (charge)

r   relative

T   overflow

0   location at which $C = C_i$

25   located 25 cm up the column

178   located 178 cm up the column

[1] But see the last paragraph in the Introduction for comments regarding the possible presence of error due to unnoticed bubble fractionation of dye or ink in dispersion experiments.

[2] Strictly speaking, the material balance requires that liquid flow rates and concentrations be on a mass basis rather than a volumetric basis. However, since the mass fraction of the solute in the liquid is so small, volumetric units (cm^3/sec and gm/cm^3) are employed for greater convenience.

REFERENCES[3]

Anderson, J. L., and Quinn, J. A. (1970). *Chem. Eng. Sci.* **25,** 373.

Cannon, K. D. (1970). M.S. thesis, Univ. of Cincinnati Library.

Cannon, K. D., and Lemlich, R. (1971). A theoretical study of adsorptive bubble fractionation, Paper no. 32b presented at Nat. Meeting Amer. Inst. Chem. Engrs., 69th, Cincinnati, Ohio.[4]

Dorman, D. C., and Lemlich, R. (1965). *Nature* **207,** 145.

Harper, D. O., and Lemlich, R. (1965). *Ind. Eng. Chem. Process Design Develop.* **4,** 13.

Harper, D. O., and Lemlich, R. (1966). *A. I. Ch. E. J.* **12,** 1220.

Kown, B. T. (1970). Rutgers, Univ., private communication.

Kown, B. T. (1971). *Water Research* **5,** 93.

Lee, S. J. (1965). M.S. thesis, Univ. of Tennessee.

Lee, S. J. (1969). Ph.D. dissertation, Univ. of Tennessee.

Lemlich, R. (1964). Prog. Rep. Research Grant WP-00161, submitted to the U.S. Public Health Service.

Lemlich, R. (1966). *A. I. Ch. E. J.* **12,** 802 ; erratum in (1967), **13,** 1017.

Lemlich, R. (1968). *In* "Progress in Separation and Purification" (E. S. Perry, ed.), Vol. I, pp. 1–56. Wiley (Interscience), New York.

Maas, K. (1969). *Separ. Sci.* **4,** 457.

Sebba, F. (1962). "Ion Flotation." American Elsevier, New York.

Shah, G. N., and Lemlich, R. (1970). *Ind. Eng. Chem. Fundamentals* **9,** 350.

[3]A videotape which includes a demonstration of bubble fractionation is cited in Chapter 3 on foam fractionation.

[4]The author anticipates the publication of this paper in *Chem. Eng. Progr. Symp. Ser.*

CHAPTER 8

SOLVENT SUBLATION

Barry L. Karger
Department of Chemistry
Northeastern University
Boston, Massachusetts

I. INTRODUCTION

Solvent sublation is a nonfoaming adsorptive bubble separation process in which enriched material on bubble surfaces is collected in immiscible liquids, rather than in foams. In a particular experiment, surface-active material will be present in a bulk aqueous phase, on top of which is placed an immiscible liquid. Gas bubbles are generated in the aqueous media and are buoyed upward into the organic phase. The bubbles selectively adsorb surface active material while in the water (as in any adsorptive bubble process) and transport this material to the nonaqueous phase. The material is either deposited in the top phase after the bubbles burst at the air–liquid interface or is dissolved during the passage of the bubble through the immiscible phase. In either case selective enrichment in the nonaqueous layer occurs. Figure 1 illustrates a simple extraction column used in solvent sublation in which gas bubbles are generated by sparging through a porous glass frit.

Sebba originated solvent sublation as an auxiliary technique to ion flotation for use in those cases in which a persistent foam existed (Sebba,

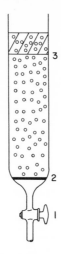

FIG. 1. Simple extraction column for solvent sublation: (1) two way stopcock, (2) porous glass frit, and (3) extraction column.

1962, 1965). The nonaqueous layer then acted both as a collection medium and as a foam breaker (actually a preventor of foam formation). The name, solvent sublation, arises from the fact that an ionic species, called the colligend, is removed by addition of a surface-active collector of opposite charge to that on the colligend. The complex formed by coulombic attraction is called the "sublate," and the process of lifting the sublate by gas bubbles "sublation."

According to Sebba, it is not necessary that the sublate dissolve fully in the organic layer, only that the salt be wetted by the solvent. Thus, both the formation of true solutions and suspensions of the sublate in the organic phase should be possible; however, most examples at the present time have dealt with the formation of solutions in the nonaqueous phase.

Caragay and Karger (1966) and Karger et al. (1967) examined in detail the solvent sublation process for the removal of two dyes, methyl orange (MO) and rhodamine B (RB), using as collector, hexadecyltrimethyl ammonium bromide (HTMAB), and 2-octanol as immiscible organic layer. The pH of the solution was adjusted so that MO was anionic and RB zwitterionic. Consequently, MO was rapidly removed from aqueous media with HTMAB (cationic collector). On the other hand, the rate of RB removal was suppressed by the added HTMAB, the collector successfully competing with RB for adsorption sites on the bubble surface. As a result of this behavior, the separation factor was roughly 100 times greater after 5 min of gas bubbling time, when compared to the results after 3 hr of bubbling. We shall return to this paper shortly.

Sheiham and Pinfold (unpublished data), Spargo and Pinfold (1970), and Pinfold and Mahne (1968) have also been active in examining solvent sublation. Sheiham and Pinfold (unpublished data) studied the sublation of hexadecyltrimethyl ammonium chloride (HTMAC) alone, following the course of its removal from aqueous media into 2-octanol by adding a ^{14}C-labeled spike of HTMAC. Spargo and Pinfold (1970) have recently examined gas bubble distribution in aqueous media during sublation, as well as other parameters; the system used was hexacyanoferrate(II) ions and dodecyl pyridinium chloride with 2-octanol as the immiscible layer. Finally, Pinfold and Mahne (1968) have used solvent sublation to remove floated material, in which suspensions of sublate occur. In this case, small amounts of surfactant are added to the organic phase prior to sublation in order to solubilize previously insoluble precipitates in the immiscible layer.

Elhanan and Karger (1969) have examined the sublation of $FeCl_4^-$ ions with tri-n-octylamine hydrochloride into anisole. Recently, Karger et al. (1970) have restudied the system of MO and HTMAB, examining in detail several parameters. A mechanism of removal of material from aqueous media by sublation has been presented.

At the present time, these publications are the only ones known to this author that deal with solvent sublation (Bittner et al., 1967, 1968).[1] With the new understanding of this separation process, there is a need to explore the general applicability of the method to separation problems both on the small and large scale. The purpose of this review is to examine solvent sublation, especially the unique features of this method. It is hoped that the review will encourage others to study the general applicability of solvent sublation.

II. EXPERIMENTAL DESIGN

A. Apparatus

Two types of sublation cells have been used to date. In the first design (Karger et al., 1967), a column of 4.5-cm diameter and 750-ml capacity, employing as sparger either a coarse- or medium-porosity glass frit, was used. Ordinarily, the volume of aqueous phase in the cell was 300 ml with 25 ml of organic solvent. The flow rate of gas was 5–30 ml/min. In the recent cell design (Spargo and Pinfold, 1970; Karger et al., 1970), the column was 9 cm in diameter and of 2-liter capacity, and contained a fine porosity glass frit with an average pore size of 10 μ. The frit from a Buechner funnel was found to be suitable. The volume of aqueous phase in this case was 1500 ml, while that of the immiscible layer was 25 ml. The flow rate was ordinarily approximately 175 ml/min.

[1]Since the writing of this manuscript, it has come to the author's attention that several ions [Fe(III), Co(II), Ni(II), Th(IV), Pa(V), and U(VI)] were removed from aqueous media by sublation of the complex phosphates, cyanides, and citrates with several collectors.

The difference between the two cells is basically one of capacity. The former contains roughly one-fifth the amount of sublate as the latter for equivalent collector and colligend concentrations in both cells. The potentially longer time necessary for sublation with the large cell is compensated by the smaller size of the gas bubbles and higher gas flow rate (that is, greater gas–liquid interface generated per unit time). Consequently, removal times are roughly comparable in the two cells. The advantage of the large cell arises from the fact that aqueous phase samples can be continuously removed from the cell during a sublation experiment in order to monitor the removal without fear of significantly altering the volume of aqueous phase. Samples can be conveniently removed by inserting a tap half way up the cell. (To obtain a representative sample of the aqueous phase, it is important that the tap not be near the porous glass frit or the liquid–liquid interface.)

Any inert gas can be used for bubbling; however, nitrogen has been found convenient. In the small cell design, the nitrogen was saturated prior to its entrance into the cell. The precaution was found to be unnecessary with the large cell, since 1500 ml of water was employed. Flow rate control can be achieved either with a flow controller (e.g., Moore flow controller) or a bulb of large capacity. The flow is measured just before the column with either a soap bubble flowmeter or a rotometer.

B. Procedure

The aqueous solution is first made up, containing the colligend and collector, at an appropriate pH. If desired, a small amount of ethanol (approximately 1 %) can be included in this solution. The ethanol acts to lower the surface tension of the aqueous phase and thus aids the formation of small bubbles (and therefore larger gas–liquid surface area for a given volume of gas). The solution is added to the cell in the appropriate volume, and then a small known amount of nonaqueous phase is slowly spread over the aqueous phase. The gas flow is commenced and dispersion of the gas throughout the aqueous phase occurs. The flow rate is set such that turbulent mixing of the aqueous and organic layers does not occur. The first few moments of sublation are not stable. The time required to reach steady conditions is significantly shorter with the large volume cell and is considered to be insignificant. For the small volume, low flow rate cell, 1–2 min are required to reach a stable and reproducible system. The removal may be followed by analysis of the organic and aqueous phase; however, because of the small volume of organic phase, it is best to follow the removal by analysis of the aqueous phase.

It can be readily seen that solvent sublation on a batch analytical scale is quite simple to perform. Therein lies one of its major advantages. Reproducibility is roughly 1 % when the system reaches steady state. This compares favorably with other extraction procedures. Finally, it should be noted that there have been attempts at making solvent sublation into a continuous process (Sebba, 1962; Bittner *et al.*, 1967, 1968). For the most part these efforts have not yet met with impressive success.

III. CHARACTERISTICS OF SOLVENT SUBLATION

A. Solvent

A wide variety of immiscible organic layers are potentially possible in solvent sublation; however, certain considerations must be kept in mind. First, the solvent must dissolve, or at least wet, the sublate. Polar solvents are therefore to be desired from this point of view. On the other hand, if the solvent is too polar, it may be too soluble in the water. Consequently, a compromise must be drawn in terms of polarity. A second consideration is solvent volatility. As gas is being passed continuously through the organic phase, it is important that the solvent have low volatility, lest significant evaporation occur. Finally, the viscosity of the organic solvent should be low. With these considerations, 2-octanol has been found to be satisfactory, as has anisole. Presumably, mixed solvents, both of which having low solubility in water, might prove useful.

B. Presaturation of Aqueous Phase

2-Octanol dissolves to some extent in water, with a saturated solution containing 10^{-2} M. This organic solvent is surface-active and will adsorb to a significant extent at the gas–liquid interface when gas bubbles pass through the water. The effect of 2-octanol on the rate of extraction can be seen in Fig. 2, which shows the removal of 10^{-5} M MO with 10^{-5} M HTMAB at pH 10.5 with the large-volume cell. It can be seen that the rate of removal is significantly lower when the aqueous phase is saturated. The removal of MO reaches a constant value after 20 min, but with saturated liquids this required 5 hr. It might be added that this behavior was independent of whether or not the organic layer was initially saturated with water. Hence, it can be assumed that the rate of sublation is based mainly on the character of the aqueous phase. Because the 2-octanol is 10^3 times more concentrated than the collector, it is clear that the slower rate with saturation conditions is due to the competition between 2-octanol and collector for the adsorption sites on the gas bubbles. Such interferences also influence the rate of removal of sublate when no 2-octanol initially exists in the aqueous phase, for the immiscible upper layer will slowly dissolve in the aqueous phase until a saturated solution is reached.

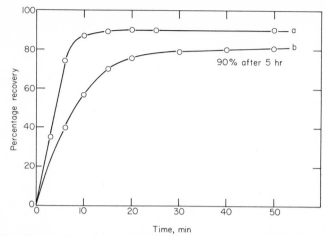

Fig. 2. The influence of 2-octanol in the aqueous phase on the rate of sublation. Conditions: 1500 ml aqueous solution of 10^{-5} M MO and HTMAB plus 10 ml EtOH; 25 ml of 2-octanol; large-cell apparatus; flow rate = 167 ml/min. (a) No 2-octanol initially present in aqueous solution, and (b) aqueous solution initially saturated with 2-octanol.

Ethanol can also slow down the rate of removal of sublate as it is surface active. Indeed in significant amounts (about 5 % by volume) a large influence on the rate is observed. However, as indicated earlier, it is desirable to add a small amount of ethanol to allow the formation of small gas bubbles in aqueous media and thus increase the gas–liquid interfacial area.

C. Bubble Size and Gas-Flow Rate

It is clear that the rate of sublation will be a function of the gas–liquid interface generated per unit time. Thus, the smaller the bubble size for a given flow rate the more rapid the removal of the sublate. In a previous study (Karget *et al.*, 1967) it has been shown that more sublate was removed when a medium-porosity frit was used rather than a coarse-porosity frit for a given flow rate and a given sublation time. It is clear that the rate of removal is related to the gas-flow rate. Thus, for a given porous glass frit, the rate of attainment of steady state is faster the faster the flow rate (Karger *et al.*, 1967; Spargo and Pinfold, 1970). Figure 3 shows a plot taken from the work of Spargo and Pinfold (1970) of the removal of hexacyanoferrate(II) with dodecyl pyridinium chloride as collector versus sublation time for three gas-flow rates. The more rapid attainment of steady state with the higher flow rates is readily seen. In addition, more sublate is removed in steady state for higher flow rates, in agreement with previous results by Karger *et al.* (1967). Thus we see the interesting conclusion that steady state is gas-flow rate dependent.

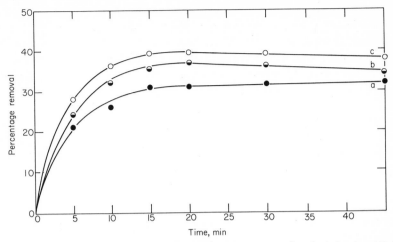

FIG. 3. Removal of 5×10^{-6} gm ion of $Fe(CN)_6^{4-}$ with 2.0×10^{-5} mole dodecyl pyridinium chloride. Conditions: 1500 ml of aqueous solution and 20 ml of 2-octanol. (a) 33 ml/min; (b) 100 ml/min; (c) 167 ml/min.

An interesting question that arises is whether the rate of solvent sublation is directly proportional to flow rate. In order for this to be so it is necessary that the bubble diameter and mass transfer across the liquid–liquid interface be independent of flow rate. In an earlier study (Karger *et al.*, 1967) it was suggested that the rate of extraction decreased as the flow rate increased when the gas volume was normalized. Thus, the product of flow rate and sublation time was maintained constant, and it was found on this basis that the percentage of removal was decreased as the flow rate increased. Recently, Spargo and Pinfold (1970) have measured bubble diameters as a function of flow rate and found that the average bubble diameter and the distribution of bubbles favored larger bubble sizes at increasingly higher flow rates.

D. Volume of Organic Phase

Sebba (1965) suggested that the amount of sublate carried into the organic layer might be independent of the volume of the organic phase. Subsequent experiments have shown this to be true for a volume range of 2-octanol of 20 (Karger *et al.*, 1967, 1970), whether the sampling is made prior to or during steady state operation. This unique characteristic of solvent sublation again points to the fact that aqueous phase conditions exert a greater influence on sublation than nonaqueous phase conditions. Mass transfer will occur mainly from the aqueous phase to the organic layer because of the unidirectional flow of gas bubbles.

The fact that the amount removed is independent of the volume of organic phase (at least as long as the concentration of the sublate does not exceed the saturation value in 2-octanol) has important consequences. First, solvent sublation can be used as a method of concentration in which a thin layer of organic phase is spread on top of a large volume of aqueous phase. The amount extracted in this case would presumably be greater than that obtained by liquid–liquid equilibrium. On the other hand, it is clear that increases in volume of the immiscible layer will eventually reach a point where liquid–liquid equilibrium removes a larger amount of the sublate than does solvent sublation. In addition, the fact that removal is independent of the volume of organic phase points strongly to the belief that solvent sublation and solvent extraction are different processes with different mechanisms of removal.

E. Comparison to Liquid -Liquid Equilibrium

The results obtained for the effect of the volume of the organic phase on the removal of sublate required a careful examination to prove that liquid–liquid equilibrium is indeed not achieved in solvent sublation. This was accomplished by Karger, Pinfold, and Palmer (1970) with a procedure originally devised by Sheiham and Pinfold (unpublished data). Solvent sublation was performed under usual conditions with the large cell using 10^{-5} M MO and HTMAB. After achievement of steady state, the flow of bubbles was stopped, and a stirrer was placed several inches below the 2-octanol layer. After four

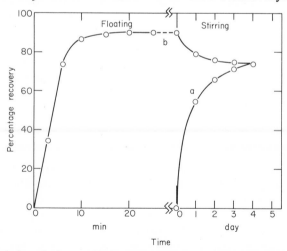

FIG. 4. The relation of solvent sublation to liquid–liquid equilibrium. Conditions: 1500 ml aqueous solution of 10^{-5} M MO and HTMAB plus 10 ml EtOH; 25 ml of 2-octanol; flow rate = 167 ml/min. (a) Attainment of liquid–liquid equilibrium by stirring. (b) Liquid–liquid equilibrium exceeded by solvent sublation but restored by stirring.

days of stirring, the removals of MO into 2-octanol had declined to the steady state value of 74%. Figure 4(b) shows this effect in detail. In another experiment, a solution of the same initial composition was not sublated but only stirred for 4 days in the sublation cell, yielding the behavior in Fig. 4(a) in which the percentage of extraction gradually reached 74%. These results clearly indicate that, using 1500 ml of aqueous solution and 25 ml of 2-octanol, the extent to which MO is removed by solvent sublation is in excess of that attained in liquid–liquid extraction. The value of solvent sublation as a method of concentration is seen in this case.

F. Other Characteristics

Several other characteristics of solvent sublation should be mentioned. First, Spargo and Pinfold (1970) and Karger *et al.* (1967) have shown that complete removal of a colligend requires a stoichiometric excess of collector over colligend. In the case of hexacyanoferrate(II) and dodecyl pyridinium chloride, it was found that more than a sixfold stoichiometric excess of collector was necessary. This result means, of course, that not all the collector transported from the aqueous to the organic phase is tied up with colligend. Second, at a given gas-flow rate, it has been found (Spargo and Pinfold, 1970) that a higher temperature results in poorer recovery. Presumably, less adsorption on the gas bubble surface occurs at the higher temperature.

IV. MECHANISM OF REMOVAL

Recently, Karger *et al.* (1970) have presented a model for the removal of sublate from aqueous media that incorporates many of the characteristics of solvent sublation. This model is not simply the direct mass transfer of enriched bubbles from the aqueous to the organic layer, as suggested by Sebba. Rather, in the light of new evidence, other factors must be taken into account.

The bubbles are enriched by adsorption in their travel through the aqueous media. Upon reaching the liquid–liquid interface, they are unable to overcome the interfacial tension immediately. Rather, coalescence must occur before bubble transfer across the interface occurs. This coalescence was noted by Sebba (1962) and confirmed by our observations. It is expected that repulsions of bubbles caused by the zeta potentials on the bubbles results in slow bubble coalescence. Consequently, a relatively stationary layer of bubbles exists below the liquid–liquid interface (this is readily observable), and the liquid trapped in this layer is effectively protected from the turbulence in the aqueous phase caused by the rising bubbles.

As the coalesced bubbles move through the liquid–liquid interface, they drag up a small amount of the liquid from the interfacial region. This entrapped liquid is considerably less than that obtained in foam separation methods. Collector and colligend carried into the organic layer readily dissolve, and the water surrounding the bubbles returns in the form of droplets of aqueous phase (which are also readily observable) after the bubbles burst. It is quite likely that liquid–liquid equilibrium is established between the water droplets and the organic layer; however, from volume considerations the amount of sublate dissolving in the water droplets should be small. This sublate is returned to the interfacial water layer, and because of the protection given by the stationary bubbles, the sublate does not enter the bulk aqueous phase. Rather, a steady state is established in which the amount of sublate travelling into the organic layer equals that carried back across the interface by returning droplets.

The fact that returning sublate from the octanol layer does not enter the bulk aqueous phase was proved experimentally by Karger *et al.* (1970). After establishing steady state for MO and HDT, a concentrated amount of MO and HDT in octanol was added to the organic layer. The gas flow rate was maintained, and analysis of samples of the aqueous phase revealed no additional sublate with time.

V. RELATIONSHIP TO SOLVENT EXTRACTION

From the above model for the removal of sublate, it is clear that solvent sublation and solvent extraction differ markedly. First, in solvent extraction, liquid–liquid equilibrium exists throughout both liquids by virtue of the intimate contact between the two phases. In solvent sublation, liquid-liquid equilibrium exists only between the organic phase and the water droplets returning to the aqueous phase. Indeed, gas-flow rates are set such that little if any disruption of the liquid–liquid interface occurs. Second, mixing in solvent extraction can cause formation of emulsions, especially if surface-active species are being extracted. The occurrence of emulsions is significantly less in the bubbling procedure used for solvent sublation. Third, the amount extracted in solvent extraction is dependent on the volume ratio of the two phases. As already pointed out, this is not true for sublation. Consequently, sublation is potentially a better concentrating tool than solvent extraction.

Further, it is important to note that in most cases of solvent extraction equilibrium is rapidly reached, even if shaking is very mild. Thus, it is difficult to use rates of extraction as a separation parameter. This is not the case in

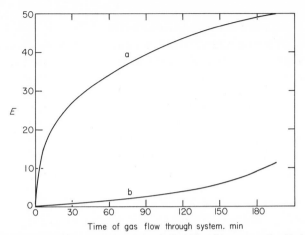

FIG. 5. Extraction coefficient as a function of time. Gas-flow rate = 5 ml/min, small cell apparatus. (a) $10^{-5}\,M$ MO, $10^{-5}\,M$ HDT, pH = 10.5, (b) $10^{-5}\,M$ RB, $10^{-5}\,M$ HDT, pH = 10.5.

solvent sublation. Indeed the rate of gas flow and the size of the cell can be useful parameters to control the rate of removal. This control can be important for separation purposes and for studying the mechanism of removal.

As an example of the use of rate of removal in separation, Karger and Caragay (1966) studied the separation of MO and RB by solvent sublation using HDT as collector. Figure 5 shows the removals of the two dyes as a function of gas bubbling time. In Figure 5(a) we see the removal of MO, which follows the usual trend in sublation of a colligend and an oppositely charged collector. RB is zwitterionic at the operating pH, so that the complex of HDT and RB is unlikely. On the other hand, RB is surface active by itself and is readily sublated into 2-octanol. Addition of HDT to the aqueous phase suppressed the rate of removal of RB, since the collector will effectively compete with the dye for adsorption sites on the gas bubble surfaces. Figure 5(b) shows the removal of RB as a function of time. It is clear from this figure that at short gas flow times, the separation factor of MO/RB will be large, while after longer time periods separation becomes poorer. For example, after 5 min the separation factor was 510, while after 180 min it dropped to only 6.5. In solvent extraction, the separation factor was only 1.7. Thus, control of the rate of removal can sometimes be useful in the separation of colligends.

VI. CONCLUSION

On the analytical scale, solvent sublation deserves more study. The process is a mild one, lending itself nicely to surface-active biochemical molecules.

The process is very simple and effective at the dilute concentration level. We have successfully worked down to the 10^{-7} M level. Relative to other adsorptive bubble methods, solvent sublation has potential for the selective removal of colligends, since very little bulk aqueous phase travels into the organic layer. In ion flotation, on the other hand, a significant amount of bulk liquid travels into the foam layer, and this factor decreases separation.

Since the rate and amount of sublate removed by solvent sublation is flow-rate dependent, careful control of conditions is required. Consequently, reproducibility may be more of a problem in this method than in solvent extraction. (The organic layer can also evaporate to some extent during a run.) In addition, separation can be understood and predicted from simple thermodynamic arguments in extraction; the same is not true for sublation. Thus more work is needed in understanding the sublation process. The greatest need at the present time, however, is to ascertain the general applicability of the method. Thus, research into the use of sublation in the removal of a variety of substances would prove very useful. Finally, while not successful at present, solvent sublation may still prove to be useful on the large scale. The process may be an inexpensive one for the concentration of colligends from large volumes of aqueous solutions into small volumes of organic solvents.

REFERENCES

Bittner, M., Mikulski, J., and Szeglowski, Z. (1967). *Nukleonika* 12, 599.
Bittner, M., Mikulski, J., and Szeglowski, Z. (1968). *Chem. Abstr.* 69, 98530.
Caragay, A. B., and Karger, B. L. (1966). *Anal. Chem.* 38, 652.
Elhanan, J., and Karger, B. L. (1969). *Anal. Chem.* 41, 671.
Karger, B. L., Caragay, A. B., and Lee, S. B. (1967). *Separ. Sci.* 2, 39.
Karger, B. L., Pinfold, T. A., and Palmer, S. E. (1970). *Separ. Sci.* 5, 603.
Pinfold, T. A., and Mahne, E. J. (1968). *J. Appl. Chem.* (*London*) 19, 188.
Sebba, F. (1962). "Ion Flotation." American Elsevier, New York.
Sebba, F. (1965). *A. I. Ch. E.—I. Chem. E., Symp. Ser.* 1, 14.
Sheiham, I. and Pinfold, T. A. Unpublished results.
Spargo, P. E., and Pinfold, T. A., *Separ. Sci.* 5, 619 (1970).

CHAPTER **9**

FOAM SEPARATION OF ENZYMES AND OTHER PROTEINS

Stanley E. Charm

New England Enzyme Center
Tufts University Medical School
Boston, Massachusetts

I. INTRODUCTION

A. History

The concentration and purification of proteins on the basis of their surface properties has been considered for the past 35 years. One of the earliest applications was the ingenious use of air to concentrate the small amount of protein in the starch wash water from industrial starch manufacturing. This protein was ultimately dried and used for animal feed (Ostwald and Siehr, 1937a).

It has long been known that proteins accumulate in the foam of their solutions and that beer foam contains more proteins and acids than the residual liquid. Beer foams have been shown, by analysis, to contain 73 % protein and 10 % water. There are a number of experiments cited in the literature describing the concentration and purification of enzyme and other proteins, (London and Hudson, 1953; Ostwald and Siehr, 1937b; Mischke, 1940; Davis et al., 1949; Nissen and Estes, 1941; Bader and Schutz, 1946; Schutz, 1937; Bader, 1944; Ostwald and Mischke, 1940; Perri and Hazel, 1946).

Enzymes are special proteins in that they exhibit biological activity in controlling the rate of specific biological reactions. The two enzymes pepsin and rennin are so difficult to separate from each other that they were at one time believed to be identical. However, when a solution of purified pepsin at pH 1.2–2.0 was frothed, the pepsin activity tended to concentrate in the froth while the rennin activity remained in solution (Andrews and Schutz, 1945). Also, gonadotropic hormones in the urine of pregnant women have been concentrated in foam (Courrier and Dognon, 1939). Urease and catalase were separated by foaming, the highest purification occurring near the isoelectric point (London et al., 1954). The greatest enrichment in the foam fractionation of bovine serum albumin also occurred at the isoelectric point with as much as a 20-fold increase in the foam (Schnepf and Gaden, 1959).

B. Protein Structure

Proteins are among the most complex molecules. They are linear polymers of 21 amino acids which may be arranged in many combinations. They are often crosslinked but not branched. To date, the arrangement of only a dozen proteins such as the enzyme lysozyme, has been established. Lysozyme attacks many bacteria by lysing or dissolving the mucopolysaccharide structure of the cell wall. It is estimated that several million enzymes exist.

The primary bonding in protein is the peptide bond between the carboxyl group and the amino group of the amino acids. The most common types of crosslinking are a covalent disulfide bridge with a bond strength of 50 cal/mole, and a weaker hydrogen bond of about 6 kcal/mole.

Protein molecules possess hydrophilic and hydrophobic sections. All of the charged polar groups of lysozyme are on its surface, as are its uncharged polar groups, with one or two exceptions. The great majority of its nonpolar hydrophobic groups are buried in the interior.

The nonpolar side chains help to hold the molecule together. When a hydrocarbon chain is in an aqueous medium, it forces the neighboring water molecules to form a cagelike structure in the immediate vicinity. This restricts the motion and number of possible arrangements of the water molecules, lowering their entropy.

For every nonpolar hydrophilic side chain of a protein that is removed from an aqueous to a nonpolar environment the protein gains an extra 4 kcal of free energy stabilization, chiefly from the entropy effect. This makes the segregation of hydrophobic side chains a powerful factor in stabilizing a protein in aqueous solution.

It is not surprising that when a protein is exposed to an interface, e.g., water–air, the hydrophobic–hydrophilic bonding is stressed. The stress may be sufficient to break a bond and denature the protein. Even the rupture of a bond remote from the active site of an enzyme may, in turn, upset the balance of forces in the structure and ultimately destroy the active site. A denatured protein cannot carry out its biological function but may be useful as a source of food.

The structure of a protein at a surface or interface may be different from that in the bulk solution. The interface stresses may cause infolding of the protein structure. Either dilute or concentrated protein films are formed. In dilute films all the molecules are in the same extensively unfolded state, while concentrated films may contain only native and unfolded molecules or molecules in many different degrees of unfolding (see Evans et al., 1970). The structure of proteins at air–water interfaces is often deduced from studies of force–area–concentration diagrams using a Langmuir trough.

Many proteins are naturally associated with surfaces, or interfaces, e.g., in cell membranes. It is also known that certain proteins are denatured upon contact with surfaces, for example, albumin and hexokinase. However, many proteins that are enzymes, contrary to general belief, do not become denatured when in contact with surfaces, for example, tripeptide synthetase, lactic acid dehydrogenase, catalase, amylase, cellulase, D-amino acid oxidase, aryl pyruvate keto-enol, and automerase. Human liver alcohol dehydrogluase suffered a 5–20 % loss in activity with a fivefold purification in foaming (Charm et al., 1966).

II. FUNDAMENTAL CONSIDERATIONS

A. Formation of Surface Phase

When a surface is presented to a protein solution, after some time has elapsed, protein molecules move to the interface, increasing the concentration there. At equilibrium, due to the differences between force fields at the interface and the bulk solution, a concentration gradient exists, as shown in Fig. 1.

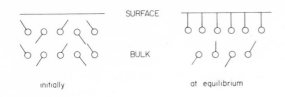

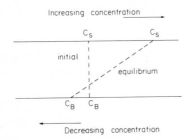

FIG. 1. Surface and bulk phase equilibrium.

The chemical potential associated with the greater concentration of protein at the interface equals the chemical potential of the bulk solution at equilibrium.

At equilibrium

$$(\mu_1)_B = (\mu_1)_S, \tag{1}$$

where $(\mu_1)_B$ is the chemical potential of protein component in bulk solution, and $(\mu_1)_S$ is the chemical potential of protein component in surface phase. The time required to achieve equilibrium for proteins may vary from minutes to hours (Evans et al., 1970).

The difference in chemical potential between the surface phase and the bulk phase supplies the energy or "driving force" for the movement of protein molecules from the bulk to the surface or interface phase. At zero time, there is no concentration gradient between the surface and the bulk, and the driving force is at its maximum; at equilibrium when a concentration gradient exists, the driving force is zero (see Fig. 1).

Studies of surface pressure versus time using a Langmuir trough indicate the unfolding time of a protein. The long-term aging effects shown by the gradual change in surface pressure in adsorbed films of lysozyme, for example, may be attributed to the slow unfolding and reorientation of native molecules in the adsorbed surface films.

B-casein solution shows no appreciable aging or unfolding effects over long periods but appears to reach a stable configuration within 25 min.

B. Driving force for Protein Movement from Bulk to Surface

The driving force for movement of protein from bulk phase to surface phase is determined by the chemical potential difference between the phases.

The chemical potential for a protein in the surface phase is

$$(\mu_1)_S = (\mu_1)_S^\circ + RT \ln(C_1)_S, \tag{2}$$

and in the bulk phase

$$(\mu_1)_B = (\mu_1)_B^\circ + RT \ln(C_1)_B, \tag{3}$$

where R is the gas constant, T is the absolute temperature, $(\mu_1)_S$ is the chemical potential of component 1 in the surface phase, $(\mu_1)_S^\circ$ is the chemical potential of component 1 in the surface phase under standard conditions, and $(C_1)_S$ is the concentration of protein in the surface phase, in moles/cm^3. The subscript B refers to the bulk phase.

The chemical potential difference, which is the driving force for transfering components between phases, is the difference between Eqs. (2) and (3):

$$(\mu_1)_B - (\mu_1)_S = (\mu^\circ)_B - (\mu^\circ)_S + RT \ln[(C_1)_B/(C_1)_S], \tag{4}$$

where $(\mu_1^\circ)_B - (\mu_1^\circ)_S$ is known as the heat of desorption.

The rate of transfer for a protein from the bulk to surface phase is

$$dN/d\theta = KA[(\mu_1^\circ)_B - (\mu_1^\circ)_S] + \ln[(C_1)_B/(C_1)_S], \qquad N/V = (C_1)_S, \tag{5}$$

$$(dC_1)_S/d\theta = KA/V\{(\mu_1^\circ)_B - (\mu_1^\circ)_S + \ln[(C_1)_B/(C_1)_S]\}, \tag{6}$$

where K is the mass transfer coefficient from bulk to surface phase, A/V is the surface area per unit volume, N is the number of moles, and θ is the time.

The term V/KA represents the resistance to protein transfer from the bulk phase to the surface.

At equilibrium, the chemical potentials of the phases are equal by Eq. (1), so that equating Eqs. (2) and (3):

$$-RT \ln[(C_1)_B/(C_1)_S] = (\mu_1^\circ)_B - (\mu_1^\circ)_S = \lambda, \tag{7}$$

where λ is the heat of desorption.

Thus, the maximum concentration of the surface active component is determined by the heat of desorption and the concentration in the bulk phase. The back-diffusion effect associated with the concentration gradient is included in λ.

C. Relationship between Heat of Desorption and Surface Tension

The heat of desorption may be related to surface tension by the Gibbs equation which also expresses a relationship between surface phase and bulk phase concentrations at equilibrium. In the Gibbs notation, the term *surface excess*, Γ^G, is employed. The Gibbs equation as stated heretofore may be derived from thermodynamic considerations (e.g., Davies and Rideal, 1966). In a two-component system, for example, protein and water, if Γ_1 is the number of moles of component 1 per unit area at the surface and Γ_2 is the number of moles of component 2 per unit area at the surface, then the Gibbs surface excess of component 1 is

$$\Gamma_1^G = \Gamma_1 - (C_1)_B \Gamma_2 / (C_2)_B. \tag{8}$$

The relationship between Γ^G and surface tension is (Gibbs, 1928)

$$\Gamma_1^G = \frac{-1}{RT} \frac{d\gamma}{d \ln(C_1)_B}, \tag{9}$$

where $d\gamma / [d \ln(C_1)_B]$ is the change of surface tension with change in the natural logarithm of the concentration.

If Eq. (9) is divided by the thickness of the surface phase t the moles per unit area may be converted to moles per unit volume, and

$$\frac{\Gamma_1^G}{t} = \frac{\Gamma_1}{t} - \frac{(C_1)_B}{(C_2)_B} \frac{\Gamma_2}{t} = (C_1)_S - \frac{(C_1)_B}{(C_2)_B} (C_2)_S, \tag{10}$$

where $(C_1)_S$ and $(C_2)_S$ are the concentration of components 1 and 2 in the surface phase, respectively.

Also

$$(C_1)_S - (C_1)_B \frac{(C_2)_S}{(C_2)_B} = \frac{-1}{tRT} \left[\frac{d\gamma}{d \ln(C_1)_B} \right] = \frac{-(C_1)_B}{tRT} \left[\frac{d\gamma}{d(C_1)_B} \right] \tag{11}$$

From Eq. (7), $(C_1)_S / (C_1)_B$ is related to the heat of desorption by

$$(C_1)_S / (C_1)_B = e^{\lambda/RT}. \tag{12}$$

Relating Eqs. (11) and (12) through $(C_1)_S / (C_1)_B$,

$$(1/tRT)[-d\gamma/d(C_1)_B] + (C_2)_S / (C_2)_B = e^{\lambda/RT}. \tag{13}$$

For dilute solutions where $(C_2)_S \cong (C_2)_B$,

$$(1/tRT)[-d\gamma/d(C_1)_B] + 1 = e^{\lambda/RT}. \tag{14}$$

The term $d\gamma/d(C_1)_B$ is determined from a surface tension–concentration diagram (see Fig. 2).

The surface phase thickness t obviously does not refer to a monolayer at the surface, but rather to that thickness through which surface forces exert their effect. It may be calculated from Eq. (14).

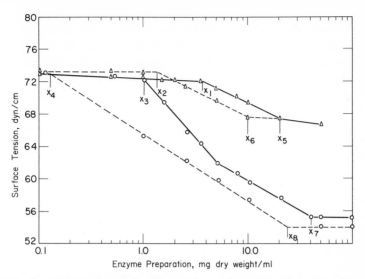

FIG. 2. Surface tension–concentration diagram for catalase and amylase: ($\triangle$ — $\triangle$) amylase in water; ($\triangle$---$\triangle$) amylase in 10% $(NH_4)_2SO_4$; ($\bigcirc$——$\bigcirc$) catalase in water; ($\bigcirc$---$\bigcirc$) catalase in 10% $(NH_4)_2SO_4$.

D. Measurement of Protein Surface Tension

The Wilhelmy method for surface tension is considered to be the best for protein surface tension measurements (Bull, 1945). This is because it is an equilibrium method in contrast to other procedures that require breaking a film, for example, the du Nouy ring method. Breaking a film of protein often becomes complicated with elasticity which interferes with the surface tension measurement unless great care is exercised. The Wilhelmy method for surface tension measures the liquid pull on a flat plate in the surface, as shown in Fig. 3.

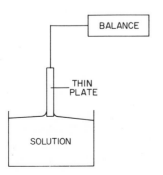

FIG. 3. Wilhelmy plate method for surface tension measurements.

The apparatus consists simply of a thin strip of glass dipping in the liquid under investigation and suspended from an arm of an analytical balance. The weight of the dry slide in air and its weight while dipping into the liquid are determined. After the buoyancy correction of the liquid displaced by the slide is added and the weight of the slide in air subtracted from the weight when dipping in the liquid, the surface tension can be calculated directly. (The net pull of the surface and the slide in grams is multiplied by the acceleration of gravity and divided by twice the length of the slide. The result is the surface tension of the liquid.) A requirement of this method is that the slide must be completely wetted by the liquid i.e., zero contact angle.

E. Surface Tension - Concentration Diagram

As shown from the Eq. (9), the surface tension–concentration diagram expresses a relationship between the bulk and surface concentration at equilibrium.

The surface tension–concentration diagram is a plot of surface tension versus the natural log of concentration. In Fig. 2, this diagram is plotted for the enzymes amylase and catalase in water and in 10% ammonium sulfate.

The general shape of the diagrams are similar. In dilute concentrations the slopes of the plots are zero, which indicates no concentration difference exists between surface and bulk phases. At a critical concentration, the slopes become negative, indicating a concentration difference between surface and bulk. Finally, at high concentrations, the slope is zero again. At this point, the protein molecules have formed micelles or aggregates and no longer behave as individual molecules.

In a mixture of two proteins, for example, catalase and amylase, where their concentrations were such that one showed a zero slope and the other a negative slope, it would be expected that one would tend to concentrate at the surface while the other would not. It is this that forms the basis of the separation and concentration of proteins through foaming.

III. FOAM SEPARATION OF PROTEINS

A. Experiment

By the passage of bubbles through a protein solution, it is possible to present a new gas surface to the solution continuously and remove those proteins from solution that have a tendency to collect at the interface. Usually this is done by collecting the foam that forms above the solution.

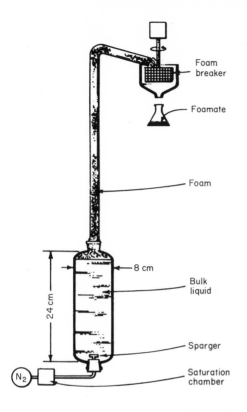

FIG. 4. Apparatus for foaming.

A simple apparatus for carrying this out is shown in Fig. 4. Saturated nitrogen is passed through a sparger into the solution forming foam which is broken after separation from the solution.

Using this technique, it has been possible to effect the separation of catalase and amylase from solution by the judicious selection of concentration as suggested by their surface tension–concentration diagrams (Fig. 2).

This was done with virtually no loss in activity. In this case amylase remained in the bulk solution while catalase was collected in the foam, as shown in Table I (Charm et al., 1966).

The employment of ammonium sulfate affects the solubility and the surface tension–concentration diagram, making it possible to carry out this separation. Varying pH in this case did not influence the surface tension–concentration diagram.

The usual problem in protein isolation is the separation of one protein from a mixture of proteins whose surface tension–concentration characteristics are unknown. As a first approach, an attempt is made to produce a concentration range where differences in the surface tension–concentration

TABLE I

FOAM SEPARATION OF MIXTURES OF CATALASE AND AMYLASE

Foaming agent	Fraction	Total vol. (ml)	Protein		Catalase (1.0 mg/ml)		Amylase (1.0 mg/ml)	
			mg/ml	Total mg	Total activity	Specific activity	Total activity	Specific activity
10% (NH$_4$)$_2$SO$_4$	Initial	970	0.207	200	80.4	40.2	23.2	116
	Foamate	53	0.406	22	49.3	222.4	1.2	58
	Residue	917	0.173	158	18.6	11.8	20.7	131
	Recovery	100%	—	90%	84.5%	—	94.5%	—
50% (NH$_4$)$_2$SO$_4$	Initial	1078	0.124	135	10.5	7.8	19.8	148
	Foamate	420	0.242	101	4.6	4.5	19.3	190
	Residue	628	0.028	18	3.7	20.3	0.3	14
	Recovery	100%	—	89.5%	78.2%	—	99%	—

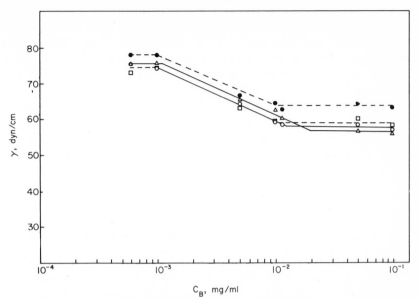

FIG. 5. Surface tension–concentration diagram for LDH mixture at various temperatures:
(●) 3°C ; (□) 20°C ; (○) 24°C ; (△) 29°C.

diagram would occur for various proteins or groups of proteins. It is in dilute
solution that great differences in slope are most likely to occur for various
proteins, as may be seen from Fig. 2.

This approach was used in the isolation of the enzyme lactic acid dehydrogenase (LDH) from chick heart. The chick heart was ground in a 0.1 M

TABLE II

CHARACTERISTICS OF FOAMATE FRACTIONS TAKEN FROM A FOAM SEPARATION EXPERIMENT[a]

Sample	LDH activity per ml solution	Protein concentration, mg/ml	Specific activity
Initial mixture	25 ± 0.6	1.91 ± 0.023	13.1 ± 1.4
Residue	23.3	0.81	28.8
Foamate 1	25	2.57	9.7
Foamate 2	23.3	2.47	9.4
Foamate 3	26.6	2.19	12.1
Foamate 4	30.0	2.12	14.2
Foamate 5	26.6	1.67	15.9
Foamate 6	26.6	1.56	17.1

[a]The foamate fractions are 20 ml.

potassium phosphate buffer of pH 7. The suspension was centrifuged and ammonium sulfate added to the supernatant to salt out an unwanted protein fraction, which then was separated by centrifugation. The supernatant from this centrifugation was subjected to foaming after dilution. The surface tension–concentration diagrams of this protein mixture is shown for various temperatures in Fig. 5.

Foamate fractions collected were assayed with the results shown in Table II. Protein in the various foamate fractions increased at first and then decreased, while the specific activity (activity units/mg protein) increased in the residue. Thus, the purification of LDH is about twofold but some loss is experienced due to some LDH passing into the foamate with the extraneous protein.

TABLE III

RESULTS OF THREE FOAMING EXPERIMENTS WITH LDH

Sample	LDH activity per ml solution	protein concentration, (mg/ml)	LDH specific activity
Initial mixture	30 ± 0.6	1.33 ± 0.023	22.6 ± 1.4
Foamate	31.7	1.235	25.6
Residue	30	0.606	49.6
Initial mixture	13.7	1.046	13.1
Foamate	13.7	1.32	10.4
Residue	13.7	0.507	27.0
Initial mixture	14.15	1.08	13.1
Foamate	14.15	1.19	11.9
Residue	14.15	0.595	23.8

TABLE IV

EFFECT OF AMMONIUM SULFATE ON LDH PURIFICATION

Initial mixture specific activity	Residue specific activity	$(NH_4)_2SO_4$ (%)
15.3	31.0	0
22.6	33.2	0
26.0	107.0	20
19.2	51.2	20
41.4	104.0	20
18.9	52.0	20
45.2	92.5	30
24.1	47.6	0

Generally, the LDH is not concentrated but rather is purified through the removal of extraneous protein. In Table III, the results of several foaming experiments are given, where it is shown that the concentration of LDH (activity/ml) remains constant but the extraneous protein is removed in the foam. Under these experimental conditions, it is possible to achieve about a twofold purification of LDH. With the addition of 20% ammonium sulfate, a fourfold increase in specific activity may be achieved, as shown in Table IV.

B. Analysis of Batch Foaming

There is a direct mathematical analogy between batch foaming and batch distillation (Grieves et al., 1963), and mathematical analysis of batch foaming may offer some insight into the factors controlling purification of proteins by this method. The analysis is basically a material balance on the foam system shown in Fig. 6.

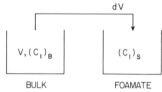

FIG. 6. Material balance for batch foaming system.

Let V be the volume of bulk phase. If differential volume dV is foamed from the bulk phase, then, by material balance,

$$(V - dV)[(C_1)_B - d(C_1)_B] + dV[(C_1)_S + d(C_1)_S] = V(C_1)_B. \quad (15)$$

Simplifying and dropping second-order differentials, this becomes

$$dV/V = d(C_1)_B/[(C_1)_S - (C_1)_B]. \quad (16)$$

The composition of the bulk and surface or foam phases is given by the equilibrium relationship of Eq. (12), i.e., $(C_1)_S = (C_1)_B e^{\lambda/RT}$.

Substituting in equation (16) for $(C_1)_S$ and integrating

$$\int_V^{V_0} \frac{dV}{V} = \int_{(C_1)_B}^{(C_1)_{B_0}} \frac{d(C_1)_B}{(C_1)_B(e^{\lambda/RT} - 1)}, \quad (17)$$

where V_0 and $(C_1)_{B_0}$ are the initial volume and bulk concentration of component 1, respectively.

Completing the integration,

$$\ln \frac{V_0}{V} = \frac{1}{(e^{\lambda/RT} - 1)} \left\{ \ln \left[\frac{(C_1)_{B_0}}{(C_1)_B} \right] \right\}. \quad (18)$$

Equation (18) is the relationship between the volume and composition of the bulk phase, assuming that equilibrium is achieved in foaming between the bulk and surface phases. This also provides a method for experimentally determining λ, the heat of desorption, provided the foaming is carried out sufficiently slowly to achieve equilibrium. A value of λ for the LDH protein mixture is estimated to be 5530 cal/mole (Potash, 1968).

C. Evaluation of the Mass Transfer Coefficient

The mass transfer coefficient K may be found by integrating Eq. (6):

$$\int_{(C_1)_{S_0}}^{(C_1)_S} \frac{d(C_1)_S}{\lambda - RT \ln (C_1)_S/(C_1)_B} = K \frac{A}{V} \int_0^\theta d\theta = K \frac{A}{V}\theta, \qquad (19)$$

where $(C_1)_{S_0}$ is the surface phase concentration at zero time and is equal to $(C_1)_{B_0}$, and θ is the time corresponding to $(C_1)_S$.

The term $(C_1)_B$ may be related to $(C_1)_S$ by a simple material balance, i.e.,

$$V_{B_0}(C_1)_{B_0} = (V_B)(C_1)_B + V_S(C_1)_S, \qquad (20)$$

where V_S is the surface phase volume.

If it is assumed, as a first approximation, that the thickness of the surface phase does not vary with the movement of component 1 into it, then the volumes of the bulk and surface phases remain constant while equilibrium is reached. $V_{B_0}/(V_{B_0} - V_B)$ is constant while equilibrium is achieved. If the surface phase thickness is known, then V_S is found from $A \times t$ and $V_S + V_B = V_{B_0}$.

Then, at any time on the way to equilibrium the concentration in the surface phase is related to the bulk phase concentration by

$$(C_1)_B = \frac{(C_1)_{B_0}(V_{B_0}) - (V_S)(C_1)_S}{(V_{B_0} - V_S)}. \qquad (21)$$

Also

$$(C_1)_{B_0} = (C_1)_{S_0}.$$

A graphical evaluation of the left side of Eq. (19) may be made by plotting $\{\lambda - RT[\ln(C_1)_S/(C_1)_B]\}^{-1}$ vs. $(C_1)_S$, as shown in Fig. 7. The initial value of $(C_1)_S$ is $(C_1)_{S_0}$ and the area under the resulting curve is $K(A/V)\theta$.

The mass transfer coefficient K may be evaluated by passing bubbles through a protein solution, collecting the first fraction of foam and a sample of residue. If the foam is the surface phase, then the concentration of protein in the foam is $(C_1)_S$ and the composition of the residue is $(C_1)_{B_0}$ or $(C_1)_{S_0}$.

The term A/V may be determined from the size and number of bubbles

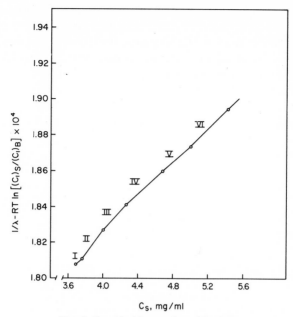

FIG. 7. Graphical integration of Eq. (19).

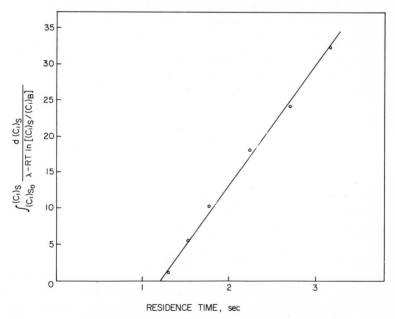

FIG. 8. Determination of the mass transfer coefficient K.

in the solution at any time, while the residence time of the surface phase θ is found from the flow rate of bubbles and the height of the foam column. With this information it is possible to solve Eq. (19) for K, as shown in Fig. 8.

Using this method, K for an LDH–protein mixture was obtained from a $(C_1)_S$ versus time experiment utilizing foam apparatus with a diameter of 2.5 cm at a gas rate of 132.5 cm^3/min (or a bubble rate of 8.5 cm/sec).

A running graphical integration of the integral in Eq. (19) was then performed (Fig. 7) and the results were plotted against the residence time (Fig. 8). From the slope of this last plot,

$$K = 2.16 \times 10^{-12} \quad \text{moles/cal cm}^2 \text{ sec.}$$

A repeat experiment yielded a K of 1.91×10^{-12} moles/cal cm^2 sec. The value of K is constant with respect to bubble flow rate (Potash, 1968) and column diameter. In these studies bubble velocity was varied from 8 to 11 cm/sec and column diameter from 2.5 to 4.0 cm.

D. Calculation of Time to Reach Equilibrium

When bubble residence time is less than 1.2 sec, there is no increase in protein concentration in the surface phase. Therefore, in calculating the time to attain equilibrium, this lag time should be added to the calculated value. The time to achieve equilibrium now may be calculated from Eq. (6).

At equilibrium $d(C_1)_S/d\theta = 0$ and λ must equal $RT \ln[(C_1)_S/(C_1)_B]$, as indicated in Eq. (7). The time to reach equilibrium may be calculated by considering Eq. (6) in an incremental form. Thus, for the first time interval

$$[(C_1)_{S_1} - (C_1)_{S_0}]/\Delta\theta = (KA/V)[\lambda - RT \ln(C_1)_{S_0}/(C_1)_{B_0}]$$

since

$$(C_1)_{S_0} = (C_1)_{B_0}, (C_1)_{S_1} = (C_1)_{S_0} + (KA/V)\lambda \, \Delta\theta.$$

For the second time interval

$$(C_1)_{S_2} = (C_1)_{S_1} + (KA/V)\Delta\theta \{\lambda - RT \ln[(C_1)_{S_1}/(C_1)_{B_1}]\}.$$

$(C_1)_{B_1}$ is determined by material balance using Eq. (20), the values of the bulk and surface phases being known.

These iterations are continued until

$$\lambda = RT \ln[(C_1)_S/(C_1)_B].$$

The sum of all the time intervals ($+ 1.2$-sec lag time) is the time to achieve equilibrium. Setting $\Delta\theta = 0.1$ sec and with $(C_1)_{B_0} = 3.69$ mg protein/ml, $\lambda = 5530$ cal/mole, it is calculated that 68 sec is the time to achieve equilibrium with the LDH protein mixture. This relatively long time to reach equilibrium may in part explain why a slow bubble velocity results in a more effective removal of protein as often noted in the literature (Karger and De Vivo, 1968).

Thus, from this analysis, the factors affecting the surface phase concentrations include λ, K, A/V, V_S, V_B, C_{B_0}, bubble diameter, surface thickness, and bubble velocity.

E. Purification by Addition of Surfactants

A number of surfactants have been tested for their capacity to effect a more efficient separation of proteins in an LDH–protein mixture.

The surfactants used over a wide range of pH included lauryl pyridinium chloride and hexadecyltrimethyl ammonium bromide, (cation ionic agents) and sodium lauryl sulfate, an anionic agent. None of these agents improved the purification of LDH.

F. Foam Fractionation of Proteins with Reflux

Fanlo and Lemlich (1965), suggested a foam fractionation with reflux. Part of the foamate was returned to the column to serve as the reflux.

When this method is used with the LDH protein mixture, the repeated foaming and foam breaking of enzyme solutions ultimately causes protein denaturation and enzyme inactivation, although a fourfold improvement in protein concentration could be achieved as compared with twofold in the batch operation.

Thus, the reflux foam column suggested by Fanlo and Lemlich does not appear to be a useful method for significantly improving separation when *active* proteins are involved.

IV. CLOSURE

In most cases, foaming improves enzyme purification two- to fourfold. On a laboratory scale, more convenient methods are available, for example, ultrafiltration, gel filtration, and ion exchange. However, on an industrial scale where economics are an important consideration, foaming may be an appropriate technique to employ.

With the increasing interest in the use of microbial protein as a food source, foaming may find a place in the large-scale separation of nucleic acids from proteins, which is one of the purification steps that must be carried out with this material.

Another area where foaming might find application is in the recovery of an economically valuable biologically active protein from a waste material. With pollution becoming such a prominent problem in processing, this technique may afford a method for reducing the cost of pollution control.

REFERENCES

Andrews, G., and Schutz, F. (1945). *Biochem. J.* **39**, 1i.
Bader, R. (1944). *Nature* **154**, 183.
Bader, R., and Schutz, F. (1937). *Trans. Faraday Soc.* **42**, 571.
Bull, H. B. (1945). "Physical Biochemistry," p. 189. Wiley, New York.
Charm, S. E., Morningstar, J., Matteo, C. C., and Paltiel, B. (1966). *Anal. Biochem.* **15**, 498.
Courrier, R., and Dognon, A. (1939). *C. R. Acad. Sci. Paris* **202**, 242.
Davies, J. T., and Rideal, E. K. (1966). "Interfacial Phenomena," p. 154. Academic Press, New York.
Davis, S. G., Fellers, C. R., and Esselen, W. B. (1949). *Food Res.* **14**, 417.
Evans, M. T. A., Mitchell, J., Mussellwhite, P. R., and Irons, L. (1970). *In* "Surface Chemistry of Biological Systems" (M. Blank, ed.), p. 1. Plenum Press, New York.
Fanlo, S. and Lemlich, R. (1965). *A. I. Ch. E.-I. Chem. E. Symp. Ser.* **9**, 75, 85.
Gibbs, J. W. (1928). "Collected Works," Vol. 1. Longmans, Green, New York.
Grieves, R. B., Kelman, S., Obermann, W. R., and Wood, R. K. (1963). *Canad. J. Chem. Eng.* **41**, 252.
Karger, B. L., and DeVivo, D. G. (1968). *Separ. Sci.* **3**, 393.
London, M., Cohen, M., and Hudson, P. B. (1954). *Biochim. Biophys. Acta* **13**, 111.
London, M., and Hudson, P. B. (1953). *Arch. Biochem. Biophys.* **46**, 141.
Mischke, W. (1940). *Wochschr. Brau.* **54**, 63.
Nissen, B. H., and Estes, C. (1941). *Amer. Soc. Brew. Chem. Proc.* **1940**, 23.
Ostwald, W., and Mischke, W. (1940). *Kolloid-Z.* **90**, 205.
Ostwald, W., and Siehr, A. (1937a). *Kolloid-Z.* **79**, 11.
Ostwald, W., and Siehr, A. (1937b). *Chem. Z.* **61**, 649.
Perri, J. M., and Hazel, F. (1946). *Ind. Eng. Chem.* **38**, 549.
Potash, M. (1968). M.S. Thesis, Dept. of Chem. Eng., Tufts Univ.
Schnepf, R. W., and Gaden, E. L., Jr. (1959). *J. Biochem. Microbil. Technol. Eng.* **1**, 1.
Schutz, F. (1937). *Nature* **154**, 629.

| CHAPTER 10 | FOAM FRACTIONATION OF SURFACTANTS, ORTHOPHOSPHATE, AND PHENOL |

Robert B. Grieves
Department of Chemical Engineering
University of Kentucky
Lexington, Kentucky

I. BATCH AND CONTINUOUS SEPARATION OF CATIONIC AND ANIONIC SURFACTANTS

A series of batch and continuous foam fractionation experiments has been carried out in an effort to establish the effects on the extent of separation of each of the following independent variables: surfactant (Grieves and Bhattacharyya, 1964, 1965b; Grieves et al., 1963, 1964), temperature (Grieves and Bhattacharyya, 1965a, b; Grieves and Wood, 1964), pH and ionic strength (Grieves and Bhattacharyya, 1964), surfactant concentration (Grieves and Bhattacharyya, 1964, 1965a, b; Grieves and Wood, 1963, 1964; Grieves et al., 1963, 1964, 1970), air rate at constant column diameter (Grieves and Bhattacharyya, 1965b, Grieves and Wood, 1963, 1964, Grieves et al., 1963, 1964, 1970), feed rate at constant column diameter (Grieves and Bhattacharyya, 1965b; Grieves and Wood, 1963, 1964, Grieves et al., 1963, 1964, 1970), solution height H_1 (Grieves and Bhattacharyya, 1965b, Grieves and Wood, 1963, 1964, Grieves et al., 1970), foam height H_f (Grieves and Bhattacharyya, 1965a, b; Grieves and Wood, 1963), and feed position P_1 (Grieves and Wood, 1964, Grieves et al., 1964).

The bubble diameter was not a controlled variable, although air diffuser porosity was held constant for a given series of experiments. Bubble diameter is generally not at the control of the design engineer, varies with time and position, and is difficult to measure experimentally (Grieves et al., 1970). The principal surfactant was the cationic ethylhexadecyldimethylammonium bromide (EHDA-Br, mol wt = 378); others were the anionic alkylbenzene-sulfonate (ABS, mol wt = 348), the anionic sodium dodecylsulfate (SDS, mol wt = 288), and the cationic hexadecylpyridinium chloride (HPy-Cl, mol wt = 340). Schematic diagrams of typical continuous and batch units are presented in Fig. 1; the nomenclature is noted in Fig. 1 and is given in a

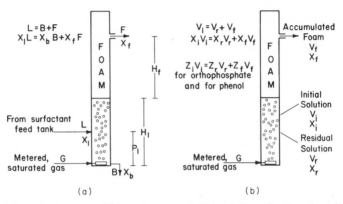

FIG. 1. Schematic diagrams of (a) continuous and (b) batch foam fractionations, indicating nomenclature and material balances.

symbol list at the end of the chapter. Material balances are also given in Fig. 1. Batch operation is desirable in the laboratory to utilize small volumes of solutions, to carry out experiments rapidly, and particularly to conduct feasibility studies (Grieves, 1968; Grieves et al., 1970). In all of the continuous experiments except those involving feed into the foam, a single, equilibrium stage separation was achieved (Grieves and Bhattacharyya, 1965b, Grieves and Wood, 1963, 1964, Grieves et al., 1970). Two parameters must be utilized as the dependent variables to completely establish the extent of separation, one a concentration and the other a quantity (Grieves, 1970). Two of the simplest are X_b (X_r for batch operation) and $X_f F$ ($X_f V_f$). A multitude of parameters has been tested, including X_b/X_1, X_f/X_b, B/X_b, F/B, B/L, $X_b B/X_1 L$, etc. Two, which indicate clearly the extent of separation and its response to the independent variables, are $1 - X_b/X_1$ ($1 - X_r/X_i$) and $X_f F/X_1 L = 1 - X_b B/X_1 L$ ($X_f V_f/X_i V_i = 1 - X_r V_r/X_i V_i$), the latter called the removal ratio; $X_f F/X_1 L \geq 1 - X_b/X_1$. For optimum operation, $1 - X_b/X_1 \to 1.0$ and $X_f F/X_1 L \to 1 - X_b/X_1$.

Initial batch experiments showed that an increase in temperature decreased $1 - X_r/X_i$ as predicted by the Gibbs equation, but decreased $X_f V_f/X_i V_i$ to a greater extent due to enhanced foam drainage (Grieves and Bhattacharyya, 1965). For EHDA-Br, increases in pH from 6.2 to 12.0 with NaOH did not alter the separation, while decreases to 1.6 with HCl and H_2SO_4 caused $1 - X_r/X_i$ and $X_f V_f/X_i V_i$ to go through maxima, principally due to the ionic strength modification by Cl^- and SO_4^{2-}. H^+ had a slightly greater effect than Na^+ (Grieves and Bhattacharyya, 1964).

Continuous foam fractionation data could be represented adequately by the following semiempirical equation (Grieves and Bhattacharyya, 1965b):

$$1 - X_b/X_1 = m(G/L)^n(X_1)^{(p/T)-1}(T)^q. \tag{1}$$

The solution height H_1 could be eliminated from the correlation because a single equilibrium stage was validated, together with the verification of a completely mixed solution (Grieves and Bhattacharyya, 1965b; Grieves and Wood, 1963, 1964). The foam height H_f for foam heights ranging from 10 to 72 cm could be eliminated from the correlation due to the independence of X_b from H_f, indicating that the liquid draining from the foam was of the same concentration as the bulk or bottoms liquid (Grieves and Bhattacharyya, 1965b; Grieves and Wood, 1963). For EHDA-Br, ABS, and SDS and for a total of 61 data points over a broad range of X_1, G, L, and for temperatures of 24–38°C, the average of the percent deviations of values calculated with Eq. (1) from experimental values was 9 % [100 × (|experimental − calculated|)/experimental]. The value of n was 0.37 for all three surfactants over the full range of the other independent variables. Correlations of the same data were attempted using the Gibbs equation plus the material balances of Fig. 1 to obtain

$$1 - X_b/X_1 = -(6/RTD_b)X_b(d\gamma/dX_b)(G/LX_1). \tag{2}$$

D_b was not measured; its effect could be taken into account with EHDA-Br by using a function $\log X_b/s$ instead of X_b, and with ABS by using $X_b{}''$ instead of X_b. Neither correlation was as accurate as Eq. (1).

To correlate the other parameter for the extent of separation, $X_f F/X_1 L$ ($X_f V_f/X_i V_i$), the batch analog of Eq. (2) was used to develop another semiempirical relation (Grieves et al., 1970). It was assumed that Γ, Gibbs' surface excess, could be related to X_r as a power function and that D_b could be related to G and X_r as power functions. Taking into account both adsorbed and entrained surfactant, the following result was obtained:

$$d(X_f V_f)/d\theta = h(G)^j(X_r)^k. \tag{3}$$

Equation (3) was tested with an extensive series of batch experiments using EHDA-Br at concentrations from 12.5 − 280 mg/liter and air rates from 0.078 to 0.85 liter/min (Grieves et al., 1970). For $10 < X_r < 45, j = 1.2$ and $k = 2.0$; for $35 < X_r < 300, j = 0.8$ and $k = 1.0$. The values of k were quite precise, as was the transition over $35 < X_r < 45$. The accuracy (average percent deviation) of Eq. (3) was about 15 %. Using the continuous analog of Eq. (3) and the values of h, j, and k determined solely from the batch experiments, excellent predictions could be made of $X_f F$ for continuous operation in the same column. The accuracy of the continuous analog of Eq. (3) with the constants obtained from batch data was 12%. Additional batch data for EHDA-Br with colloidal ferric oxide and polynucleated complexed cyanide also yielded a value of $k = 1.0$ in Eq. (3) (Grieves et al., 1970). Similarly, application of the continuous analog of Eq. (3) to the data for EHDA-Br discussed in conjunction with Eq. (1) and (2) yielded a value of $k = 1.0$ and an accuracy of 18% for 19 points.

Variation of the position of the feed to a continuous foam fractionation process from the solution, as indicated in Fig. 1, to the column of foam yielded multiple separation stages and increases in the values of $1 - X_b/X_1$ and $X_f F/X_1 L$, both for EHDA-Br (Grieves and Wood, 1964) and for ABS (Grieves et al., 1964). Optimum position for the feed was at the midpoint of the column of foam, yielding a high value of $1 - X_b/X_1$ and, more important, a value of $X_f F/X_1 L \to 1 - X_b/X_1$ (Grieves and Wood, 1964, Grieves et al., 1964).

II. BATCH SEPARATION OF ORTHOPHOSPHATE AND OF PHENOL

Orthophosphate (Grieves and Bhattacharyya, 1966a, b) and phenol (Grieves and Aronica, 1965, 1966) have been foam fractionated as soluble species in two series of batch experiments [see Fig. 1(b)]. Primary standard KH_2PO_4 or analytical reagent grade C_6H_5OH were dissolved in distilled water and contacted with the cationic EHDA-Br, the pH was adjusted with KOH or NaOH, and the solutions were foam fractionated at a nitrogen rate of 0.35–0.40 liter/min for periods of time corresponding to no further foam formation. The effects of pH are indicated in Fig. 2(a) at initial orthophosphate and phenol concentrations of about 0.25 mM, and an EHDA-Br concentration of 400 mg/liter (1.06 mM) with orthophosphate and 200 mg/liter with phenol. The maxima with orthophosphate were produced by the conversion of $H_2PO_4^- \to HPO_4^{2-}$ as the pH was elevated (with the higher charged species being more easily foam fractionated) and by incipient competition with OH^- above pH 11 (Grieves and Bhattacharyya, 1966). The maximum with phenol was brought about by ionization of phenol to the more readily foam fractionated phenolate $(C_6H_5O^-)$ as the pH was elevated, together with competition with OH^- at pH values above 11 (Grieves and Aronica, 1966). The effects of ionic strength ("anionic strength" $= \mu^* = (1000/2)[Cl^-] + (2)(1000)[SO_4^{2-}]$) are indicated in Fig. 2(b) and (c). It is clear that the addition of sulfate had a much more pronounced influence on $1 - Z_r/Z_i$ for orthophosphate than for phenol (note the contrasting ranges of the abscissa scales) and that the addition of chloride had a somewhat more pronounced influence. Also, the reduced foam fractionation of both orthophosphate and phenol was produced not solely by an "ionic strength effect"; sulfate was obviously more

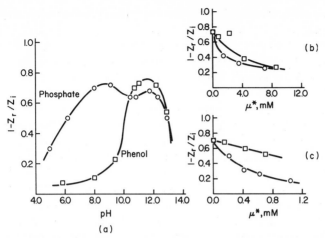

FIG. 2. (a) Effects of pH on foam fractionation of orthophosphate and of phenol. Effect of "anionic strength" (b) with phenol at pH 11.7 and (c) with orthophosphate at pH 8.0: ($\square$) Cl, ($\bigcirc$) SO$_4$.

detrimental than chloride, due primarily to competition with HPO$_4^{2-}$ (the pH was 8.0), and C$_6$H$_5$O$^-$ (the pH was 11.7) for the monovalent EHDA cations. Both sulfate and chloride provided significant competition with the non-surface-active orthophosphate, but less competition with the surface-active phenol.

For orthophosphate at pH = 8.0, with no added sulfate or chloride, with Z_i (initial orthophosphate concentration) ranging from 0.13 to 1.58 mM, and with X_i (initial surfactant) ranging from 50 to 600 mg/liter (0.13–1.59 mM), the following equation gave a good fit (Grieves and Bhattacharyya, 1966a):

$$1 - Z_r/Z_i = aX_i + cX_i^{0.5}/Z_i. \qquad (4)$$

The accuracy (average percent deviation) of Eq. (4) was 9% for 14 points. The collapsed foam volume V_f was independent of Z_i and was directly proportional to $X_i^{1.07}$. To give an idea of the separation achievable, $Z_f V_f/Z_i V_i = 0.99$ for $Z_i = 0.20$ mM with $X_i = 1.06$ mM and for $Z_i = 0.25$ mM with $X_i = 1.59$ mM.

For phenol at pH 12.9, with no added sulfate or chloride, with Z_i (initial phenol concentration) ranging from 0.12 to 7.15 mM, and with X_i ranging from 50 to 600 mg/liter, a power relation gave a good fit (Grieves and Aronica, 1965):

$$1 - Z_r/Z_i = wX_iZ_i^{-0.51}. \qquad (5)$$

The accuracy of Eq. (5) was 12% for 29 points. The collapsed foam volume V_f was independent of Z_i and was directly proportional to X_i. Typical separations were $Z_f V_f / Z_i V_i = 0.99$ for $Z_i = 0.19$ mM with $X_i = 1.06$ mM, or for $Z_i = 0.43$ mM with $X_i = 1.59$ mM. With a given value of X_i, better separations were always achieved for phenol than for orthophosphate, except at very low values of Z_i.

A "relative fractionation" parameter was introduced to give the ratio of EHDA$^+$ to HPO$_4^{2-}$ or $C_6H_5O^-$ at the gas–solution interfaces associated with the bubbles as the foam left the foam fractionation column. The parameter eliminated the effect of entrained bulk liquid carried with the foam. Values of the relative fractionation parameter ranged from 2.6 to 7.6 mole EHDA$^+$/mole HPO$_4^{2-}$, compared to the theoretical value of 2.0 [(EHDA)$_2$-HPO$_4$] corresponding to no competition from other anions such as bromide which was the counterion to the surfactant. Values ranged from 0.62 to 7.2 mole EHDA$^+$/mole $C_6H_5O^-$, compared to the theoretical value of 1.0 (EHDA-C_6H_5O). Competition with bromide (and also with hydroxide in the case of phenol) clearly decreased the extent of separation for both orthophosphate and phenol.

SYMBOLS

B flow rate of bottoms or effluent stream, continuous process, liters/min

D_b bubble diameter, cm

F flow rate of foam (collapsed, as liquid) stream, continuous process, liters/min

G gas rate, liters/min

H_f height of column of foam above foam–bulk solution interface, cm

H_1 height of bulk solution in column above base of column, cm

L flow rate of feed stream, continuous process, liters/min

P_1 height at which feed is introduced to bulk solution in column above base of column, cm

R gas constant, dyn cm/mg °K

T temperature, °K

V_f volume of foam (collapsed, as liquid) accumulated in θ min, batch process, liters

V_i volume of initial solution, batch process, liters

V_r volume of residual solution remaining after θ min, batch process, liters

X_b surfactant concentration in bottoms or effluent stream, continuous process, mg/liter

X_f surfactant concentration in foam stream or in accumulated foam, mg/liter

X_i surfactant concentration in initial solution, batch process, mg/liter

X_1 surfactant concentration in feed stream, continuous process, mg/liter

X_r surfactant concentration in residual solution, batch process, mg/liter

Z_b concentration of dissolved or particulate species of interest in bottoms or effluent stream, continuous process, mM

Z_f orthophosphate or phenol (or other dissolved or particulate species of interest) concentration in accumulated foam, or in foam stream, mM

Z_i orthophosphate or phenol (or other dissolved or particulate species of interest) concentration in initial solution, batch process, mM

Z_1 concentration of dissolved or particulate species of interest in feed stream, continuous process, mM

Z_r orthophosphate or phenol (or other dissolved or particulate species of interest) concentration in residual solution, batch process, mM

$a, b, c, g, h, j, k, m, n, p, q, s, u$ empirical constants

γ surface tension, dyn/cm

θ foaming time in batch process, min

REFERENCES

Grieves, R. B. (1968). *Brit. Chem. Eng.* **13**, 77.

Grieves, R. B. (1970). *J. Water Pollut. Contr. Fed.* **42** (part 2), R336.

Grieves, R. B., and Aronica, R. C. (1965). *Int. J. Air. Water Pollut.* **10**, 31.

Grieves, R. B., and Aronica, R. C. (1966). *Nature* **210**, 901.

Grieves, R. B., and Bhattacharyya, D. (1964). *Nature* **204**, 441.

Grieves, R. B., and Bhattacharyya, D. (1965a). *J. Amer. Oil Chem. Soc.* **42**, 174.

Grieves, R. B., and Bhattacharyya, D. (1965b). *J. Water Pollut. Contr. Fed.* **37**, 980.

Grieves, R. B., and Bhattacharyya, D. (1966a). *Separ. Sci.* **1**, 81.

Grieves, R. B., and Bhattacharyya, D. (1966b). *J. Amer. Oil Chem. Soc.* **43**, 529.

Grieves, R. B., and Wood, R. K. (1963). *Nature* **200**, 332.

Grieves, R. B., and Wood, R. K. (1964). *A. I. Ch. E. J.* **10**, 456.

Grieves, R. B., Kelman, S. Obermann, W. R., and Wood, R. K. (1963). *Canad. J. Chem. Eng.* **41**, 252.

Grieves, R. B., Crandall, C. J., and Wood, R. K. (1964). *Int. J. Air Water Pollut.* **8**, 501.

Grieves, R. B., Ogbu, U. I., Bhattacharyya, D., and Conger, W. L. (1970). *Separ. Sci.* **5**, 583.

CHAPTER **11**   ION, COLLOID, AND PRECIPITATE
FLOTATION OF INORGANIC ANIONS[1]

Robert B. Grieves and Dibakar Bhattacharyya
Department of Chemical Engineering
University of Kentucky
Lexington, Kentucky

I. BATCH ION FLOTATION OF CHROMIUM(VI)

Upon addition of a quaternary ammonium surfactant, ethylhexadecyl-dimethylammonium bromide (EHDA-Br), to a dilute solution of $K_2Cr_2O_7$ (0.46 mM as $HCrO_4^-$), minute orange crystals were formed (about 1 μ in diameter) which tended to agglomerate upon standing. Filtration experiments indicated the presence of 0.96 mole $EHDA^+$/mole $HCrO_4^-$ in the particulates. A series of batch ion flotation studies was made at pH 3.6, with initial distilled water solutions 0.046–0.93 mM in $HCrO_4^-$ (Grieves and Wilson, 1965; Grieves et al., 1965). For each experiment, the surfactant (EHDA-Br) was added in a series of dosages generally ranging from 2 to 4 but in some experiments as many as 16. An increase in the number of dosages (at constant total surfactant concentration) consistently decreased V_f, the collapsed foam volume (see Fig. 1 of Chapter 10), because the ion flotation was being carried out from more dilute solutions and entrainment was decreased. V_f increased linearly with X_i and decreased linearly with Z_i. The chromium removal ratio, $Z_f V_f/Z_i V_i$, ranged from 0.81 to 0.92 at the approximately stoichiometric feed ratio of 1.14 mole $EHDA^+$/mole $HCrO_4^-$.

[1]Symbols are listed and defined in the previous chapter.

A second series of experiments was conducted, using a lower foam height, lower air rate, and a single surfactant dosage (Grieves et al., 1969c). With initial solutions 0.93 mM in $HCrO_4^-$ at pH 4.2, $Z_f V_f/Z_i V_i = 0.96$ at $X_i/Z_i = 1.14$. Changing from distilled water to tap water (for which the conductivity was about 400 μmho/cm at 23°C) decreased $Z_f V_f/Z_i V_i$ to 0.80, but the addition of 5 mg/liter Dow N-12, a nonionic polymer, increased the removal ratio to 0.93. EHDA-Br was considerably superior to other surfactants. In tap water with no polymer the removal ratios for several surfactants at $X_i/Z_i \approx 1.0$ were as follows: cetyldimethylbenzylammonium chloride, 0.61; hexadecylpyridinium chloride, 0.54; hexadecylamine, 0.42; and octadecylamine, 0.46. The addition of polymer provided some improvement in the flotation with the amines.

With a single dosage of EHDA-Br, the relative fractionation parameter ranged from 1.0 to 1.1 moles $EHDA^+$/mole $HCrO_4^-$ at the bubble interfaces as long as the average concentration ratio in the column was less than 1.1 (Grieves et al., 1968).

II. CONTINUOUS ION FLOTATION OF CHROMIUM(VI)

Continuous ion flotations were carried out with distilled water feed solutions 0.092–0.70 mM in $HCrO_4^-$ at pH 5.2. A number of operating variables were investigated, including air rate, column height per unit feed rate, feed $HCrO_4^-$ concentration, feed $EHDA^+/HCrO_4^-$ ratio, multistage operation, foam height, and feed position (Grieves and Schwartz, 1966a, b). At a ratio, $X_1/Z_1 = 0.86$, the concentration removal could be related as follows:

$$1 - Z_b/Z_1 = 1 - aZ_1^{-0.10}(H_1/L)^{-0.18}. \tag{1}$$

The average of the percent deviations of values calculated with Eq. (1) from experimental values was 11% for 21 points. For experiments with $X_1/Z_1 = 0.86–1.14$, $1 - Z_b/Z_1$ was practically independent of the feed surfactant concentration. The removal ratio, $Z_f F/Z_1 L$, went through a maximum with Z_1 at constant X_1/Z_1: $Z_f F/Z_1 L = 0.90$ at $Z_1 = 0.58$ mM for $X_1/Z_1 = 1.14$, and $Z_f F/Z_1 L = 0.62$ at $Z_1 = 0.48$ mM for $X_1/Z_1 = 0.86$. The maxima were produced by minima in B/L and could be explained on the basis of surfactant-particle effects on foaminess (Grieves and Schwartz, 1966b). Air rate had a very slight effect, again in contrast to foam fractionations. The effect of H_1/L and lack of effect of air rate may have been produced by the rate-limiting step being the reaction between $HCrO_4^-$ and $EHDA^+$ in the flotation column. The behavior was in direct contrast to the continuous foam fractionation experiments described in the previous chapter.

The performance of two or more ion flotation columns operated in series, with surfactant added in each column, could be calculated (Grieves, 1966). For a feed 0.93 mM in $HCrO_4^-$, four columns in series would yield $1 - Z_b/Z_1 \approx 0.97$ and $Z_f F/Z_1 L = 0.98$, with a total X_1/Z_1 of 1.16.

The effects of foam height and feed position were established for single-column operation (Grieves and Schwartz, 1966b; Grieves, 1966). Of several combinations tested, optimum operation was achieved with the foam height increased from 15 to 42 cm and feed at the midpoint of the column of foam. Compared to feed into the solution and to the lower foam height utilized for all experiments described above, $1 - Z_b/Z_1$ was increased and $Z_f F/Z_1 L$ was closer to $1 - Z_b/Z_1$.

III. CONTINUOUS DISSOLVED-AIR ION FLOTATION OF CHROMIUM(VI)

As an alternative to the generation of gas–liquid interfacial area by dispersing air through a sintered glass, porous metal, or ceramic diffuser, part of the effluent stream from a continuous ion flotation process may be recycled, with air dissolution at high pressure followed by throttling and precipitation of fine bubbles. A dissolved-air unit was designed to handle from 8 to 32 gal/hr of feed; the ion flotation column was 28 cm in diameter, with $H_1 = 121$ cm. Two contrasts with dispersed-air operation were immediately obvious. First, a scum or froth was formed instead of a foam and the scum had to be removed from the column by a vacuum trap. Second, also due to the extremely fine air bubbles formed with dissolved air, the nonionic polymer, Dow N-12, had to be added to the feed to aggregate the EHDA-HCrO$_4$ particles (Grieves and Ettelt, 1967).

Ion flotation data were submitted to multiple regression analysis, generating an equation relating $1 - Z_b/Z_1$ to five independent variables with about 4% accuracy (Grieves et al., 1969a). The conditions at optimum operation were as follows: polymer dosage 1.5–2.3% of the feed concentration of Cr (mg Cr/liter), detention time of 85 min, recycle rate 170% of the feed rate, $X_1/Z_1 = 1.05$, and $Z_1 = 0.92$ mM. For these conditions, $1 - Z_b/Z_1 = 0.94$ and the froth contained 231 mM HCrO$_4^-$.

The chemical costs of treating an acid chromate solution by ion flotation were about 80¢/lb Cr treated, including the recovery and reuse of 90% of the surfactant. A feasible recovery process was established (Grieves et al., 1969b), including pH depression to 2–3, HCrO$_4^-$ reduction with NaHSO$_3$, extraction with 1:1 isopropanol–chloroform, and vacuum distillation of the solvents. Redissolution of EHDA–Br in water and reuse in batch experiments with HCrO$_4^-$ plus polymer in tap water indicated no impairment of its ion flotation capability.

IV. BATCH AND CONTINUOUS PRECIPITATE FLOTATION OF CHROMIUM(III)

Instead of ion-floating soluble $HCrO_4^-$, a series of experiments was carried out to study the flotation of precipitated chromium(III) hydroxide formed by lowering the pH of an aqueous $Na_2Cr_2O_7 \cdot 2H_2O$ solution to 2.5–3.0 with HCl, reducing the Cr with $NaHSO_3$, and precipitating the Cr(III) with NaOH (Bhattacharyya et al., 1970). The effect of pH (at essentially constant ionic strength) on the batch precipitate flotation of Cr(III) hydroxide is given in Fig. 1. Sodium dodecylsulfate (SDS) was the principal surfactant

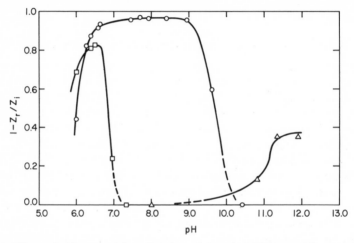

Fig. 1. Effects of pH on precipitate flotation of chromium(III) hydroxide, at two initial chromium concentrations and with two surfactants: ($\bigcirc$) $Z_i = 0.93$ mM, SDS; ($\square$) $Z_i = 1.86$ mM, SDS; ($\triangle$) $Z_i = 0.93$ mM, EHDA-Br.

that was used, at a molar feed ratio of 0.093 mole DS^-/mole Cr. First, considering the 0.93 mM Cr suspensions, the rapid decrease in $1 - Z_r/Z_i$ below pH 6.5 was produced by incomplete precipitation of Cr(III) hydroxide plus incomplete flotation of the soluble $Cr(OH)_2^+$ and/or $Cr(OH)^{+2}$ due to their stoichiometric surfactant demand. Above pH 9.0 the charge on the precipitate was reduced, and the charge was reversed from positive to negative at pH 9.7, as indicated by surface potential measurements. This was shown further by enhanced flotation with the cationic surfactant EHDA–Br above pH 10, at a feed ratio, X_i/Z_i, of 0.080. The removal ratio remained less than 0.4, principally due to competition by $Cr(OH)_4^-$ and OH^- for the cationic surfactant. For the 1.86-mM Cr suspensions, the optimum pH range for efficient flotation was reduced and narrowed in comparison with 0.93-mM suspensions (Fig. 1). Above pH 7.0, the surface charge was decreasing but remained positive, indicating other modifications in particle surface area characteristics and in particle–surfactant interactions.

Batch experiments were also used to study the effects of X_i/Z_i, which included surfactant adsorption measurements, and the effect of ionic strength, particularly including Na^+ and Ca^{2+}. Compared to ion flotation, precipitate flotation resulted in a reduction in surfactant requirement of 90%.

Continuous precipitate flotation experiments were conducted with Cr(III) hydroxide suspensions, establishing the effect of several operating variables (Grieves, 1970). For two columns operated in series, with no surfactant added in the second column, a 1.86-mM Cr suspension could be treated with $X_1/Z_1 = 0.10$, yielding $1 - Z_b/Z_1 = 0.98$ and $Z_f F/Z_1 L \approx 0.98$, with chemical costs of 62¢/lb Cr treated.

V. ION AND COLLOID FLOTATION OF CYANIDE COMPLEXED BY IRON

Cyanide (CN) readily reacts in aqueous solution with Fe^{2+} and its hydrolyzed forms to produce (1) $Fe(CN)_6^{4-}$ and $Fe(CN)_5H_2O^{3-}$, (2) $FeFe(CN)_6^-$ and $FeFe(CN)_6^{2-}$, called soluble Prussian blue, the former including Fe^{3+} formed by air oxidation of Fe^{2+}, and (3) $Fe(II)[Fe(CN)_6Fe]$ and $Fe(II)[Fe(CN)_6Fe]_2$, precipitated species. The species produced depend on the initial Fe/CN molar ratio. Foam separation of the first is ion flotation, foam separation of Prussian blue with the complexed cyanide forming colloidal polynucleated species is colloid flotation, and foam separation of the third is precipitate flotation. Figure 2(a) indicates the effect of pH on the batch flotation of complexed cyanide with EHDA-Br at initial Fe/CN

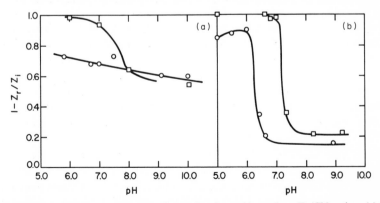

FIG. 2. Effects of pH on the flotation of complexed cyanide at three Fe/CN ratios: (a) (○) Fe/CN = 0.21, X_i/Z_i = 0.43 ; (□) Fe/CN = 0.35, X_i/Z_i = 0.41. (b) Fe/CN = 0.56, (○) X_i/Z_i = 0.042, (□) X_i/Z_i = 0.10.

ratios of 0.21 and 0.35. The concentration removal, $1 - Z_r/Z_i$, is expressed only in terms of complexed CN (as CN), as is the feed ratio, X_i/Z_i. For all of the experiments described in this section, the free noncomplexed cyanide was in the range of 0.30–0.35 mM at Fe/CN $= 0.21$ (for total cyanide from 1.54–3.08 mM), and in the range 0.22–0.30 mM at Fe/CN $= 0.35$ (Grieves and Bhattacharyya, 1968, 1969a). The merging of the two curves in Fig. 2(a) corresponds to the conversion at Fe/CN $= 0.35$ from Prussian blue to soluble $Fe(CN)_6^{4-}$.

For two series of experiments at Fe/CN $= 0.21$ and 0.35 with initial total cyanide concentrations ranging from 1.54 to 3.08 mM (complexed cyanide from about 1.2 mM to 2.8 mM), the removal of complexed cyanide, $Z_f V_f/Z_i V_i$, was a linear function of X_i/Z_i for X_i/Z_i from 0.20 to 0.65 (Grieves and Bhattacharyya, 1969a). From these results and also from the relative fractionation parameter (Grieves et al., 1968), the predominant species at Fe/CN $= 0.21$ was $Fe(CN)_6^{4-}$, and that at Fe/CN $= 0.35$ was polynucleated $FeFe(CN)_6^{2-}$.

All of the above experiments were conducted with a foaming time of 25 min, corresponding to the cessation of foam formation. An additional rate study was conducted at Fe/CN $= 0.35$ with $Z_i = 1.3$ and 2.4 mM, and the following equation was established:

$$-d(Z_r V_r)/d\theta = b(Z_r V_r - Z_r V_{r\infty})^{0.93}, \tag{2}$$

in which $Z_r V_{r\infty}$ is the residual quantity of complexed cyanide at foam cessation. The average percent deviation of the integrated equation was 22 % for 17 points.

VI. PRECIPITATE FLOTATION OF CYANIDE COMPLEXED BY IRON

Figure 2(b) shows the effect of pH on the flotation of cyanide complexed and precipitated by Fe^{2+} at Fe/CN $= 0.56$, at two values of X_i/Z_i (Grieves and Bhattacharyya, 1969b). The abrupt decreases in flotation over pH 6.0–6.5 and 7.0–7.5 were produced by greater demands for the surfactant by Fe(II) and Fe(III), precipitated as the hydroxide and/or carbonate. Thus, less surfactant was free to act as a collector for the precipitated complexed cyanide and to act as a frother. Also, pH elevation may have modified the particle surface area characteristics.

Other variables that were studied in the batch precipitate flotation work included FeCN and FeCN–surfactant mixing times, ionic strength, and initial CN and EHDA-Br concentrations. Ionic strength in terms of increased concentrations of Cl^- and SO_4^{2-} produced a distinct decrease in flotation. At pH 6.0 and over the range of initial complexed cyanide concentrations of 1.3–2.8 mM, and at an initial surfactant to complexed cyanide ratio of 0.042, the removal of complexed cyanide, $Z_f V_f / Z_i V_i$, ranged from 0.91 to 0.95. At $Fe/CN = 0.35$ (flotation of polynucleated species) a ratio $X_i / Z_i = 0.35$ would be required for essentially complete flotation. Thus, similarly to chromium, a 90% reduction in surfactant requirement would be realized with precipitate flotation. The chemical cost of precipitate flotation would be 37¢/lb CN treated versus 97¢/lb at $Fe/CN = 0.35$. These figures include 70% surfactant recovery and alkaline chlorination of the residual 0.3-mM noncomplexed cyanide.

REFERENCES

Bhattacharyya, D., Carlton, J. A., and Grieves, R. B. (1970). *A. I. Ch. E. J.* **17,** 419.

Grieves, R. B. (1966). *Separ. Sci.* **1,** 395.

Grieves, R. B. (1970). *J. Water Pollut. Contr. Fed.* **42** (part 2), R336.

Grieves, R. B., and Bhattacharyya, D. (1968). *Separ. Sci.* **3,** 185.

Grieves, R. B., and Bhattacharyya, D. (1969a). *J. Appl. Chem.* **19,** 115.

Grieves, R. B., and Bhattacharyya, D. (1969b). *Separ. Sci.* **4,** 301.

Grieves, R. B., and Ettelt, G. A. (1967). *A. I. Ch. E. J.* **13,** 1167.

Grieves, R. B., and Schwartz, S. M. (1966a). *J. Appl. Chem.* **16,** 14.

Grieves, R. B., and Schwartz, S. M. (1966b). *A. I. Ch. E. J.* **12,** 746.

Grieves, R. B., and Wilson, T. E. (1965). *Nature* **205,** 1066.

Grieves, R. B., Wilson T. E., and Shih, K. Y. (1965). *A. I. Ch. E. J.* **11,** 820.

Grieves, R. B., Ghosal, J. K., and Bhattacharyya, D. (1968). *J. Amer. Oil Chem. Soc.* **45,** 591.

Grieves, R. B., Ettelt, G. A., Schrodt, J. T., and Bhattacharyya, D. (1969a). *J. Sanit. Eng. Div., Amer. Soc. Civil Eng.* **95,** 515.

Grieves, R. B., Sickles, J. E. II, Ghosal, J. K., and Bhattacharyya, D. (1969b). *Separ. Sci.* **4,** 425.

Grieves, R. B., Bhattacharyya, D., and Conger, W. L. (1969c). *Chem. Eng. Progr. Symp. Ser. No. 91* **65,** 29.

Robert B. Grieves

Department of Chemical Engineering
University of Kentucky
Lexington, Kentucky

I. COLLOIDAL FERRIC OXIDE

In many particulate–solution systems, the initial charge of the particles plays an important role in the adsorption of a surfactant on the particles, although coordination effects may also be significant. The colloidal or precipitated particles (for example, precipitated Cr(III) hydroxide and precipitated $Fe(II)[Fe(CN)_6Fe]_2$ discussed in Chapter 11) acquire a charge by desorption of ionic species from the solid or by adsorption of ions from solution, preferentially constituent ions. The surfactant added as a collector-frother in the flotation process serves three functions, as follows. The adsorption of the surfactant on the surfaces of the particles may cause destabilization of the colloid and partial aggregation and makes the particles suitable for bubble attachment; interaction between the particles plus their adsorbed surfactant and "free" surfactant adsorbed at the gas–liquid interfaces associated with generated gas bubbles may produce bubble attachment of the particles; and "free" surfactant acts as a frother producing a stable foam which may be further stabilized by the presence of particulates.

An initial batch study was conducted with a nondialyzed Fe(III) oxide sol prepared by the dropwise hydrolysis of $FeCl_3$ (Grieves and Bhattacharyya, 1965a). The foam separation experiments were carried out at pH 5.0 with the

[1] Symbols are listed and defined in Chapter 10.

anionic surfactant SDS. Surfactant addition to the sol produced partial aggregation of the particulates, which was readily determined by turbidity measurements. With initial suspensions of 0.45–2.69 mM in Fe and with X_i/Z_i that ranged from 0.024 to 0.073, the concentration removal could be fitted by the following:

$$1 - Z_r/Z_i = 1 - (a/Z_i) \exp(-bX_i/Z_i). \tag{1}$$

The accuracy of Eq. (1) was 26 % for 12 points. Typical results were $Z_f V_f/Z_i V_i = 0.99$ at $Z_i = 1.89$ mM and $X_i/Z_i = 0.046$, at $Z_i = 0.70$ mM and $X_i/Z_i = 0.074$, and at $Z_i = 0.38$ mM and $X_i/Z_i = 0.092$. Both $1 - Z_r/Z_i$ and $Z_f V_f/Z_i V_i$ increased with X_i/Z_i at constant Z_i, with Z_i at constant X_i/Z_i, and with Z_i at constant X_i. The foam was stabilized by the presence of particles, with V_f increasing in the same fashion as $Z_f V_f/Z_i V_i$. These results could be explained qualitatively in terms of a simple model involving adsorbed and "free" surfactant, and involving particles at the bubble inter-faces and those carried in the entrained bulk liquid (Grieves et al., 1967).

A series of experiments involving pH variation was conducted at a shorter (5 min) foaming time and a lower foam height (Grieves and Bhattacharyya, 1967). The effect of pH for an anionic and a cationic surfactant is given in Fig. 1. The point of charge reversal of the colloidal suspension is shown

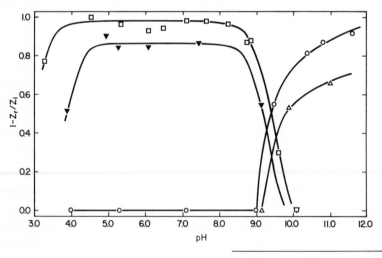

Symbol	Surfactant	X_i/Z_i
□	SDS	0.070
▼	SDS	0.041
○	EHDA-Br	0.070
△	EHDA-Br	0.041

FIG. 1. Effect of pH on the foam separation of colloidal ferric oxide using a cationic and an anionic surfactant. $Z_i = 1.67$ mM Fe.

clearly. The poor flotation with SDS below pH 4.0 was produced by solubilization of a fraction of the Fe(III) and/or by an increase in the required surfactant to Fe(III) ratios for polynucleated hydrolyzed iron species. Solubilization of Fe(III) above pH 10 provided the incomplete (<0.8) flotation with EHDA-Br.

Extensive rate studies were made at pH 5.8 with SDS and at pH 10.8 with EHDA-Br ($Z_i = 1.67$ and 2.38 mM, and X_i/Z_i from 0.015 to 0.075). The removal rates could be fitted by the following (Grieves and Bhattacharyya, 1968):

$$d(Z_f V_f)/d\theta = [c(X_r/Z_r)](Z_r V_r)^{1.72} \qquad \text{with EHDA-Br,} \qquad (2)$$

$$d(Z_f V_f)/d\theta = [\exp(g X_i)](X_r/Z_r)(Z_r V_r)^{1.89} \qquad \text{with SDS.} \qquad (3)$$

The data could not be fitted without including the surfactant concentration. The utility of $Z_r V_r$, instead of Z_r, was to permit integration of Eqs. (2) and (3), replacing the residual surfactant concentration by an exponential function of time. It is significant that the interfacial area did not have to be included in the correlations, which were fairly accurate for this type of data (24 % for 47 points). In the rate studies, the relative fractionation parameter ranged from 0.032 to 0.053 mole EHDA$^+$/mole Fe and from 0.016 to 0.029 mole DS$^-$/mole Fe.

II. SIX SPECIES OF BACTERIA

Bacterial cultures in aqueous media behave in many respects as colloids, possessing characteristics of both hydrophilic and hydrophobic colloids, and having a net negative charge at neutral pH. Batch foam separation experiments were carried out with six species: *Escherichia coli, Serratia marcescens, Proteus vulgaris, Bacillus cereus, Pseudomonas fluorescens,* and *Bacillus subtilis* var. *niger,* all with cationic EHDA-Br at pH 7.0 (Bretz *et al.,* 1966; Grieves and Wang, 1966, 1967a, b). The following effects on the flotation (based on total cell count as determined by a membrane filtration technique) were established: bacterial species, initial cell concentration, initial surfactant concentration, foaming time, air rate, foam height, and the presence, mode, and time of contact of several inorganic salts. Results indicated an interaction between cells and surfactant that was clearly different than that between a hydrophobic colloid [such as Fe(III) oxide] and a surfactant. There was no evidence of partial aggregation of the cells by the surfactant, and the foam produced was determined by the "free" surfactant not bound by the cells, analogous to $HCrO_4^-$, polynucleated $FeFe(CN)_6^{2-}$, precipitated Cr(III) hydroxide, and precipitated $Fe(II)[Fe(CN)_6 Fe]_2$, with the cells having a neutral or possibly negative effect on foam stability. The EHDA cations were adsorbed at oppositely charged surface groupings or within the cytoplasm, were partly dissolved in the surface, or reacted chemically with surface molecules.

Experiments with *Escherichia coli* at an initial cell count of 2.5×10^7 cells/ml and at $X_i = 0.053, 0.079$, and 0.106 mM indicated approximately first-order removal: $Z_f V_f / Z_i V_i = 1 - he^{-j\theta}$, after a short time during which a higher removal rate was observed. At $\theta = 20$ min, $Z_f V_f / Z_i V_i = 0.99976$. Both $Z_f V_f / Z_i V_i$ and $1 - Z_r/Z_i$ were not significantly dependent on X_i. At $\theta = 10$ min, $1 - Z_r/Z_i = 0.983$ at $X_i = 0.079$ mM for $Z_i = 8.3 \times 10^5$ to 1.0×10^8 cells/ml; lower concentrations of surfactant provided increases in $1 - Z_r/Z_i$ and made $1 - Z_r/Z_i$ dependent on Z_i. In direct contrast to colloidal Fe(III) oxide, the foam volume decreased markedly with an increase in Z_i at constant X_i. At constant $\theta = 10$ min, $Z_i = 2.5 \times 10^7$ cells/ml, and $X_i = 0.079$ mM, increases in the gas rate and decreases in the foam height consistently produced increases in $1 - Z_r/Z_i$ and $Z_f V_f / Z_i V_i$. The dependence of $1 - Z_r/Z_i$ on foam height indicated substantial foam breakage in the rising column of foam.

For the six bacterial species and based on an average of five experiments for each species at the single set of experimental conditions that was employed, $Z_f V_f / Z_i V_i$ ranged from 0.753 for *Serratia marcescens* to 0.998 for *Bacillus subtilis* var. *niger*. The concentration removal, $1 - Z_r/Z_i$, responded similarly. The foam volume was a strong function of the species.

For *Bacillus subtilis* var. *niger*, the presence of 5.0 mEq/liter of any of eight Cl or SO_4 salts of Na, K, Ca, or Mg increased the residual cell concentration by an order of magnitude or more, irrespective of the sequence of contact of salt, surfactant, and cells. For *Pseudomonas fluorescens*, 5.0 mEq/liter of $MgSO_4$ increased the residual cell concentration by two orders of magnitude; however, the exceptional influence could be overcome by contacting cells with surfactant before the salt. $MgSO_4$ also had an unusually adverse effect on the flotation of *Escherichia coli*. The addition of surfactant in several doses during a foaming experiment (in contrast to a single dose) provided a substantial reduction in V_f when salts were present, but had little effect on $1 - Z_r/Z_i$.

III. ACTIVE CARBON WITH ADSORBED PHENOL

A batch foam separation study was made of powdered active carbon equilibrated with phenol and a surfactant: EHDA-Br, SDS, or nonionic Triton X-100 (alkyl phenoxy polyethoxy ethanol) (Grieves and Chouinard, 1969). The carbon was Aqua Nuchar "A" (Westvaco), with a minimum of

90% passage through a U.S. Series 325 screen and a particle size range from 1.4 to 42 μ. Initial carbon concentrations ranged from 100 to 800 mg/liter, with 0.10–0.60 mM phenol, giving equilibrium concentrations of non-adsorbed phenol of 0.05–0.10 mM. A series of experiments at pH 3.0, 6.0 or 7.0, and 10.0 indicated best results with the cationic EHDA-Br at pH 6 or 10, with the anionic SDS at pH 3, and with the nonionic Triton X-100 at pH 7. For a constant X_i/Z_i, the best flotation, considering surfactant and pH, was achieved with EHDA–Br at pH 6. For example, at pH 6–7 with $Z_i = 200$ mg carbon/liter, and $X_i \approx 0.12$ mM: $1 - Z_r/Z_i = 0.84$ using EHDA-Br, $1 - Z_r/Z_i = 0.65$ using SDS, and $1 - Z_r/Z_i = 0.68$ with Triton X-100; with $Z_i = 800$ mg carbon/liter, $1 - Z_r/Z_i = 0.96$ using 0.37-mM EHDA-Br, and $1 - Z_r/Z_i = 0.53$ using 0.44 mM SDS. The pH behavior and advantage of the cationic surfactant could be predicted from the surface charge behavior of the carbon.

The foam production was a direct function of the "free" nonadsorbed surfactant, which was monitored in the experiments. Flotation was impaired by excessive quantities of surfactant in the initial suspensions, due to the formation on the bubble surfaces of stable hydrated envelopes of surfactant ions or perhaps by the formation of hydrated micelle coatings on the particle surfaces. This effect was also evident in the foam separation of polynucleated $FeFe(CN)_6^{2-}$, *Escherichia coli*, and $HCrO_4^-$.

IV. CLAYS AND CLAY SEDIMENTS: CLARIFICATION OF TURBID WATERS

A rather unique application of foam separation is to water clarification, replacing the commonly employed coagulation–sedimentation-based process by flotation of the negatively charged, colloidal (hydrophilic), clay particulates. A research program was established to test the feasibility of the process for batch operation and for continuous flow operation for water supplies up to 1000 gal/day, using a cationic surfactant as the collector–frother. Initial batch studies (Grieves, 1966; Grieves and Crandall, 1966) were carried out with "synthetic" waters, built from distilled water, salts, clays, etc. The effects of the following were established on the flotation process: pH, several mono- and divalent cations and anions, the presence of organics, the mode of sur-factant addition, air rate, temperature, and the use of bentonite clay as a flotation aid to overcome interferences from Fe(III), Al, and phosphate. Six individual clays (kaolinites, illites, and montmorillonites) were foam separated from distilled water suspensions over the range pH 2–12, in both the absence and presence of Fe(II) and Fe(III) (Crandall and Grieves, 1968). The results were interpreted in terms of a simple model involving the interaction of clay particles, surfactant cations, and soluble and/or particulate iron.

Batch experiments were also carried out with natural waters sampled from over 15 sources (Grieves, 1967 ; Grieves and Conger, 1969). Five surfactants were tested, generally added in multiple doses; the clarification effected by any dose was found to be uninfluenced by previous doses. For water samples with turbidities (optical densities at 400 mμ with 10-mm light path) ranging from 0.04 to 0.80 (suspended solids up to 500 mg/liter), three doses of cetyl-dimethylbenzylammonium chloride (Cetol) totaling 45 mg/liter or three doses of trimethyltallowammonium chloride (Adogen 471) totaling 35 mg/liter produced effluents with turbidities all less than 0.01 (suspended solids < 10 mg/liter). The air rate was 3.0 liter/min at 10 psig and the maximum foaming time was 30 min for a 3-liter sample.

Continuous foam separation was evaluated for the clarification of "synthetic" waters (Grieves and Schwartz, 1966). The effects of air rate, temperature, organics, and pH (as long as the pH was < 9.2) were slight. On the other hand, increases in the concentrations of Fe(III) and Al in the feed water from 0 to 0.5 mEq/liter (of both) had a substantial negative effect which could be offset by the addition of up to 70 mg/liter of bentonite clay to the feed, and to a lesser extent by an increased feed surfactant concentration and column height per unit feed rate.

Utilizing a 40-gal/day unit, natural waters sampled from the Kentucky River were clarified by continuous foam separation. The unit was designed for dispersed-air or dissolved-air operation with secondary carbon adsorption and terminal disinfection stages (Grieves and Conger, 1969). The effluent turbidity was related by multiple regression analysis to the following independent variables: feed surfactant concentration, air rate, column height per unit feed rate, and feed water turbidity (Grieves et al., 1970). The resultant equation, involving a cubic term in X_1, in Z_1, and in H_1/L and a linear term in G, was accurate to about 25 %, which was quite satisfactory from a consideration of the low-effluent turbidities which were correlated. The equation was used to obtain an operating cost estimate for a 1000-gal/day unit, using constant X_1 and approximately constant G/L and G per column cross section as the basis for the scaleup. Operating costs ranged from 27 to 71¢/1000 gal, depending on the turbidity of the feed water. The effects of the use of ethanol as a frothing agent and of variable foam height were also established. Foam separation is a most promising process for the clarification of very turbid waters supplied in small quantities.

V. EFFECT OF PARTICULATES ON FOAM SEPARATION OF SURFACTANTS

Throughout the experiments detailed in this chapter and in Chapter 11 the effluent or residual concentrations of the surfactant were carefully monitored, together with the collapsed foam volume generated in each foam separation process. Comparisons could be made with foam separation experiments involving solutions of the surfactants only. In the cases of the hydrophobic colloids Fe(III) oxide and Sn(IV) oxide, the presence of the particulates produced decreases in the residual surfactant concentrations and substantial increases in the collapsed foam volumes (Grieves and Bhattacharyya, 1965b, c). In the cases of bacterial species; active carbon; kaolinite clays and natural sediments; precipitated Cr(III) hydroxide and $Fe(II)[Fe(CN)_6Fe]_2$; and $HCrO_4^-$, $Fe(CN)_6^{4-}$, and polynucleated $FeFe(CN)_6^{2-}$, which form particulates upon reaction with the surfactant, the presence of particulates produced reduced collapsed foam volumes due to the adsorption or binding of a portion of the surfactant. Somewhat reduced residual surfactant concentrations were produced due to ionic strength effects and due to the more efficient removal of surfactant adsorbed on or associated with particulates.

REFERENCES

Bretz, H. W., Wang, S. L., and Grieves, R. B. (1966). *Appl. Microbiol.* **14,** 778.

Crandall, C. J. and Grieves, R. B. (1968). *Water Res.* **2,** 817.

Grieves, R. B. (1966). *J. Sanit. Eng. Div. Amer. Soc. Civil Eng.* **95,** 515.

Grieves, R. B. (1967). *J. Amer. Water Works Assoc.* **59,** 859.

Grieves, R. B., and Bhattacharyya, D. (1965a). *Canad. J. Chem. Eng.* **43,** 286.

Grieves, R. B., and Bhattacharyya, D. (1965b). *A. I. Ch. E. J.* **11,** 274.

Grieves, R. B., and Bhattacharyya, D. (1965c). *Nature* **207,** 476.

Grieves, R. B., and Bhattacharyya, D. (1967). *J. Amer. Oil Chem. Soc.* **44,** 498.

Grieves, R. B., and Bhattacharyya, D. (1968). *J. Appl. Chem.* **18,** 149.

Grieves, R. B., and Chouinard, E. F. (1969). *J. Appl. Chem.* **19,** 60.

Grieves, R. B., and Conger, W. L. (1969). *Chem. Eng. Progr. Symp. Ser. No. 97* **65,** 200.

Grieves, R. B., and Crandall, C. J. (1966). *Water Sewage Works* **113,** 432.

Grieves, R. B., and Schwartz, S. M. (1966). *J. Amer. Water Works Assoc.* **58,** 1129.

Grieves, R. B., and Wang, S. L. (1966). *Biotechnol. Bioeng.* **8,** 323.

Grieves, R. B., and Wang, S. L. (1967a). *Biotechnol. Bioeng.* **9,** 187.

Grieves, R. B., and Wang, S. L. (1967b). *Appl. Microbiol.* **15,** 76.

Grieves, R. B., Bhattacharyya, D., and Crandall, C. J. (1967). *J. Appl. Chem.* **17,** 163.

Grieves, R. B., Conger, W. L., and Malone, D. P. (1970). *J. Amer. Water Works Assoc.* **62,** 304.

CHAPTER **13**

REMOVAL AND USE OF HYDROLYZABLE METALS IN FOAM SEPARATIONS

Alan J. Rubin

Water Resources Center
College of Engineering
Ohio State University
Columbus, Ohio

I. INTRODUCTION

Metals dispersed in a liquid may be removed from solution or dispersion by a number of adsorptive bubble separation techniques. With foam separations the exact mechanistic regime of the transport from the bulk to the foam phase depends primarily on the state of dispersion or subdivision of the metal both before and after adding the collector surfactant. The sublate, that is, the surface-active product of the interaction between the metal and the collector, may be removed by flotation or partition. The latter results when soluble sublates are separated, in which case the foam separation is known as foam fractionation. Flotation occurs whenever the components to be removed or its sublates are insoluble. For one reason or another, it is not always possible to distinguish between these two transport mechanisms. This is particularly true with colloidal materials or substances of only moderate solubility.

The foam separation of metal ions is a foam fractionation process when metal and collector are attracted by simple ion-pair formation or by weak coordination. In such cases the sublate is soluble, and stoichiometric or greater amounts of collector are required. When the metal ion and collector react to form an insoluble product, the process is ion flotation. Again, a stoichiometric or greater concentration of surfactant is used and must be added in such a way that it exists as simple ions and not as micelles. Condensation of the metal by pH adjustment or other means prior to the addition of collector results in precipitate flotation. With this process only a very small amount of collector is needed, the lower limit being based on the surfactant requirements for a stable surface (foam) phase. It should be noted that ion flotation can be viewed as a special case of precipitate flotation, as can microflotation. The mechanism by which the particles are transported from the bulk to the foam phase is identical with all three techniques. The major differences amongst the various foam separation processes, and especially these three, lie in the nature of the interaction between the collector and the surface inactive substance to be removed.

With hydrolyzable metals, such as aluminum(III), lead(II), zinc(II), and the transition metals, the mechanism, then, depends to a large extent upon the pH of the dispersion medium as well as the nature and concentration of the metal. Other parameters which are very important are the ionic strength and the type and concentration of collector. Additional variables which should be considered in order to understand the controlling mechanisms include temperature and gas flow rate.

Microflotation is a special technique for removing microorganisms and other colloids, and is not to be confused with "colloid flotation" or "foam flotation." Individual colloidal particles are removed by the latter while microflation is a true flotation process. This is accomplished by enmeshing the ordinarily small particles in the flocculent precipitate of a hydrolyzable metal. Usually the floc is produced by hydroxide precipitation; aluminum sulfate has been found to very effectively "gather" or "enmesh" organic and inorganic sols over a broad pH range. The resultant floc has the effect of increasing the rate and degree of removal by decreasing the stability of the particles and by increasing their size and providing a new surface for the efficient adsorption of both anionic and cationic collectors. Strongly ionized soluble collectors may be used since the hydroxide floc stabilizes the surface foam phase. With a stable foam only a very low rate of gas flow is required for rapid clarification. As with precipitate flotation, this prevents redispersion of the float and increases the volume reduction by decreasing the amount of bulk liquid transported to the foam. The major advantages of microflotation over conventional settling or foam separation techniques are the high rates of efficient removals possible and low selectivity with regard to the nature of the colloidal particles that are to be removed. As expected, solution pH is the most fundamental parameter controlling the process.

II. EXPERIMENTAL METHODS AND MATERIALS

A schematic of the experimental apparatus used in the author's work is shown in Fig. 1. Nitrogen gas was passed from a cylinder to a foam column

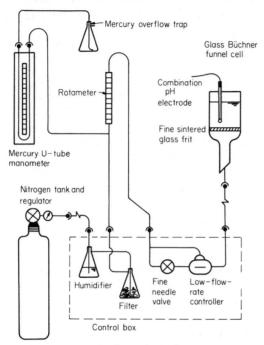

FIG. 1. Schematic of experimental apparatus.

through a gas humidifier, glass-wool filter, Manostat rotameter (model FM 1042B) and Moore low-flow-rate controller (model 63 BU-L). Glass Büchner funnels of 600-ml capacity, having diameters of 10 cm and fine sintered-glass frits, served as the foam columns. The line gas pressure was monitored with an open mercury U-tube manometer and maintained at 30 in. upstream of the controller. The gas flow was adjusted with a fine Nupro needle valve, reported in milliliters per minute at average atmospheric pressure and room temperature.

The surfactant collectors were used without special purification, and absolute alcohol was used as the frother to refine further the small bubbles supplied by the fine frit. The collector solution was prepared with distilled water so that the desired concentration of collector and 1 ml of alcohol could be added simultaneously in a single injection. The collector concentration is expressed as the "collector ratio" S, which is defined as the molar ratio of collector to metal. Solutions of reagent grade sodium hydroxide, sodium perchlorate, and nitric or hydrochloric acid were used to adjust the ionic strength and/or pH. Stock solutions of the metals were prepared from the reagent grade nitrate or perchlorate salt. The concentrations remaining after foaming were determined colorimetrically. The details of these analyses and many aspects of the experimental techniques are discussed in the appropriate papers as cited in the text.

Suspensions of the bacterium *Bacillus cereus* or the clay illite served as the dispersed phases for the microflotation studies. Suspension turbidities as determined by absorption measurements were used to estimate concentration. Reagent grade aluminium sulfate (alum) was used as the hydrolyzing metal coagulant and the applied doses are reported as mg/liter of $Al_2(SO_4)_3 \cdot 18H_2O$.

Experiments were batch type using 400 ml of the test solution or suspension. After adjustment of the gas-flow rate, pH and ionic strength, collector–frother solution was added to the foam column; samples of the bulk were taken at predetermined intervals. The foam phase was undisturbed during experimentation and was not removed for analysis. A pH meter and combination electrode were used to monitor pH during each experiment.

III. FOAM SEPARATION OF HYDROLYZABLE METALS

A. Effect of Hydrogen Ion Concentration

1. Hydrolysis and Foam Separation of Lead(II)

When a soluble metal salt is added to water, it will dissociate and the ions will become hydrated. In addition, if the metal is capable of undergoing hydrolysis, it will also react with water thereby lowering the pH and forming mononuclear and perhaps polynuclear hydroxo species in a stepwise series. The exact distribution of the various species formed, and indeed their nature, will depend primarily on the solution pH, and to a lesser extent on the temperature and the type and concentration of solution components as well as other physical and chemical characteristics of the system. It would be expected then that varying the hydrogen ion concentration of a solution of such a metal would have a profound effect on its removal by foam separation.

A case in point is the effect of pH on the foam separation of lead using the anionic surfactant sodium lauryl sulfate (NaLS) as the collector (Rubin and Lapp, 1969). Typical curves showing the removal of lead(II) as a function of time at collector ratio 2 ($S = 2$) are illustrated in Fig. 2. At a gas flow rate of about 25 ml/min the maximum or steady state removals are reached in

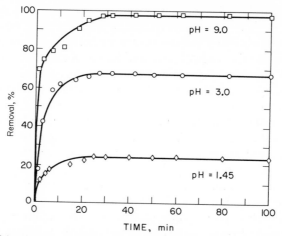

Fig. 2. Typical data for foam separation of 0.1 mM lead(II) as a function of time at different pH; 0.2 mM sodium lauryl sulfate served as the collector. [From A. J. Rubin and W. L. Lapp, *Anal. Chem.* **41**, 1133 (1969).]

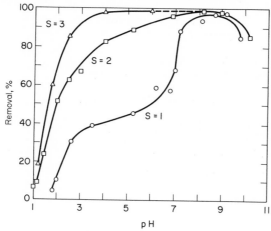

Fig. 3. Removal of 0.1 mM lead(II) as a function of pH at different sodium lauryl sulfate concentrations. [From A. J. Rubin and W. L. Lapp, *Anal. Chem.* **41**, 1133 (1969).]

20–40 min at all pH values. As a reminder to the reader, these experiments were performed in a short column and the foam phase was not removed or recycled during experimentation. Steady state removals are estimated by the 100-min removals from data such as these. The removal of 0.2 mM lead(II) at minimum ionic strength is summarized in Fig. 3 as a function of pH at three collector ratios.

At this concentration of lead there is no evidence of precipitate flotation of insoluble hydroxide at any pH. Note also that the effective pH range of removal increases on the acid side with increasing collector concentration. This range, of course, is characteristic of the particular metal, reflecting competition between H^+ and metal ions for collector at low pH and competition between OH^- and collector ions for metal at higher pH. At collector ratios 1 and 2 there is a discrete maximum in removal at pH 8.2. In general, removals were less sensitive to collector concentration at this pH. At the lowest collector concentration there is also an uncharacteristic plateau in removals near pH 5. This plateau was also observed in similar experiments at collector ratio 2 when run at higher ionic strength. Removals decreased as the pH was increased above 10.2, as expected, but were highly scattered and therefore are not shown in this figure. This scatter in the results may have been due to the presence of small amounts of insoluble hydroxide formed by localized precipitation during pH adjustment.

The results in Fig. 3 can best be interpreted by examining the hydrolysis of lead(II). The hydrolytic reactions and related equilibrium constants (Fuerstenau and Atak, 1965) are summarized below:

$$Pb^{2+} + H_2O \rightleftharpoons PbOH^+ + H^+, \qquad pK_1 = 6.17, \tag{1}$$

$$PbOH^+ + H_2O \rightleftharpoons Pb(OH)_2 \ (aq) + H^+, \qquad pK_2 = 10.90, \tag{2}$$

$$Pb(OH)_2 \ (aq) \rightleftharpoons HPbO_2^- + H^+, \qquad pK_3 = 10.92. \tag{3}$$

For simplicity the waters of hydration are not shown.

In dilute solutions lead hydroxide is soluble:

$$Pb(OH)_2 \ (c) \rightleftharpoons Pb(OH)_2 \ (aq), \qquad pK_d = 3.44. \tag{4}$$

Therefore precipitate flotation would not be expected in solutions less concentrated than $3.65 \times 10^{-4} \ M$.

The reactions listed above are adequate to define a $10^{-4} \ M$ lead(II) solution since at this low concentration the formation of polynuclear species is negligible. A nitrate complex of lead(II) is also known (Gilbert, 1964) but its concentration is negligible in dilute solutions using chloride or perchlorate salts to adjust the pH and ionic strength. Equations (1)–(3) are summarized in Fig. 4, which shows the distribution of each species in percent as a function of pH.

Assuming that only positively charged species are foam separated by an anionic collector, the maximum concentration removed in the presence of an excess of collector (collector ratio of 2 or greater) is given by

$$M_2 = [Pb^{2+}] + [PbOH^+]. \tag{5}$$

Similarly, at a collector ratio of 1 the maximum concentration removed is given by

$$M_1 = \tfrac{1}{2}[Pb^{2+}] + [PbOH^+]. \tag{6}$$

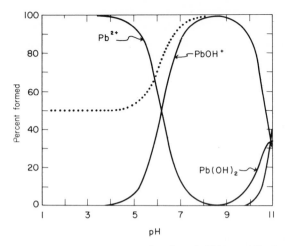

FIG. 4. Distribution of lead(II) species as a function of pH. Dotted line is the maximum that can be removed at a collector ratio of unity. [From A. J. Rubin and W. L. Lapp, *Anal. Chem.* **41,** 1133 (1969).]

The latter function is indicated in Fig. 4 by the dotted line, and explains the plateaus in removal at $S = 1$. Increasing the ionic strength also has the effect of reducing the effective collector concentration.

Differences between removals calculated from the above and experimental results can be accounted for by H^+ competition at low pH and instability of the lead–lauryl sulfate complex. The removals are less sensitive to collector and reach a maximum at pH 8.2 because stoichiometry of collector to metal is most favorable in the presence of the $PbOH^+$ ion. Removals decrease above pH 8.2 upon formation of the soluble lead hydroxide and plumbite ion. It would be expected that the latter would be removed by a cationic collector.

In summary, the results demonstrate the applicability of hydrolysis data to estimating removals by foam separation. Such data alone, however, are not sufficient to completely characterize the system. Other factors which must be considered include the stability of the metal–collector complex, the type of collector used, and the effect of competing cations. Before examining some of these and other variables it is instructive to consider the effect of pH on the foam separation of a metal in which its hydroxide is insoluble, and the success of hydrolysis data in predicting precipitate flotation.

2. Precipitate Flotation of Iron(III) and Zinc(II)

Iron(III) and zinc(II) also hydrolyze in solution but, unlike lead, form insoluble hydroxides in the approximate 10^{-4} M concentration range. If it were assumed that these hydroxide precipitates would be completely removed from dispersion, say by precipitate flotation, then a "theoretical" precipitate flotation curve could be drawn based on calculations from hydrolysis data (Rubin and Johnson, 1967; Rubin and Lapp, 1971). For iron(III) at room temperature, neglecting polynuclear species, the distribution of soluble iron species C_a is given by

$$C_a = [Fe^{3+}] + [FeOH^{2+}] + [Fe(OH)_2^+] + [Fe(OH)_4^-] \qquad (7)$$

or as a function of pH in acid solution:

$$C_a = K_s[H^+]^3 + K_sK_1[H^+]^2 + K_sK_1K_2[H^+], \qquad (8)$$

where K_s is the dissociation or "solubility" constant $[Fe^{3+}]/[H^+]^3$, K_1 and K_2 are the first and second hydrolysis constants for iron(III), respectively, and $[H^+]$ is the hydrogen ion concentration calculated directly from pH measurements. Assuming 100% efficiency, the percentage of removal of the metal precipitate is calculated by substituting C_a into $\% R = 100(1 - C_a/C_0)$, where C_0 is the total (analytical) concentration of iron added. Such a curve compared to experimental data is shown in Fig. 5. Further details about its derivation are given elsewhere (Rubin, 1968).

The calculated precipitate formation curve for 0.2 mM iron(III) predicts that the metal is completely soluble up to pH 2.67 (the precipitation point).

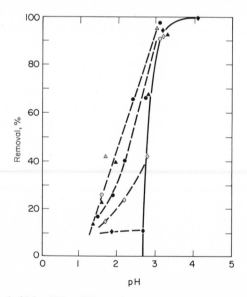

FIG. 5. Removal of 0.2 mM iron(III) as a function of pH at different gas flow rates G (ml/min) and concentrations C (mM) of sodium lauryl sulfate. Solid line is the calculated precipitate flotation removal curve. (◆) $C = 0.2$, $G = 11$; (●) $C = 0.4$, $G = 11$; (▲) $C = 0.6$, $G = 11$; (◇) $C = 0.2$, $G = 19$; (○) $C = 0.4$, $G = 19$; (△) $C = 0.6$, $G = 19$. [From A. J. Rubin and J. D. Johnson, *Anal. Chem.* **39**, 298 (1967).]

At the precipitation point and above, ferric ions at this concentration exist in both soluble and insoluble forms. However, at pH 4 only an insignificant fraction of the metal remains in solution. The removals shown in Fig. 5 at C mM concentration of sodium lauryl sulfate and gas-flow rates of G ml/min fall on the curve or to its left. The removals at pH values less than pH 2.67 must be due to the foam separation of dissolved iron(III) species. Removals on the curve, then, are due to precipitate flotation, while removals between the curve and the precipitation point are due both to ion removal and precipitate flotation. The removals at the higher collector concentrations and gas flow rates increase with pH. At the lowest gas rates and sodium lauryl sulfate concentrations, the removals are lower but increase sharply at pH values at and above the precipitation point, in excellent agreement with the calculated curve.

The results, that is, the pattern of removal with pH, are similar for both iron(III) and aluminum(III) (Rubin, unpublished paper). Both of these metals form polynuclear ions in dilute solution. Since no one soluble species predominates below the precipitation point, the removal curves do not demonstrate a plateau in removal as does lead. Examination of the hydrolytic reactions of zinc(II) (Kanzelmeyer, 1961), on the other hand, reveals that the free unhydrolyzed Zn^{2+} ion is the species in greatest concentration up to a pH of about 8; $ZnOH^+$ is formed but is almost completely negligible in concentration. Also at the concentration examined, 0.1 mM zinc(II), there was no evidence of soluble polynuclears. Above pH 8.5 and below pH 11 the insoluble hydroxide is predominant, with soluble $Zn(OH)_2$ molecules making up much less than 3% of the total. Thus, the total concentration of soluble zinc is given by

$$C_a = [Zn^{2+}] + [Zn(OH)_3^-]. \tag{9}$$

The predicted precipitate removal curve is compared with experimental data in Fig. 6, showing the removal of 0.1 mM zinc(II) with various concentrations of sodium lauryl sulfate at a gas-flow rate of about 25 ml/min.

The removals are greatest, even at very low collector concentrations, in the pH region of hydroxide precipitation. Agreement between calculated and observed removals at the lower pH limit of hydroxide formation is excellent. Removal above pH 10 decreases upon formation of negatively charged zinc species; however, this occurs at lower pH than predicted by the solubility limits of the precipitate on its basic side. There is the possibility that the precipitate has acquired a negative charge at the higher pH due to adsorption of hydroxide ions and thus has a lowered affinity for the collector.

Although the removals are relatively insensitive to collector concentration above pH 8, at lower pH the removal of Zn^{2+} is quite dependent upon this parameter. At a stoichiometric collector ratio of 2, the removals averaged about 92% within the pH range of 3–8, forming a plateau in the removal

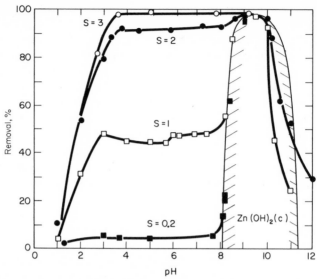

Fig. 6. Removal of 0.1 mM zinc(II) with sodium lauryl sulfate as a function of pH. Experimental results are compared with a calculated precipitate flotation curve. (From Rubin and Lapp, 1971.)

curve. This plateau is indicative of a single predominant ionic species. The difference between the observed removal and the theoretical removal of 100% is due to the instability of the zinc–lauryl sulfate complex. Below pH 3 removals drop off sharply because of protonation of the collector.

3. Studies of the Foam Separation of Copper(II)

Soluble and condensed copper(II) are also efficiently removed by foam separation with NaLS. The removal pattern with pH is similar in many respects to those for zinc and lead. Unlike iron and lead, which readily co-ordinate with oxygen atoms, there is very little interaction between copper and lauryl sulfate. The sublate is completely soluble and thus the mechanism under which copper is removed at low pH is unambiguously a foam fractionation. For these reasons copper(II) is ideal for comparing the various metals and collectors, as well as offering a simple comparison between partition and flotation processes by merely changing pH (Rubin *et al.*, 1966).

In practice, of course, foam fractionation requires tall columns with provisions for reflux and foam drainage to maximize the separation and to reduce the amount of liquid carried over with the foam. For the studies outlined here, a short column was used since this is experimentally convenient and permits the controlling variables to be studied individually and independently of one another. During experimentation the foam phase was not disrupted and only a single experiment was performed on each single batch.

B. Comparison of Ion and Precipitate Removal

1. Interactions between Solution Components

The nature of the collector surfactant is very important. Figure 5 suggests that sodium lauryl sulfate, a strongly ionized anionic surfactant, is capable of removing dissolved, colloidal, and condensed iron(III). Soluble ferric species,

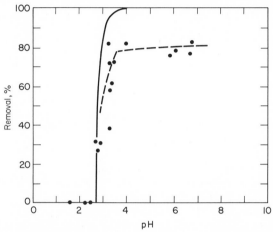

FIG. 7. Removal of 0.2 mM iron(III) with stearylamine as a function of pH. Experimental results are compared with calculated precipitate flotation curve. [From A. J. Rubin and J. D. Johnson, *Anal. Chem.* **39**, 298 (1967).]

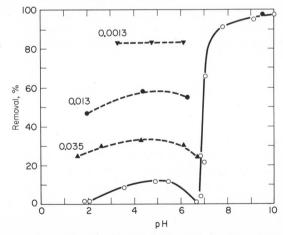

FIG. 8. Foam separation of 0.3 mM copper(II) with stearylamine (open points and solid line) and NaLS at different ionic strengths (blackened points and dashed line): (▼) 0.0013, (●) 0.013, (△) 0.035. Removals as a function of pH. [From A. J. Rubin and J. D. Johnson, *Anal. Chem.* **39**, 298 (1967).]

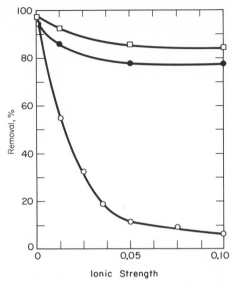

FIG. 9. Effect of ionic strength on the foam separation of 0.1 mM zinc(II) with sodium lauryl sulfate. Removal of ions at pH 5 and of precipitate at pH 9. ($\square$) $S = 1$, pH $= 9.0$; ($\bullet$) $S = 0.2$; pH $= 9.0$; ($\bigcirc$) $S = 2$, pH $= 5.0$. (From Rubin and Lapp, 1971.)

except at high pH, are positively charged and therefore their removal from solution by a cationic collector would not be expected. Iron precipitates, however, are known to be floated by amine collectors. These effects, the non-removal of soluble iron and the floatability of its hydroxide precipitate using stearylamine as the collector, are shown in Fig. 7. This surfactant is a weakly basic primary amine, being positively charged only in acid solutions. Iron is not removed by the amine until the precipitation point and at higher pH the removals are less than predicted. The scattered and nonreproducible results shown in this figure are due to the variable character of iron hydroxide and because iron and stearylamine form an unstable complex. This may be contrasted with the removal of copper by the same amine as shown in Fig. 8. The removals form a smooth curve and approach 100 % because copper is a strong nitrogen coordinator and interacts strongly with amines. Because of this its ions although positively charged are also partially removed.

It is apparent that the removal of hydrolyzable metals in soluble systems involves a competition between collector and the various ions in solution. Where coordination between collector and the metal is a minimum, electrostatic effects predominate. In these cases the sublate is soluble and removals are sensitive to the other components of the solution such as sodium and other simple ions. This is the "ionic strength" effect and is illustrated in Fig. 8

for copper(II) and in Fig. 9 for zinc(II), both using sodium lauryl sulfate as the collector. The ionic strengths were adjusted with sodium perchlorate.

Removals of 0.3 mM copper(II) after foaming for 40 min are shown as a function of pH at different ionic strengths, all other conditions being held constant. Points on the extreme left in Fig. 8 are for removals at the lowest pH possible at each ionic strength. Shown also in this figure is a single point on the precipitate flotation curve at an intermediate ionic strength. That zinc removal is similarly affected is shown in Fig. 9. Experiments were conducted at pH 5, where the Zn^{2+} ion predominates and at pH 9 where essentially all of the metal is present as the precipitate. These figures demonstrate a fundamental difference between precipitate flotation and the foam fractionation of metal ions. Thus, although there may be a special affinity between collector and hydrolyzable metal, the probability of surfactant entering the foam as an ion pair with a neutral salt increases with an increase in the concentration of the latter. On the other hand, precipitate removal is nonstoichiometric and surfactant need only react with ions on the surface of the condensed phase to effect its removal. This requires very small amounts of the collector, the excess being used to stabilize the surface phase.

Differences in ion and precipitate removal with regard to surfactant requirements will be examined further in the next section. In addition to altering the ionic competition in a system, reductions in removals with increasing ionic strength may be attributable in part to decreases in activity of the soluble species. The salt content of the solution may also influence removals by changing the critical micelle concentration of the collector and by altering foam stability. Note then that selectivity in some foam separation systems could be improved by adjusting the concentration of neutral salts as well as the pH.

2. Effect of Metal and Collector Concentrations

To a large extent the effects of collector concentration on both ion and precipitate removal are shown in Figs. 3, 5, and 6. As indicated by these figures and in Fig. 10 for 0.3 mM copper(II) at pH 6.3 (Rubin et al., 1966), ion removal is strongly dependent on collector concentration, requiring stoichiometric or greater concentrations for complete removal. That precipitate flotation is considerably less affected is demonstrated in Fig. 11 for the same concentration of copper at pH 9.5. Neither the rate nor the total removal of copper hydroxide is affected at collector ratios of 0.2 or greater. At very low amounts of surfactant the rate of removal is still high but the surface phase is unstable, and after a few minutes of foaming the float is redispersed and removals correspondingly decrease. This effect is evident when using alcohol

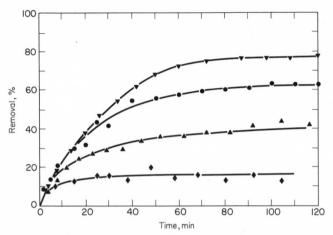

FIG. 10. Effect of sodium lauryl sulfate concentration on the foam fractionation of 0.3 mM copper(II). Ionic strength 0.013 at pH 6.3. Collector ratio : ($\triangledown$) 3, ($\bullet$) 2, ($\triangle$) 1, ($\diamondsuit$) 0.5. [From A. J. Rubin et al., Ind. Eng. Chem. Process Des. Develop. **5**, 368 (1966).]

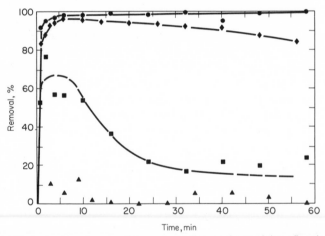

FIG. 11. Effect of sodium lauryl sulfate concentration on the precipitate flotation of 0.3 mM copper(II). Ionic strength 0.013 at pH 9.5. Collector ratio : ($\bullet$) 1.0, ($\blacklozenge$) 0.01, ($\blacksquare$) 0.001, ($\blacktriangle$) 0. [From A. J. Rubin et al., Ind. Eng. Chem. Process Des. Develop. **5**, 368 (1966).]

frother but without collector, as shown in the lower portion of Fig. 11. Partial removal of the precipitate is observed since ethanol has some surface-active properties. However, without collector redispersion occurs very rapidly.

Since ion removal is strongly dependent upon the stoichiometry of collector to metal, an increase in the concentration of the latter would require a corresponding increase in surfactant. The amount removed at steady state is dependent only on the collector ratio; however, the rate is altered by the metal concentration. Conversely, neither the rate nor degree of precipitate flotation is affected by the initial concentration of the metal, provided enough surfactant remains after reaction with the surface of the condensed phase to produce a stable foam.

3. Effect of Temperature and Gas-Flow Rate

Two parameters that might be significant in foam separations are temperature and gas-flow rate. Both are important considerations not only in the way in which they might affect removals but also in what they tell us about the mechanisms of the process. In general, for example, temperature has been found to have very little affect (Rubin *et al.*, 1966; Schoen and Mazella, 1961), provided that foam structure is not destroyed thermally. Therefore, it is apparent that foam separations are not controlled by such temperature sensitive parameters as diffusion and surface tension.

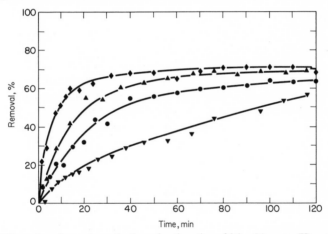

Fig. 12. Effect of gas flow rate on the foam fractionation of 0.3 mM copper(II) with 0.6 mM sodium lauryl sulfate. Ionic strength 0.013 at pH 6.3. Gas rates (ml/min): (◆) 29.8, (▲) 19.0, (●) 10.8, (▼) 5.2. [From A. J. Rubin *et al.*, *Ind. Eng. Chem. Process Des. Develop.* **5**, 368 (1966).]

Gas-flow rate, on the other hand, as shown in Fig. 12, strongly affects the rate of removal of soluble sublates without significantly affecting the steady state removals. The effect is not as great with insoluble systems; apparently, the larger the particle being removed, the less the rate of removal is determined by gas-flow rate. The difference between ion and precipitate removal is reflected in these observations. The removal of soluble sublates involves their distribution or partition between gaseous and aqueous phases. The greater the interfacial area, as occurs with increasing gas rate, the greater is the removal at any given time although the total removal depends upon stoichiometric relationships. Precipitates, on the other hand, are levitated out with bubbles that serve to reduce the effective density of the particles. Interfacial area is not important and the process is not controlled by adsorption at the bubble surface. In fact, because gelatinous precipitates are easily broken up, and once redispersed in the bulk are extremely hard to refloat, an increased gas rate may be detrimental to precipitate and microflotation. A further disadvantage of high gas rates, especially with soluble systems is that the amount of bulk liquid entrained in the foam may be very high, resulting in a smaller volume reduction.

IV. MICROFLOTATION FOR REMOVAL OF COLLOIDS

A. Studies with *Bacillus cereus*

When individual colloidally dispersed particles are removed by foaming, the exact mechanistic regime of the process may be ambiguous, appearing to operate by partition. Generally, the rate of removal is rather slow and the particles may be selectively removed. Unless a chromatographic separation is desired, the process would be more rapid and less selective if the particles were increased in size and if a new surface was produced for the efficient adsorption of collector. Both are accomplished by the addition of hydrolyzing metal and pH control prior to foaming. The resultant process is known as "microflotation." Microflotation was developed as a consequence of the author's work with hydrolyzable metals (Rubin and Cassell, 1965) and has now been applied to both organic and inorganic colloids.

Applications and the effects of several parameters on the process have been described in detail elsewhere (see Rubin *et al.*, 1966a). In general, microflotation is sensitive to the same variables as precipitate flotation. What is unique is the use of a hydrolyzing metal to aid the removal of sols. This aspect will be discussed here.

The first example is the microflotation of the gram positive bacterium *Bacillus cereus* (Rubin and Lackey, 1968). Removals of the organism within 20 min of foaming using lauric acid as the collector are shown in Fig. 13. Without aluminum sulfate (alum) the removals are irregular; high removals occur only in acid solutions. Even with this poor collector the addition of alum results in rapid and essentially complete separation near pH 7. Flotation of the organism has also been examined without collector or frother, both with and without alum. There was no removal at any pH. Aggregation alone does not result in flotation when collector and frother are not used.

Figure 14 shows that removal in such systems is a combination of colloid flotation of free particles and precipitate flotation of sol trapped in an alum floc. Bacterial cells are removed at pH 4.5 and below when alum is not used.

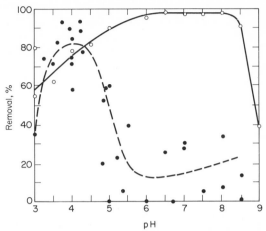

FIG. 13. Foam separation of *B. cereus* with and without the presence of aluminum sulfate using lauric acid as the collector. This compares microflotation with colloid flotation. (○) 150 mg/liter alum ; (●) no alum. [From A. J. Rubin and S. C. Lackey, *J. Amer. Water Works Assoc.* **60,** 1156 (1968).]

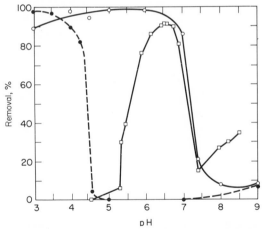

FIG. 14. Removal of *B. cereus* with sodium lauryl sulfate, comparing colloid flotation and settling with microflotation. (○) 50 mg/liter alum ; (●) no alum ; (□) settling 50 mg/liter alum. [From A. J. Rubin and S. C. Lackey, *J. Amer. Water Works Assoc.* **60,** 1156 (1968).]

This is one pH unit less than observed with lauric acid ; however, the removals with sodium lauryl sulfate are greater and form a smooth curve. At low pH the organisms are coagulated by hydrogen ion, that is, their negative charges are reduced so that an anionic collector is adsorbed. Lauric acid, on the other hand, is uncharged in acid solution and is thus capable of hydrogen bonding with the cell surfaces at slightly higher pH. In the pH range between

5 and 7 the cells can be removed from dispersion in the presence of alum by settling or flotation. This is the "sweep zone" of aluminum hydroxide precipitation and enmeshment of the organisms in the resultant floc (Rubin and Hanna, 1968). It should be noted that the ability of the organisms to be collected is sharply reduced once the optimum pH is exceeded. At this boundary the hydroxide floc goes back into solution as the negatively charged aluminate ion, $Al(OH)_4^-$.

B. Studies with the Clay Illite

Microflotation was originally introduced as a removal technique for microorganisms and other organic sols. Recently, it has been demonstrated

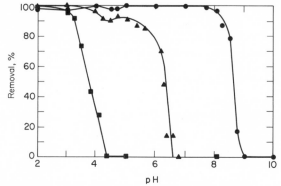

FIG. 15. Effect of aluminum sulfate concentration on the foam separation of illite with sodium lauryl sulfate. (■) No alum; (▲) 6.25 mg/liter alum; (●) 50 mg/liter alum. [From A. J. Rubin and S. F. Erickson, *Water Res.* **4** (on press).]

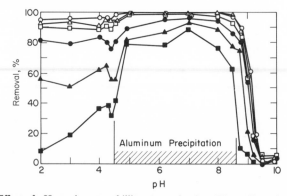

FIG. 16. Effect of pH on the rate of illite removal using 100 mg/liter alum and 30 mg/liter sodium lauryl sulfate. Removals at a given time are shown and the range of aluminum hydroxide precipitation is defined. Time (min): (○) 16, (△) 8, (□) 4, (●) 2, (▲) 1, (□) 0.5. [From A. J. Rubin and S. F. Erickson, *Water Res.* **4** (on press).]

that the process is just as effective with inorganic systems and that removal is subject to the same variables. An example is the flotation of the clay illite (Rubin and Erickson, 1971).

Figure 15 shows the removals of illite obtained at 16 min as a function of pH using about 30 mg/liter sodium lauryl sulfate. As with *Bacillus cereus*, the clay is removed only at low pH when alum is not used. Upon adding the flocculating agent, the pH range of removal is widened with increasing concentration. As demonstrated in Fig. 16, the addition of alum also increases the rate of clarification; this occurs in the pH regions of aluminum hydroxide precipitation. Here the sols are enmeshed, or gathered, in the flocculent precipitate and the floc itself provides the sites for the adsorption of collector. In general, because the hydroxide is formed over a wider pH range with increasing metal concentration, the pH range of removal also increases on the alkaline side. The pH range of removal for many sols in acid solutions is independent of the alum concentration and the rates of removal are considerably less. Changes in collector concentration also do not significantly affect the removal rate.

In summary, the removal of sols can be facilitated by the addition of a suitable hydrolyzable metal salt. The resultant process, microflotation, possesses the same advantages as precipitate flotation. The particles are removed more rapidly and less selectively than would be possible without the addition of the hydrolyzable metal.

REFERENCES

Fuerstenau, M. C., and Atak, S. (1965). *Trans. A. I. M. E.* **232,** 24.

Gilbert, T. W. (1964). *In* "Treatise on Analytical Chemistry" (I. M. Kolthoff, P. J. Elving, and E. B. Sandell, eds.), Vol. 6, Part II. Wiley (Interscience), New York.

Kanzelmeyer, J. H. (1961). *In* "Treatise on Analytical Chemistry" (I. M. Kolthoff, P. J. Elving, and E. B. Sandell, eds.), Vol. 3, Part II. Wiley (Interscience), New York.

Rubin, A. J. (1968). *J. Amer. Water Works Assoc.* **60,** 832.

Rubin, A. J. (unpublished). Precipitate flotation of aluminum(III) and the mechanism of microflotation, Ohio State Univ., Columbus.

Rubin, A. J., and Erickson, S. F. (1971). *Water Res.* **4** (on press).

Rubin, A. J., and Hanna, G. P. (1968). *Environ. Sci. Technol.* **2,** 358.

Rubin, A. J., and Johnson, J. D. (1967). *Anal. Chem.* **39,** 298.

Rubin, A. J., and Lackey, S. C. (1968). *J. Amer. Water Works Assoc.* **60,** 1156.

Rubin, A. J., and Lapp, W. L. (1969). *Anal. Chem.* **41,** 1133.

Rubin, A. J., and Lapp, W. L. (1971). *Separ. Sci.* **6,** 357.

Rubin, A. J., and Cassell, E. A. (1965). *Proc. Southern Water Resources Pollut. Contr. Conf.* **14,** 222.

Rubin, A. J., Cassell, E. A., Henderson, O., Johnson, J. D., and Lamb, J. C. (1966a). *Biotechnol. Bioeng.* **8,** 135.

Rubin, A. J., Johnson, J. D., and Lamb, J. C. (1966b). *Ind. Eng. Chem. Process Des. Develop.* **5,** 368.

Schoen, H. M., and Mazzella, G. (1961). *Ind. Water Wastes* **6,** 71.

CHAPTER 14

APPLICATION OF ADSORPTIVE BUBBLE SEPARATION TECHNIQUES TO WASTEWATER TREATMENT

David Jenkins
Department of Sanitary Engineering
University of California
Berkeley, California

Jan Scherfig
Department of Sanitary Engineering
University of California
Irvine, California

David W. Eckhoff
H. F. Ludwig and Associates
New York, New York

I. INTRODUCTION

The application of techniques involving adsorptive bubble separation to wastewater treatment can best be discussed on a unit operation basis under the separate categories of municipal and industrial wastewater treatment. The objectives of a wastewater treatment scheme should vary depending on the quality criteria of the receiving water, and these quality criteria should be

TABLE I

EFFECTS OF TREATMENT PROCESSES ON TYPICAL MUNICIPAL WASTEWATER COMPOSITION

	Composition (mg/liter)						
	BOD[a]	COD[b]	SS[c]	MPN[d]	TDS[e]	Total N	Total P
Raw wastewater	170	300	250	10^9	500	40	10
Wastewater after:							
Primary sedimentation	120	200	100	5×10^8	500	30	9
Secondary treatment							
(activated sludge)	20	50	20	5×10^5	500	25	8
+ Chlorination	20	50	20	10–100	500	25	8
Tertiary treatment:							
Lime precipitation							
and sedimentation	10	30	10	1–10	500	25	1
+ Filtration	5	20	<5	<10	500	25	0.5
+ Carbon adsorption	<5	<10	<1	<10	500	25	0.5
+ Reverse osmosis	<1	<5	<1	0	<100	<5	<0.5

[a]5-day biochemical oxygen demand—an index of the degradable organic matter.
[b]Chemical oxygen demand—an index of the total organic matter.
[c]Suspended solids.
[d]Most probable number of coliform organisms per liter.
[e]Total dissolved solids.

based on the specific beneficial uses of the receiving water. It is therefore wise to preface any discussion of "typical" wastewater treatment process flow sheets with a statement that their universal application to specific situations is not justified. For example, the traditional domestic wastewater schemes outlined below are probably not justifiable for the treatment of sewage in coastal communities prior to ocean disposal. Nor are these schemes always the correct way to meet some of the more recent waste treatment objectives such as removal of nutrients, toxic compounds, and floating and slick-causing materials. With such a qualification in mind it has been possible and traditional to divide waste treatment processes into three broad categories whose descriptive titles "primary," "secondary," and "tertiary" describe the sequence in which these processes treat wastewater.

Primary treatment processes largely involve physical–chemical separation of particulate or dispersed material by processes such as sedimentation and flotation, possibly aided by precipitation–coagulation and flocculation. The treatment and disposal of the concentrated overflow or underflow of these processes is attained by a variety of physical–chemical and biological sludge treatment and disposal processes.

TABLE II

ROLE OF ADSORPTIVE BUBBLE SEPARATION IN TYPICAL MUNICIPAL WASTEWATER TREATMENT PROCESSES

Process	Purpose	Method	Role of adsorptive bubble separation
Screening	Remove large (>1.5–3 in.) solid material	Interception of material with screens placed in waste flow	None
Grit removal	Remove sand and grit (>2 mm-diam, sp. gr. 2.65)	Sedimentation in channels or aerated chambers	Aeration *may* aid in flotation of grease particles
Preaeration	Aid later removal of particulate material by sedimentation	Aeration in basin with approx. 0.1 ft³ air/gal sewage	Aeration *may* break up agglomerates of grease and settleable particles allowing later flotation of grease and sedimentation of particles
Vacuum flotation	Remove floatable (especially grease) and settleable particles	Apply vacuum to aerated sewage. Skim floating particles. Remove settled material as underflow.	Particles removed by adsorptive bubble separation. Bubbles generated by application of vacuum.
Pressure flotation	Remove floatable and settlable organic particles.	Aerate wastewater under pressure. Release pressure to generate bubbles. Skim floating particles. Remove settled material as underflow.	Particles removed by adsorptive bubble separation. Bubbles generated by pressurization of wastewater followed by release of pressure.
Primary sedimentation (with skimming)	Remove settlable solid particles and floating solids	Provide quiescent sedimentation (1–2 hr). Remove settled particles as primary sludge underflow.	Gas bubbles *may* cause flotation of particles (especially grease).

TABLE II (continued)

Process	Purpose	Method	Role of adsorptive bubble separation
Biological treatment	Remove suspended matter and biodegradable soluble organic matter.	Grow enrichment culture of microorganisms in (1) aerated basins (activated sludge), (2) on surface of solid medium (biofiltration), (3) in pond or lagoon (oxidation pond aerated lagoon).	Probably none.
Secondary sedimentation	Remove and partially concentrate microbial culture from effluent of secondary treatment device.	Provide quiescent sedimentation (0.5–1.5 hr). Remove settled biomass for sludge treatment (biofiltration) or for sludge treatment and return to reactor (activated sludge).	Deleterious: Gas bubbles attached to sludge particles either by aeration in activated sludge basin or by release of nitrogen gas by denitrification in anoxic sludge blanket at bottom of secondary clarifier may float sludge to surface ("rising sludge") of secondary clarifier and cause deterioration of effluent quality.
Foam fractionation	Remove undegraded detergent and other surface-active materials and floatable particles.	Aerate to generate foam. Separate, collapse, and dispose of foam.	Foam is enriched with surface active materials. Foaming provides opportunity for flotation of particulates by adsorptive bubble separation.
Sludge thickening	Increase suspended matter (solids) content of sludges wasted from biological treatment processes.	Pressurize settled sewage with air. Mix with waste biological sludge. Collect concentrated float and settled materials.	Sludge particles floated by adsorptive bubble separation.

Secondary treatment processes involve the growth of an enrichment culture of predominantly aerobic organisms using the various organic (and inorganic) compounds of the wastewater as a substrate, and growing in such a form that they can be physically separated from the treated wastewater. The excess microbial mass produced in these processes is treated by the previously mentioned sludge treatment and disposal processes.

Tertiary treatment processes encompass a wide variety of processes that have been "tacked on" to the traditional primary–secondary process flow sheet and designed to improve removals of various materials greater than that effected by the primary–secondary treatment train. Physical–chemical processes such as precipitation–coagulation, froth flotation, flocculation, filtration, adsorption on active carbon, ion exchange, electrodialysis, and reverse osmosis have been employed or proposed as tertiary treatment processes. Biological processes such as algal oxidation ponds and biological denitrification are also in this group of processes.

The effects of the various treatment processes on the composition of a typical domestic wastewater are illustrated in Table I.

The established or postulated role of adsorptive bubble separation in each of these processes is summarized in Table II.

II. APPLICATIONS TO DOMESTIC WASTEWATER

A. Replacement for Primary Sedimentation

As a replacement for primary sedimentation, adsorptive bubble separation has been used only sparingly and for the treatment of domestic sewage–industrial waste mixtures containing large amounts of flotable, greasy, or fibrous solids, that is, mixtures of domestic sewage and food processing or cannery wastes. However, many instances can be cited where adsorptive bubble separation plays an incidental role in primary treatment such as the floating of grease particles which are then removed by skimming devices. There appears to be renewed interest in the application of adsorptive bubble separation for floatable matter removal from raw domestic wastewater prior to its discharge to the marine or estuarine environment or from storm sewer overflows.

Peck (1920) holds a U.S. patent which claims that the organic material in waste liquors could be coagulated to form particles having selective attraction for gas bubbles and then be separated by a flotation operation which may be of substantially the same nature as is practiced in the mineral flotation art. No data were presented to support this claim and only limited application of processes of this nature have been made in the treatment of domestic sewage. Eliassen (1943) cites a flotation process developed at the Easterly plant of Cleveland, Ohio, by Hawley in which waste lubricating oil was used as a "flotation agent." The oil was added to the sewage and the mixture aerated to produce a scum which was skimmed off and burned. While "high suspended solids removals" were claimed, the process does not operate currently and indeed there would be significant deterrents to the addition of waste lubricating oil to sewage in present times. Laboratory studies by Hansen and Gotaas (1943) demonstrated that a "heteropolar lauryl amine hydrochloride, DP243" was an affective "flotation coagulant" agent for the flotation of solids from raw domestic sewage but Eliassen (1943), in discussing this work, pointed out that the cost of $200–300/million gal for this coagulant made the application of the process to waste treatment impractical.

Osterman (1955) described the "Colloidair process"—a pressure flotation process in which air at 25 lb/in^2 was dissolved in the entire sewage flow. No performance data were given nor have any subsequent reports on the application of this process appeared. Hay (1956) reports on pilot plant studies of pressure flotation on the Racine, Wisconsin, grit chamber effluent. At optimum performance the pressure flotation unit reduced suspended solids in raw sewage by 40–60% to produce an effluent with 50–80 mg/liter suspended solids and a skimmed sludge with 8–9% solids. The pressurization, to recommended levels of 30–40 lb/in.^{2} was carried out on a portion (20–50%) of recirculated effluent. It was found that 40% air saturation was required to obtain good performance.

Commercially produced equipment designed to use adsorptive bubble separation as a replacement for primary sedimentation has not been common and has almost completely been confined to one proprietary unit—the *Vacuator*, which is a vacuum flotation unit manufactured by Dorr-Oliver Inc. The major applications of the Vacuator has been in the primary clarification of wastes with high grease loads, wastes from vegetable and fruit processing, and wastes with high concentrations of H_2S (in which case foul gases can be collected in the enclosed system and then burned). The Vacuator is a covered cylindrical concrete tank into which sewage is introduced by a central draft tube and maintained at a vacuum of about 9 in. of mercury by a wet vacuum pump. Air is introduced into the sewage prior to application of vacuum by a short period of mechanical aeration or by aerating a portion of recycled Vacuator effluent using centrifugal pumps and an eductor.

The floated solids are skimmed from the surface of the sewage, and scrapers on the Vacuator floor collect grit and other solids that settle to the bottom of the unit. Design overflow rates for Vacuators are between 5000 and 8000 gal/ft^2 day and nominal detention times of about 20 min are employed. Units installed in the field vary in diameter from 12 to 70 ft.

Published operating data for Vacuators is sparse—there being two reports for field installations, both of which operated at overflow rates much lower than the reported design rates of 5000–8000 gal/ft^2 day. Logan (1949) reports the operation of a Vacuator at Palo Alto, California on domestic wastewater in admixture with apricot, peach, pear, and tomato canning wastes at overflow rates that average 2860 gal/ft^2 day, a value that is roughly 50 % of the design value of 5000 gal/ft^2 day. Removals of suspended solids average 41 % from 285 to 169 mg/liter, while settlable solids removals were 57.5 % from 181 to 77 mg/liter. Mays (1953) presents data on the performance of a lightly loaded Vacuator (approx. 2000 gal/ft^2 day) in the treatment of domestic sewage and mixtures of tomato waste and domestic sewage. Removals of suspended solids depended on the type of aeration device and amount of air introduced into the sewage prior to vacuum flotation. An eductor seemed to be the best aeration device tested, and with aeration rates of 2.3 ft^3/min per million gal/ day a 57 % suspended solids reduction (450–195 mg/liter) was obtained. When no aeration was practiced prior to vacuum flotation, reductions in suspended solids of 34 % were accomplished (330–220 mg/liter).

B. Floatables Removal Prior to Marine or Estuarine Disposal of Liquid Wastes

A review of waste disposal practices in countries such as the United States, which includes major cities both inland and along the seaboard, shows conclusively that the costs of waste disposal can be considerably less in areas where it is feasible to discharge waste effluents into open marine waters. The treatment of wastewaters for discharge to confined bodies of water has become increasingly costly because of the need to meet contemporary water quality standards for the protection of the beneficial uses of the receiving waters, particularly for public water supply and fish and wildlife procreation. In contrast, disposal in the ocean requires a much lesser degree of treatment, provided that is is possible to construct a submarine outfall conduit so that the wastewaters can be discharged at sufficient depths and at sufficient distances from the shore in unconfined marine waters where the ocean waters themselves serve as an efficient treatment system.

A comprehensive survey and evaluation of waste discharge effects on the marine environment made by the State of California (1961) concluded that for disposal to unconfined marine waters with adequately designed outfalls treatment can often be focused on the removal of floatable material. [Floatable matter is defined operationally using the "flotation funnel" technique of Scherfig and Ludwig (1967).] Ludwig and Onodere (1964) concluded that marine wastewater outfalls designed to maintain low coliform organism counts in the surf and littoral zones would also provide satisfactory dispersion of soluble and suspended matter, especially since dilution factors of 1000 or more are characteristic within a short distance of the outfall terminus. Settlable solids appear to be transported seaward in innocuous concentrations (Ludwig and Carter, 1961 ; State of California, 1961). Floatable matter, however, because of prevalent on-shore winds in most areas, tends to work shoreward and appear on beaches and in near-shore waters.

Provisions have been made in the design of marine wastewater disposal systems for several metropolitan areas, including Rio de Janeiro (Brazil) and Accra (Ghana) for floatable removal prior to discharge. For example, pilot plant flotation studies at the Rio de Janeiro site showed that quiescent flotation and sparged air flotation were unsatisfactory for removing floatable material. The process of choice was dissolved-air flotation at hydraulic loading rates of 6000–8000 gal/ft^2 day with direct pressurization of the entire waste stream to approximately 40 lb/in.^{2} and with a specific air release of 0.5–1.0% by volume. Subsequent treatment of the skimmed float by mixing with nine volumes of primary effluent, with comminution and gravity sedimentation at overflow rates of 1000 gal/ft^2 day, was found to be necessary to effect a 90% efficient separation of true floatables from other nonfloating particulates that were also removed. Disposal of the latter will be by return to the effluent for ocean discharge while the floatables will be incinerated.

Pressure flotation is also a rational treatment method for combined sewer overflows discharging into estuarine receiving waters. (Combined sewers convey sanitary wastewaters to treatment facilities and during periods of rainfall they also transport the storm flows to points of diversion, where the mixtures of sanitary sewage and stormwater overflow.)

For example, a study of the pollution resulting from combined sewer overflows in San Francisco, California revealed that:

1. Receiving water coliform counts were 4–9 times greater during periods of stormwater discharge than in dry weather.
2. Receiving water grease concentrations were 1.2 mg/m^2 in wet weather and 0.1 mg/m^2 in dry weather.
3. Near a combined sewer overflow, the grease concentration in the intertidal beach area had a wet weather maximum of 23 mg/m^2 whereas the dry weather maximum was only 0.2 mg/m^2.
4. After rainfall, on-shore winds were prevalent in several recreational beach areas.

Since no other significant pollution could be detected, treatment of storm water overflows by chlorination for disinfection, and dissolved air pressure flotation for floatables and oil and grease removal, was recommended (City of San Francisco, 1967). A flotation plant, scheduled for completion in Fall, 1970 will include the following design variables: hydraulic loading rate, 6000 gal/ft^2 day ; recycle stream, 10% of design flow ; recycle pressurization, 50 lb/in.^{2} The removed float will be temporarily stored at the facility and will be fluidized and pumped to a dry-weather flow treatment facility following the storm.

C. Concentration of Sludges

The use of dissolved-air flotation thickening, especially for the concentration of secondary treatment sludges but also for mixed primary- and secondary-treatment sludges, has significantly increased in the last two decades. This parallels the increase in the amount of sewage solids that must be disposed of from the nation's waste treatment plants. The first and generally most difficult step in the ultimate disposal of sewage solids collected in primary and secondary gravity sedimentation units (which without chemical aids rarely exceed a 2% solids concentration) is water removal. Since a sludge solids increase of from 1.5 to 5% is equivalent to a water removal of 70%, sludge thickening results in an increase in the capacity of ultimate disposal units and a reduction in unit disposal costs. Ettelt and Kennedy (1966) show that general savings of about $10/ton of solids will accrue if solids concentration is increased from 3 to 6% prior to ultimate treatment and disposal.

Another potentially important but not yet widely used application of adsorptive bubble separation is as a replacement for the gravity separation of activated sludge in a secondary sedimentation basin prior to return of the activated sludge to the aeration basin. Dissolved-air flotation separation of activated sludge requires less time than gravity separation and therefore provides less opportunity for the release of organic and inorganic materials from the sludge into the liquid phase. In addition, dissolved-air flotation separation may allow activated sludge plants to be operated at higher than standard rates where the poor settling characteristics of the activated sludge have, in the past, made the process inoperable.

The design and operating variables that govern the performance of dissolved-air sludge thickeners are similar to those which govern the performance of other dissolved-air processes, viz., pressure, recycle volume, temperature, ratio between air release and solids, solids and hydraulic loading rates, feed solids concentration, and solids detention period. In addition to the standard parameters, there are two parameters of special significance in sludge flotation: the type of solids and the type of surface active compounds present in the feed sludge.

In the evaluation of sludge thickening processes it is desirable to combine the three parameters pressure, recycle volume, and temperature into a single parameter, namely, the amount of released air per unit volume of influent. While the most commonly used design parameter is the air–solids ratio (pound air per pound solids), Ettelt (1965) has indicated that this parameter cannot predict performance except under the limited conditions of either a low air input or a high solids loading. The widely variable solids loadings encountered in dissolved-air sludge thickening units make the usefulness of the air–solids ratio parameter somewhat limited.

In Howe's (1958) model,

$$C_o = C_i \exp\left(-\frac{V_r}{Q/A_h}\right), \tag{1}$$

where V_r is the rising velocity of the sludge–air particle, Q/A_h is the surface loading (flow rate per unit surface area), C_o is the effluent suspended solids concentration, and C_i is the influent suspended solids concentration, attempted to over come of these difficulties by considering the hydraulics of the treatment unit.

Investigations by Mulbarger and Huffman (1969) have led to the following design equation for activated sludge solids separation units:

$$C_i = C_o \exp\left(\frac{\text{air/solids}}{\mu Q/A_h}\right). \tag{2}$$

V_r in Eq. (1), which is difficult to evaluate in practice, is replaced by the air/solids ratio on the basis of finding by Katz (1958) that V_r is proportional to this ratio. Temperature effects are accounted for by dividing the exponent in Eq. (1) by the dynamic fluid viscosity μ.

Systems were found to operate in three distinct zones: failure, unstable, and stable, the stable zone being characterized by values of $[(\text{air/solids})/\mu(Q/A_h)]$ ranging from 10.0 to 24.0 $\times$ 10^{-5} ft/lb.

Equation (2) does not account for two important design variables, namely, surface tension and the nature of the solid surfaces. Mathematical evaluation of surface tension effects is not yet possible because of the limited information available on the nature and concentration of surfactants. Katz (1964) showed that the addition of the surfactant ABS to activated sludge mixed liquor at levels of 5, 10, and 40 mg/liter increased the rise velocity from 0.11 to 0.23–0.25 ft/min, possibly because of a reduction in bubble size so that more bubbles in the optimum range were present. Increasing ABS concentration to 100 mg/liter reduced rise velocity to 0.11 ft/min, possibly because of dispersal of activated sludge particles and alteration of particle surfaces which made bubble attachment more difficult. Katz (1964) and Mulbarger and Huffman (1969) have observed that highly dispersed sludges with unsatisfactory settling characteristics in general also exhibit unsatisfactory flotation thickening characteristics. The use of chemicals to improve these characteristics has been tested in many plants, but no general results are available and chemicals must be tested for each specific use. It has also been observed that both the method of activated sludge aeration, especially the shear from mechanical aerators, influences the performance of air flotation units more significantly than do the basic operational variables such as the air–solids ratio, and solids and hydraulic loadings.

Recent advances in the design of air flotation thickeners have overcome most of the difficulties experienced in early designs, the most common of which was the disintegration of sludge particles in the recycle system to create small hard-to-remove particles. The use of flotation aids (Jones, 1968) and polyelectrolytes (Garwood, 1967) has alleviated these difficulties. The design of the inlet mixing chamber and the point of addition of flotation aids are very significant, the best results being obtained when flotation aids are introduced in the recycle line immediately before the pressure release valve.

Sludge thickening occurs in the floating sludge blanket, which is normally between 8 and 24 in. thick. The air bubbles associated with the sludge particles reduce the overall density of the air–solids matrix and force part of the blanket above the water level so that a significant drainage of water can occur. The lower point of the sludge blanket may have a solids concentration of only 1 %, while solids concentrations of 4–7 % have been observed in the upper part of the blanket (Katz and Geinopolos, 1967).

Two methods of air addition have commonly been used. These are (1) pressurization of a special dilution flow stream (often recycled effluent or primary sewage), and (2) pressurization of the combined dilution and feed flow. Pressurization of the total flow [as practiced by Peixoto (1970) for raw sewage flotation] for activated sludge thickening might be expected to cause breakup of the fragile activated sludge particles following pressure release through a pressure-reducing valve. However, owing to the much lower pressures required to dissolve a given amount of air when using total flow pressurization, this effect is not as significant as might be expected from a comparison with partial flow pressurization systems. Moreover, pressurization of the total flow requires the use of pressure-reducing valves of large aperture so that the shearing of activated sludge particles is reduced. Indeed, Ettelt (1965) concluded, on the basis of pilot plant studies, that total flow pressurization was the best process for activated sludge thickening.

Solids removal from the thickened float blanket is usually accomplished by a skimmer blade that moves the sludge into a trough. Skimmers, however, tend to produce turbulence in the lower regions of the floating blanket and reduce the solids concentration attainable. A device consisting of a rotating cylinder that lifts solids from the surface of the liquid using the tendency of sludge to stick to the collector surface has been developed by Geinopolos and Katz (1964) to overcome the disadvantages of a skimmer blade. The sludge collected by the rotating cylinder is removed by a Doctor blade and discharged into a trough.

Sludge collection devices must be operated so as to provide adequate sludge blanket thicknesses for dewatering to take place in the blanket. Blanket depths should not be allowed to increase to the point where they occupy the entire thickening zone and cause large amounts of solids to be carried out in the underflow.

Extensive laboratory and pilot plant activated sludge thickening experiments conducted by the Metropolitan Sanitary District of Greater Chicago have demonstrated that dissolved-air flotation thickening is superior to gravity thickening. Typical pilot plant results reported by Ettelt and Kennedy (1966) are presented in Table III.

Recycle-pressurization systems normally operate at 60–70 lb/in.^{2} and common solids loadings are in the range of 2–3 lb/ft^2 hr when flotation aids are not used and to 5 lb/ft^2 hr when these aids are employed. Thickened sludge solids concentrations of 4 % can be readily achieved and concentrations of up to 6 % have been reported with use of floatation aids. Thickened sludge solids concentration decreases with increasing net mass loading [Katz (1964); see Fig. 1] and increases with increasing flotation aid dose. An overall solids

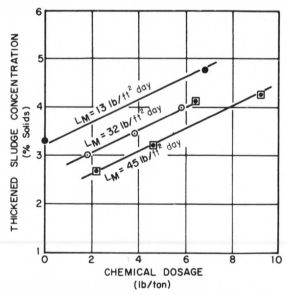

FIG. 1. Effect of solids loading L_M and chemical dose on solids concentration in thickened activated sludge (Katz, 1964).

removal efficiency of 95 % is readily achieved with flotation aids, and up to 98 % removal can be expected (Jones, 1968).

The performance of field-scale dissolved air flotation thickeners with and without flotation aids is summarized in Tables IV and V (Katz and Geinopolos, 1967). The use of dissolved air flotation thickening at these plants has

TABLE III

AVERAGE RESULTS OF FLOTATION UNITS AT CHICAGO[a, b]

Run no.	Feed sludge	Number of days	Floated plus settled solids			
			Ratio	lb/ft² day	Concentration (%)	Removal (%)
1	Activated[c] (SVI = 81)	3	0.42	11.4	3.2	97
2	Activated (SVI = 84)	6	0.67	13.8	3.5	96
3	Activated (SVI = 101)	5	0.43	12.5	3.1	98
4	15% Primary; 85% activated	10	0.70	19.3	4.5	98
5	Activated[d] (SVI = 104)	1	0.65	17.2	3.5	98
6	Activated[d] (SVI = 107)	1	0.97	10.0	4.0	99

[a] From Ettelt, G. A. and Kennedy, T. J. (1966). *J. Water Pollut. Contr. Fed.* **38**, 248.

[b] Optimum recycle: activated sludge—250% of feed flow, activated + primary sludge mixtures—300% of feed flow. SVI = sludge volume index, ml sludge settled after 30 min/gm dry wt. sludge.

[c] 50% air solubility; all other runs, 75–82%.

[d] Polymer dosage of 12 lb/ton.

TABLE IV

RESULTS OF SLUDGE THICKENING BY FLOTATION FROM INSTALLATIONS NOT USING FLOTATION AIDS[a]

Installation	No. of tanks	Type sludge[b] treated	Capacity (tons/day)				Feed sludge conc. (%)			Thickened sludge conc. (%)				95% conf. limits	Volatile solids (%)			Solids recovery (%)		
			Max	Avg	Min	Rated	Max	Avg	Min	Max	Avg	Min	Rated		Max	Avg	Min	Max	Avg	Min
Dalton, Ga.	1	AP	—	4.55	—	3.86	—	1.29	—	7.8	6.1	4.8	3.5	5.88–6.32	86	77	64	—	—	—
Atlanta, Ga.	2	A, P, AP	57.3	34.2	16.7	15.6	2.7	1.90	1.1	8.3	7.4	6.4	5.0	5.80–8.80	—	—	—	—	—	—
Nassau, N.Y.	2	A	17.1	9.6	4.0	8.3	1.25	0.81	0.55	7.2	4.9	3.3	4.0	4.51–5.23	82	70	65	98	85	60
Wayne Cnty, Mich.	6	A	49.5	28.7	15.9	48.0	1.00	0.77	0.48	4.75	3.7	2.82	4.0	3.50–3.94	83	81	79	100	99	98
San Jose, Calif.	1	A	0.59	0.57	0.54	0.33	0.50	0.45	0.34	4.9	4.6	4.4	4.0	2.5–6.7	—	61	—	89	83.4	75
(noncanning season)	12	AP	620	458	310	391	2.79	2.30	1.79	8.10	7.1	5.87	5.0	6.82–7.34	—	—	—	97.0	94.4	92.2
Canning season		A	517	400	275	391	2.06	1.77	1.52	6.14	5.3	4.55	5.0	5.11–5.53	—	—	—	99.4	88.0	72.2
Coors Brewery		A	—	3.6	—	6.7	—	0.77	—	—	4.1	—	3.5	—	—	—	—	—	90.0	—
Levittown, Pa.	1	A	—	3.8	—	2.5	—	0.80	—	—	6.5	—	4.0	—	—	—	—	—	93.0	—
		AP	—	7.0	—	4.9	—	0.64	—	—	8.6	—	7.5	—	—	—	—	—	91.0	—
Boise, Ida.	1	A	4.4	4.0	1.5	4.0	0.50	0.46	0.40	4.10	4.0	3.60	3.5	3.65–4.27	85	81	76	92.4	88.0	80.4

[a] From Katz, W. J. (1964). J. Water Pollut. Contr. Fed. 36, 407.

[b] A = activated; P = primary; AP = combination of activated and primary.

TABLE V

RESULTS OF SLUDGE THICKENING BY FLOTATION FROM INSTALLATIONS USING ORGANIC POLYELECTROLYTE AIDS[a]

Installation	Type sludge treated	Quantity of sludge treated (tons/day)			Feed sludge conc. (%)			Chemical dosage (lb/ton solids)		Thickened sludge conc. (%)			95% conf. limits	Volatile solids (%)			Solids recovery (%)		
		Max	Avg	Min	Max	Avg	Min	Max	Min	Max	Avg	Min		Max	Avg	Min	Max	Avg	Min
Nassau County, N.Y.	Activated	12.3	—	4.0	1.00	0.77	0.48	26.1	2.9	4.69	4.05	3.43	3.89–4.21	83	81	79	100.0	99.9	99.9
Boise, Idaho	Activated	4.2	4.0	3.9	0.53	0.45	0.37	12	6	4.2	—	3.8	—	—	81	—	—	99	—

[a] From Katz, W. J. (1964). J. Water Pollut. Contr. Fed. 36, 407.

resulted in (1) an increase in anaerobic sludge digester capacity or in increased digester residence time, (2) a reduction in the amount of grit pumped to the digester which minimizes the need for expensive digester cleaning, and (3) a reduction in raw sludge heating requirements prior to digestion.

D. Foam Fractionation and Froth Flotation

Foaming of wastewaters has classically been regarded as a nuisance. Its occurrence on activated sludge aeration basins at startup or under lightly loaded conditions has been reported widely. With the commercial introduction in the middle 1940's of synthetic detergents based on aryl alkyl sulfonates, foaming in sewage treatment plants—especially on activated sludge aeration basins, in secondary effluents, and on the watercourses into which sewage and sewage effluents discharged—became a problem of major magnitude (Degens, 1954; Anon., 1950). A survey conducted at this time showed that 50 out of 104 activated sludge plants surveyed throughout the nation experienced foaming as a daily occurrence (Polkowski et al., 1959). It was shown that the generation of froth was largely due to the presence of synthetic detergent residues that, unlike other foam-producing components of sewage, were not degraded during the waste treatment processes. Considerable research established that the then most common surfactant, alkyl benzene sulfonate (ABS), was degraded 50–65% by typical primary and secondary treatment processes, so that with raw sewage concentrations of 10 mg ABS/liter, about 4–5 mg ABS/liter (which is well above the foaming threshold concentration of 0.5 mg/liter) commonly remained in secondary effluents. In sewage treatment plants the foaming nuisance was controlled by spraying secondary effluent on to aeration tanks and effluent channels to destroy the foam by jet action. In some locations, foam control was achieved by drip feeding an antifoam (for example, Mobilpar W) to the aeration basin in concentrations of about 1 mg/liter. These measures eliminated the immediate foaming nuisance but ABS continued to pass through treatment processes partially undegraded. In the 1950's and early 1960's ABS residues began to appear and foaming began to occur in some well waters, indicating the presence of water of sewage origin. By 1965, however, the detergent industry had ceased production of ABS and other branched-chain alkyl aryl sulfonates for the domestic market and replaced them with more readily degradeable surfactant materials. The major foaming problems disappeared as detergent residues in secondary effluents, receiving waters and ground waters decreased. Parenthetically it might be added that while ABS residues and the foaming problems that advertised their presence were eliminated, the sewage effluents that accompanied them were and are still present though now unadvertised.

The detergent-caused foaming of sewage stimulated the development of foaming processes for ABS removal. The original proposers of such a process (McGauhey et al., 1959) showed, in continuous pilot plant studies, that 80 % removal of ABS from settled sewage or activated sludge effluent could be achieved by aeration with 1 ft^3 air per gallon of liquid. The addition of a foaming agent (2-octanol) increased ABS removal but resulted in a wet, dilute foam stream (foamate). McGauhey and Klein (1963) found that activated sludge effluents ceased to foam when the ABS concentration reached about 1 mg/liter. However, they concluded that an ABS concentration of 1–2 mg/liter would appear to be a value more practically attainable by field-scale foam fractionation units. Pilot plant foaming studies by Rubin and Everett (1963) on a high rate activated sludge effluent indicated that 0.4 mg/liter was the minimum ABS concentration attainable by foaming. This low ABS residual, however, does not appear to have been attained by any other pilot plant or plant-scale foam separation units.

Many observations on the composition of foams produced by the forced aeration of sewage and sewage effluents have confirmed the early observations of Donaldson (1952) that dissolved and particulate materials other than ABS were concentrated into sewage-derived foams. Donaldson (1952) observed that only 12 % of the total solids of aeration tank froth was ABS. Harkness and Jenkins (1960) found that 70–80 % of the solids in foams produced in sewage treatment plants and on polluted rivers was organic, and that these foams had a high grease content. Studies by Jenkins (1964) showed that foams produced from settled sewage and activated sludge effluents were highly enriched in ABS and in suspended solids, and to a lesser extent in dissolved organic matter other than ABS. These observations support the idea that froth flotation is important in the foaming of sewage and sewage effluents, and that foam separation may lead to a polishing of activated sludge effluents by removal of suspended and dissolved matter in addition to ABS. Relatively few plant-scale foam fractionation units were constructed because the conception of the process in response to stringent ABS effluent criteria only preceded the replacement of ABS by about 1½ years.

The first plant-scale foam fractionation unit to be constructed was installed at the Barstow, California, Maintenance Yard of the Atcheson, Topeka and Santa Fe Railroad Co. and was part of a treatment scheme for the removal of oils and detergents from the wastes produced by washing railroad cars (Ludwig, 1962). The foam fractionation unit followed a holding basin coagulation–flotation system designed to remove oil and grease, and consisted of an enclosed rectangular basin containing air diffusers designed to provide approximately 10 min nominal residence time and air/liquid ratios of about 1 ft^3/gal. Pilot plant tests demonstrated that this unit could decrease ABS concentration from 6 to 2 mg/liter. Reduced ABS removal efficiency was found to occur when the upstream oil removal system did not function properly. Foamate was discharged to a holding basin where, after collapsing, it was transported to a remote desert area for disposal by dumping.

The Los Angeles County Sanitation District operated two foam fractionation units: a 12 million gal/day plant-scale unit at the Whittier Narrows Water Reclamation Plant and a 300 gal/min experimental foamer pilot plant at the Pomona Water Reclamation Plant. The Whittier Narrows foamer was constructed to meet an effluent criterion of 2 mg/liter ABS prior to recharge of the effluent to the ground. It was a longitudinal flow unit, 50 ft long by 14.5 ft wide by 11 ft deep, and was constructed by modifying an existing chlorine contact chamber. Two rows of Saran tube diffusers were located 4 ft and 11 ft below the liquid surface. The top of the unit was enclosed by a flexible air-tight plastic cover and the foam produced spilled over a freeboard into a well where it was collapsed by spraying with secondary effluent. Activated sludge effluent detention time varied between 6.5 and 7.3 min.

The Pomona foamer was in the form of a "column on its side." Foam was produced by aerating a 5-ft depth of activated sludge effluent with Saran tube diffusers located at the bottom of the unit along its entire 15-ft length. Foam spilled over a side weir that ran the length of the unit and was collapsed by spraying with already collapsed foam. Typical operating parameters were 5 min nominal liquid residence time and an air/liquid ratio of 0.6 ft^3/gal. The foamate flow was 6–12% of the liquid flow.

A foaming unit to treat 0.5 million gal/day of activated sludge effluent was constructed at the Valley Community Services Districts (VCSD) San Ramon, California Water Reclamation Plant to meet an effluent ABS criterion of 1 mg/liter ABS. The unit consisted of a metal hood 8 ft long by 9 ft wide by 13 ft deep that sat on the surface of a chlorine contact chamber. Air was supplied through cylindrical carborundum diffusers located at the bottom of the unit. At the liquid surface a layer of "Hexel" expanded metal separated the liquid from the foam layer. Foam spilled over a weir running the width of the unit and was collapsed by spraying with already collapsed foam.

All the foaming units operating on activated sludge effluents reduced ABS concentration to less than 2 mg/liter and two of the units (Pomona and VCSD) achieved ABS residuals of 1 mg/liter or less. Referring to Table VI, significant froth flotation is indicated by the removals of particulate matter (measured as suspended solids or suspended COD). Other soluble organic matter (soluble COD other than ABS) is only removed to a limited extent.

Foaming of activated sludge effluents generates a new, fairly concentrated stream—the foamate. The volume (and hence the strength) of this stream depends largely on the amount of foam drainage that is allowed to take place, and this in turn is determined by the freeboard between the liquid surface and the foam overflow weir. Generally the foamate flow is 1% or less of the influent flow, depending on the amount of foam reflux allowed. While foamate

TABLE VI

TYPICAL OPERATING RESULTS FROM FOAMING UNITS[a]

Constituent	Whittier Narrows			Pomona			Valley community services district			Barstow		
	Influent	Effluent	Foamate	Influent	Effluent	Foamate	Influent	Effluent	Foamate	Influent	Effluent	Foamate
ABS	3.0	1.6	30	3.7	0.92	34	2.0	0.79	18	6	2	—
Total COD	40	29	170	64	43	260	49	37	170	—	—	—
Suspended COD	13	8	80	19	10	110	8	4	64	—	—	—
Soluble COD other than ABS	19	18	24	36	30	73	36	31	63	—	—	—
Suspended solids	7	4	43	15	10	90	9	3	110	—	—	—

[a]Units are milligrams per liter.

volume was found to be important in early studies (McGauhey and Klein, 1963) in which foamate disposal by burning was attempted, in neither of the two foaming units that have been installed in operating activated sludge plants is foamate volume a critical parameter. The Whittier Narrows Plant discharged the foamate to a trunk sewer, while the VCSD plant used the foam recycle process to dispose of foamate.

The foam recycle process was conceived by Sharman and Kyriacou (1962) and consists of recycling the collapsed foam back through the activated sludge aeration basin. The process is predicated on two assumptions: (1) that the rate of ABS degradation by activated sludge organisms is proportional to its concentration, and (2) that there are no ABS isomers totally resistant to biodegradation. In a one-year laboratory scale test of the process, Sharman et al. (1964) reported that no continued ABS buildup took place in the activated sludge aeration basin as a result of foamate recycle. With an ABS concentration of 10–11 mg/liter in the primary sewage influent and a final effluent from the foamer containing about 1 mg/liter ABS, the aeration basin ABS concentration fluctuated between 18 and 25 mg/liter. All indications from this experiment pointed to an acclimatization of the activated sludge microflora to the biodegradation of the more resistant ABS isomers. The unit employing foam recycle thus removed approximately 90% of the influent ABS as compared with a 50–75% ABS removal achieved by a control activated sludge unit.

The foam recycle process was used with success at the VCSD Water Reclamation Plant, and a study by Jenkins (1966) at this plant demonstrated its success under admittedly lightly loaded circumstances (Fig. 2). Of the average 31.9 lb ABS/day entering the plant, 29.2 lb ABS/day were biologically removed. In a more critical test of the foam recycle process, Jenkins (1966) operated a foaming column in conjunction with an activated sludge plant of 600-gal aeration capacity (standard rate). Prior to foam recycle the activated sludge plant removed 55% of the influent ABS to produce a residual ABS of 4 mg/liter. During the month-long foam recycle experiment, effluent ABS concentration averaged 0.8 mg/liter for a total ABS removal of 94%. No buildup of ABS was detectable in the aeration basin. Foaming of the secondary effluent also produced improvements (reductions) in effluent suspended matter content from an average of 57–24 mg/liter.

It would therefore appear that the capability exists, by a combination of activated sludge treatment, secondary effluent foaming, and foam recycle to reduce the ABS content of sewage effluents (and provide polishing from a suspended solids standpoint) to a level that will not cause persistant foaming. While such a waste treatment objective can be achieved without resort to tertiary foaming processes where readily biodegradeable detergents are used, there are many areas of the world where "hard" synthetic detergents are still

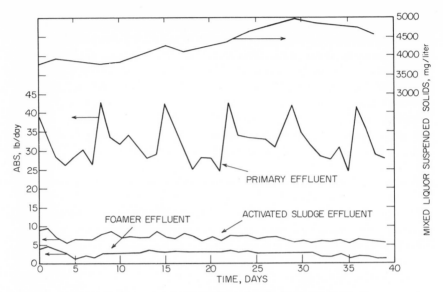

FIG. 2. Performance of VCSD plant using the foam recycle process.

used and in which processes such as those described here are applicable for the prevention of persistant foaming of sewage effluents.

III. APPLICATIONS TO INDUSTRIAL WASTES

A. Introduction

Dispersed-air, dissolved-air, and electro-air flotation have been employed in the treatment of industrial wastes. Dispersed air flotation, where bubbles are generated by the shearing action of propellers or passage of gas through a porous solid into the liquid, has found only limited application because the strong shearing action required for bubble formation tends to break up the relatively fragile suspended solids encountered in wastewaters. Pav and Gleeson's (1967) modification of dispersed-air flotation in which an oscillator is used for air dispersion may circumvent this problem and increase the use of dispersed-air flotation.

Dissolved-air flotation for solid–liquid separation is applicable to many industrial wastes containing finely dispersed material (for example, oil, grease, fibers, and small suspended particles) with a density close to water, which cannot be removed readily by the more common solid–liquid separation processes such as centrifugation, sedimentation, or quiescent flotation.

A fairly recent development in treatment of industrial wastes is the electro-air flotation process described by Gerasimov (1962) and by Matov (1966).

Gas bubbles are produced by direct electrolysis of the waste, and aluminum or iron electrodes can sometimes be used to produce Al(III) or Fe(III), which act as coagulant and flotation aids.

The design of a dissolved-air flotation process should be based on laboratory tests on the specific wastewater. A convenient testing system consists of a pressure chamber and a large cylinder (Rohlich, 1954) or a pressure chamber connected directly to the bottom of the flotation funnel described by Scherfig and Ludwig (1967). Wastewater, saturated (or partially saturated) with air in the pressure chamber, is introduced into the flotation funnel and the height of the solid–liquid interface is recorded at intervals. The height of the sludge–liquid interface provides the necessary rise velocity information and the total amount of floated material can be measured after the rise of the suspended particles is completed. The amount of air released from solution is determined in a parallel test using an effluent that does not contain solids.

B. Oil

The earliest and still the most common wastewater treatment application of dissolved-air flotation is to oil refinery effluents. Most oil refineries have a separate system for the collection of oily process and storm waters. The first treatment step, normally gravity separation in an American Petroleum Institute (API) separator (American Petroleum Institute Standards, 1963), is usually followed by chemical flocculation with Al(III) or Fe(III) chlorides (and more recently aided by polyelectrolytes) as flocculating agents and emulsion breakers. Dissolved-air flotation can follow chemical treatment to increase the rate of oil separation, and this type of process chain has been shown to be capable of producing effluents with less than 10 mg/liter oil (Simonsen, 1962a). Alternatively, it is often possible to redesign conventional API separators to air flotation units (Simonsen, 1962b). Quigley and Hoffman (1967) report that air flotation by itself removes 70–80% oil from refinery effluents; further treatment with 50 mg/liter alum improved oil removal to 80–85% and an additional 100 mg/liter lime dose increased oil removal to 85–90%.

Soluble oil wastes from cold reduction steel mills are treated by gravity separation, followed by the addition of ferrous sulfate and lime to break emulsions, and then dissolved air flotation.

Paint-containing wastes, which are an important waste stream produced by the automobile industry, can be treated economically by chemical addition for emulsion breaking, followed by air flotation to produce effluents with suspended solids concentrations of about 20–25 mg/liter.

C. Grease

Slaughterhouse wastes, either with manure or manure free, can be given preliminary treatment by air flotation prior to either discharge to a municipal waste system or to in-plant biological treatment.

Suspended solids and grease removals greater than 50% can be achieved at hydraulic loadings of about 2000 gal/ft^2 day for wastes containing initial concentrations of solids in the range 700–1700 mg/liter and grease in the range 270–1370 mg/liter (Johnson, 1965). By comparison, gravity sedimentation units operating at hydraulic loadings of only 500 gal/ft^2 day reduce suspended matter by 50–80% (Johnson, 1965).

Scherfig and Ludwig (1967) have reported that vacuum flotation is effective in the removal of grease from mixtures of domestic sewage and slaughterhouse waste.

Dispersed-air flocculation was used by van Vuuren et al. to treat slaughterhouse wastes after chemical treatment with sulfuric acid and ferrous sulfate. Chemical oxygen demand reductions of 77% were obtained in a laboratory-scale unit.

D. Fibers

Most of the fibers and fillers (clays and $CaCO_3$) that make up the bulk of the suspended solids in pulp and paper mill wastes are recovered in-plant and reused. Most commonly, recovery devices have consisted of filtration and sedimentation units but more recently flotation units have been incorporated as a final step in the recovery process.

Fiber and filler removal efficiency from paper mill white waters can be improved by using oils with a specific gravity less than 1.0 as flotation agents. The oil flotation agent must be separated from the fibers in a subsequent sedimentation process.

Improved separation of fibers has been obtained at Deutsche Gold und Silber-Scheideanstalt (1962) by the use of a mixture of a peroxide, sodium silicate, and a nonionic wetting agent as a flotation aid. Rose and Sebald (1968) have noted that the removal of surfactants and other organic chemicals occurs by foam fractionation during the air flotation of paper mill wastes.

IV. CONCLUDING REMARKS

In the past, the use of adsorptive bubble separation techniques for solid–liquid separations in wastewater treatment has been largely confined to sludge concentration and to somewhat specialized solids removal, e.g., grease, oil, fiber, and detergent removal. However, it now appears that adsorptive bubble separation dissolved-air flotation is about to take its place among the array of processes commonly used for municipal wastewater treatment. This development appears imminent because of the identification of floatables removal as an important waste treatment objective for wastes discharged to marine and estuarine environments.

REFERENCES

American Petroleum Institute (1963). "Manual on Refinery Wastes," 7th ed., Vol. VI.

Anon. (1950). *J. Water Pollut. Contr. Fed.* **22,** 362.

California State Water Pollution Control Board (1961). Report on Collation, Evaluation, and Presentation of Scientific and Technical Data Relative to the Marine Disposal of Liquid Wastes, Engineering Science, Inc.

City of San Francisco (1967). Characterization and Treatment of Combined Sewer Overflows— San Francisco. Report prepared by Engineering Science Inc., Arcadia, California.

Degens, P. N. (1954). *J. Water Pollut. Contr. Fed.* **26,** 1494.

Deutsche Gold und Silber-Scheideanstalt (1962). Belg. Pat. 619,737.

Donaldson, W. (1952). *J. Water Pollut. Contr. Fed.* **24,** 1033.

Eliassen, R. (1943). *Sewage Works J.* **15,** 252.

Ettelt, G. A. (1965). *Proc. Purdue Ind. Waste Conf. Eng.* p. 210.
    Extension Series, Purdue Univ., Lafayette, Indiana.

Ettelt, G. A., and Kennedy, T. J. (1966). *J. Water Pollut. Contr. Fed.* **38,** 248.

Garwood, J. (1967). *Effluent Water Treatment J.* **7,** 380.

Geinopolos, A., and Katz, W. J. (1964). *J. Water Pollut. Contr. Fed.* **36,** 712.

Gerasimov, I. (1962). *Horostic. Heft i Gaz Tekn Neftepererabotka i Heftekhin* **12.**

Hansen, C. A., and Gotaas, H. B. (1943). *Sewage Works J.* **15,** 242.

Harkness, N., and Jenkins, S. H. (1960). *Water Waste Treatment J.,* 170, (Nov./Dec.).

Hay, T. T. (1956). *J. Water Pollut. Contr. Fed.* **28,** 100.

Howe, R. H. L. (1958). *In* "Biological Treatment of Sewage and Industrial Wastes" (J. McCabe and W. W. Eckenfelder, Jr., eds.), Vol. II, p. 241. Van Nostrand (Reinhold), Princeton, New Jersey.

Jenkins, D. (1964). Application of Foam Fractionation to Sewage Treatment. Part II: Foam Fractionation of Sewage and Sewage Effluents, Berkley Sanit. Eng. Res. Lab. Univ. of Calif. Rep. No. 64–10.

Jenkins, D. (1966). *J. Water Pollut. Contr. Fed.* **38,** 1737.

Johnson, A. (1965). *In* "Industrial Wastewater Control" (C. T. Gurnham, ed.), p. 25. Academic Press, New York.

Jones, W. H. (1968). *Water Sewage Works* **115,** R177.

Katz, W. J. (1958). *Public Works* **89,** 12.

Katz, W. J. (1964). *J. Water Pollut. Contr. Fed.* **36,** 407.

Katz, W. J., and Geinopolos, A. (1967). *J. Water Pollut. Contr. Fed.* **39,** 946.

Logan, R. P. (1949). *Sewage Ind. Wastes* **21,** 799.

Ludwig, H. F. (1962). *Ind. Water and Wastes* **7,** 44.

Ludwig, H. F., and Carter, R. C. (1961). *J. Water Pollut. Contr. Fed.* **33,** 1123.

Ludwig, H. F., and Onodera, B. (1964). *Advan. Water Pollut. Res.* **3,** 37.

McGauhey, P. H., and Klein, S. A. (1963). *J. Water Pollut. Contr. Fed.* **35,** 100.

McGauhey, P. H., Klein, S. A., and Palmer, P. B. (1959). A Study of Operating Variables as They Affect ABS Removal by Sewage Treatment Plants. Berkley, Sanit. Eng. Res. Lab., Univ. of Calif.

Matov, B. M. (1966). *Elek. Obrab. Muter Akad. Hauk Mold SSR (USSR)* **4**, 94; (1967) *Chem. Abstr.* **67**, 76145.

Mays, T. J. (1953). *Sewage Ind. Wastes* **25**, 1229.

Mulbarger, M. C., and Huffman, D. D. (1969). Mixed liquor solids separation by flotation. Report of Municipal Treatment Research Program, U. S. Dept. of the Interior, June, 1969.

Osterman, J. (1955). *J. Water Pollut. Contr. Fed.* **27**, 20.

Pav, J., and Gleeson, M. A. (1967). Brit. Pat. 930,951 ; *Water Pollut. Abstr.* **116**, 1967.

Peck, C. L. (1920). *In* Eliassen, R. (1943) *Sewage Works J.* **15**, 252.

Peixoto, E. C. (1970). Paper presented at Int. Conf. Water Pollut. Res., 5th, San Francisco.

Polkowski, L. B., Rohlich, G. A., and Simpson, J. R. (1959). *Sewage Ind. Wastes* **31**, 1004.

Quigley, R. E., and Hoffman, E. L. (1967). *Ind. Water Eng.* **4**, 22.

Rohlich, G. A. (1954). *Ind. Eng. Chem.* **46**, 304.

Rose, J. L., and Sebald, J. E. (1968). *Tappi* **51**, 314.

Rubin, E., and Everett, R., Jr., (1963). *Ind. Eng. Chem.* **55**, 44.

Scherfig, J., and Ludwig, H. F. (1967). *Proc. Int. Conf. Water Pollut. Res., 3rd* **2**, 217.

Sharman, S. H., and Kyriacou, D. (1962). A novel method for removal of detergent from sewage. Paper presented at 43rd Ann. Conf. of Calif. Sect. Amer. Water Works Assoc.

Sharman, S. H., Kyriacou, D., Schuett, W. R., and Sweeney, W. A. (1964). Foam recycle : a method for improved removal of detergents from sewage, paper presented at Amer. Chem. Soc. Div. of Water and Waste Chem., April 1964.

Simonsen, R. N. (1962a). *Oil Gas J.* **60**, 146.

Simonsen, R. N. (1962b). *Hydrocarbon Process. Pet. Refiner.* **41**.

CHAPTER **15**

SEPARATION OF SURFACTANTS AND
METALLIC IONS BY FOAMING:
STUDIES IN FRANCE

J. Arod

Service d'Analyse et de Chimie Appliquée
Centre d'Études Nucléaires de Cadarache,
France

I. INTRODUCTION

Foam separation has been applied in France to the decontamination of radioactive wastes of medium-level activity. These solutions, which are generally acid, have a high percentage of sodium nitrate (0.1–1 M) after a neutralization with caustic soda, and their concentration in calcium and magnesium ions is between 100 and 200 mg/liter. The principal radioisotopes present are the following: strontium-90, cesium-137, and cerium-144.

A preliminary study, which was carried out in an equilibrium column and in a countercurrent column (at a flow rate of 0.6 liter/hr), familiarized the author with the process of foam separation. It was then necessary to adapt this technique to the treatment of the waste, to search for a surfactant which would be effective even in a highly saline medium, and to devise a chemical pretreatment for the waste so as to make it compatible with the possibilities of the process.

The result of these studies was the construction and operation of a pilot plant which can treat a flow rate of 100 liters/hr.

II. PRELIMINARY STUDY

In the light of some studies which were already published (Weinstock *et al.*, 1961; Rubin and Gaden, 1962; Leonard and Lemlich, 1965), experiments

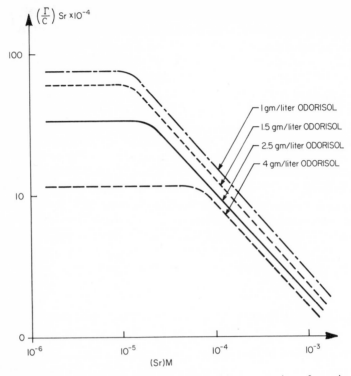

FIG. 1. Variation of the strontium distribution factor with concentrations of strontium ion and surfactant.

TABLE I

EXTRACTION OF STRONTIUM IONS IN A COUNTER-CURRENT COLUMN:
INFLUENCE OF FOAM COLUMN HEIGHT

Height (cm)				Extraction
Liquid column	Foam column	DF^a	VR^b	(%)
125	0	10.5	4.3	90.5
19	0	11.8	3.6	91.5
19	17	185	4.6	99.46
19	27	294	5.7	99.66
19	55	254	5.8	99.61
19	107	560	6.8	99.82
19	157	356	6.3	99.73

^{a}DF: decontamination factor = ($[Sr^{2+}]$ in the initial solution)/($[Sr^{2+}]$ in the final solution).
^{b}VR: volume reduction = (volume of the treated solution)/(volume of broken foam which is collected).

were conducted for the extraction of strontium ions (in a dilute aqueous solution) with a commercial surfactant, Odorisol SPS (sodium lauryl sulfate with 27% active material).

Using an equilibrium column, values of the distribution factor Γ/C for strontium ions and its variations in terms of different parameters were determined (Arod et al., 1964a).

Figure 1 shows some of the results obtained and also shows the effect that the concentration in strontium ions and in surfactant has upon the distribution factor. It was also noticed that the reduction of the distribution factor was due to the presence of sodium and calcium ions in the solution.

Trials for extraction which were carried out afterwards in a counter-current column (Arod et al., 1964b) showed that it was possible to have several stages of mass transfer in the column of foam, and that the liquid pool at the bottom of the column behaved like a single stage regardless of its height. As can be seen in Table I, the extraction of strontium ions is greatly improved by the presence of a column of foam, but above a certain column height the extraction output remains nearly constant.

III. ADAPTATION OF FOAM SEPARATION TO WASTE TREATMENT

A. Influence of Sodium and Calcium Ions

Due to the chemical composition of wastes of medium-level activity, a search was conducted for a surfactant which would have a great affinity for strontium, cesium, and cerium ions in solutions at high concentration of sodium nitrate. About thirty products were tested. They were either commercial products or products synthesized upon request, in accordance with

TABLE II

EXTRACTION OF STRONTIUM IONS BY DBDTTA

$[Sr^{2+}]$ (M)	$[NaNO_3]$ (M)	pH	$[Ca^{2+}]$ (mg/liter)	DF	VR
10^{-4}	1×10^{-1}	12	0	208	216
10^{-4}	5×10^{-1}	12	0	286	72
10^{-4}	1	12	0	172	90
10^{-4}	3×10^{-1}	7	0	68	116
10^{-4}	3×10^{-1}	9	0	165	302
10^{-4}	3×10^{-1}	12	0	352	284
10^{-4}	3×10^{-1}	11	4	183	107
10^{-4}	3×10^{-1}	11	8	73	86
10^{-4}	3×10^{-1}	11	80	1.6	100

effects of selectivity observed with certain solvents or certain ion exchange resins.

For solutions with 3×10^{-1} M sodium nitrate the only interesting results were found with the chelatant surfactant DBDTTA (sodium salt of the dodecylbenzyldiethylenetriamine tetraacetic acid) (Schonfeld *et al.*, 1960). The extraction of strontium ions is carried out with an output which is over 99 %; that of the cerium ions with an output which is over 98 %; while by contrast, that of the cesium ions remains low (10–20 %).

This surfactant was then used for all ulterior runs (Arod, 1966). Table II shows the effect of various chemical parameters, especially calcium ions, on the extraction of strontium ions.

B. Chemical Pretreatment of Waste

As it is impossible to fix cesium ions directly on the molecule of a surfactant, and as it is obligatory to reduce the concentration of calcium ions, it is necessary to apply a chemical pretreatment to the waste.

The cesium ion decontamination is done by adsorption on a colloidal precipitate of nickel ferrocyanide (Pushkarev *et al.*, 1960), made by adding 50 mg/liter ferrocyanide ions to 10 mg/liter nickel ions. Calcium ions are eliminated by precipitation of calcium oxalate.

The addition of the reagents is carried out in two mixed tanks connected in series. The precipitates which appear and the solution are then put into a settling tank in which 60–80 % of the solid particles settle by sedimentation. The solution and the solid particles which do not settle finally enter the foam column. Strontium and cerium ions are picked up by the molecules of surfactant. The solid particles are adsorbed at the gas–liquid interface that constitute the foam bubbles; the phenomenon is called flotation.

For solutions containing strontium, cerium, and cesium ions, decontamination factors of 200 and a volume reduction of 50 have been achieved.

IV. EXPLOITATION OF A 100-LITER/HR PILOT PLANT

Operation of the pilot plant was adjusted with respect to the hydrodynamic parameters (such as air rate, bubble diameter, degree of concentration in DBDTTA, etc.) and the mechanical devices (air disperser, feeding of the solution, the setting of packing inside the column). All this made for a very stable operation of the pilot plant and good reproductibility of results. Continuous 24-hr trials have shown that a decontamination factor of 250 could be reached and maintained for strontium ions, with a volume reduction of 50 (Arod and Rosset, 1967). The installation is shown in Fig. 2.

FIG. 2. Decontamination of radioactive wastes by foam separation in a pilot plant with a capacity of 100 liters/hr.

V. CONCLUSION

The studies of foam separation provided (in the case of synthetic solutions) a good knowledge of the extraction mechanisms of the radioactive cations, strontium, cesium, and cerium.

Mastery was obtained of the hydrodynamic phenomena that occur in a countercurrent column of foam at a flow rate of 100 liters/hr.

Among the different surfactants that were studied, sodium dodecylbenzyl-diethylenetriamine tetraacetate (DBDTTA) gave the best results because of its good chelating qualities.

The application of the process to the decontamination of a real waste, although it is limited, is not very encouraging. This is not because of low decontamination factors, but because of the great chemical and, above all, hydrodynamic interferences that could be noticed. The corresponding modifications of the mechanical behavior and of the wetness of the foam require the consideration of each type of waste to be treated as a specific problem, and, consequently, the restudy of the hydrodynamics of the foam column before each treatment. This indeed is not very compatible with industrial exploitation.

Because of the lack of a field of application specific to foam separation in the area of wastes, the aforementioned research was discontinued.

However, it is not unthinkable to consider the application of this method to the recovery and separation of traces of expensive nuclear materials for which the restrictions of cost are less severe, and the flow rates and the constancy of the solutions more compatible with the possibilities of foam separation.

REFERENCES

Arod, J. (1966). Extraction–concentration de cations métalliques en solution diluée à l'aide d'un agent tensio-actif chelatant—Essais d'application à la décontamination des effluents radio-actifs. SECA 085.

Arod, J., and Rosset, F. (1967). Extraction–concentration de cations métalliques en solutions diluées à l'aide d'agents tensio-actifs; Exploitation d'un pilote-mousse 100 l/hr; Recherche d'agents tensio-actifs sélectifs, SECA 115.

Arod, J., Fould, H., and Wormser, G. (1964a). Extraction-concentration de cations métalliques en solutions diluée à l'aide d'agents tensio-actifs—Essais d'applications à la décontamination d'effluents radio-actifs. SC 64-049.

Arod, J., Fould, H., and Wormser, G. (1964b). Extraction–concentration de cations métalliques à l'aide d'agents tensio-actifs. SECA 063.

Leonard, R. A., and Lemlich, R. (1965). A. I. Ch. E. J. 11, 18–28.

Pushkarev, V. V., Skrylev, L. D., and Bagretsov, V. F. (1960). Zh. Prikl. Khim. SSSR 33, 1, 81–85.

Rubin, E., and Gaden, E. L. New Chemical Engineering Separation Techniques. In "Foam Separation" (H. M. Schoen Ed.) Chapter 5, p. 318–385. Wiley (Interscience), New York.

Schonfeld, E., Sandford, R., Mazella, G., Ghosh, D., and Mook, S. (1960). The removal of Sr and Cs from nuclear waste solutions by foam separation. U.S. At. Energy Comm. Rep. NYO 9577.

Weinstock, J., Schonfeld, E., Rubin, E., and Mook, S. (1961). Foam separation. RAI 100, 101, 104.

CHAPTER **16**   SEPARATION OF SURFACTANTS AND
METALLIC IONS BY FOAMING:
STUDIES IN ISRAEL

Eliezer Rubin

Department of Chemical Engineering
Technion—Israel Institute of Technology
Haifa, Israel

I. INTRODUCTION

This chapter reviews and summarizes the experimental and theoretical work on foam separation done in Israel. Practically all the work reported here was conducted during recent years at the Technion, Israel Institute of Technology.

Until recently many authors, including the present one, have used the names "foam separation" and "foam fractionation" synonymously to indicate separation of solutes from solution by foaming. However, it was suggested recently (Lemlich, 1968b) that foam separation be called the general branch of adsorptive bubble separation dealing with the separation by foaming of two- and three-phase (gas–liquid–solid) systems. Foam fractionation is then the special case of foam separation where only two phases, gas and liquid, are involved.

Foam fractionation may be defined as a separation method utilizing foam, as a medium of large specific interfacial area, for partial separation of components of a solution containing surface-active solutes. Thus, foam fractionation deals with pure solutions, in contrast to the various flotation techniques in which solid phases are also present.

Most of the work conducted at the Technion is connected with foam fractionation. Moreover, all experimental work involved only stable foams.

The experimental and theoretical work involving foam fractionation may be divided into five categories: (1) adsorption to gas–liquid interfaces, (2) properties of foams, (3) equipment design and performance, (4) foam breaking, (5) practical applications.

To date, the work done in Israel has been of a basic nature and covered the first four categories. It was aimed at understanding fundamental aspects of foam fractionation, that is, basic theory, limitations and advantages, possible equipment design configurations, and so on. No specific attempt was made to explore practical situations for which foam fractionation may have advantages over other separation techniques. However, in view of the results of the basic research, some work directed towards the practical applications is planned for the very near future.

The work done on the first three categories mentioned above is summarized and reviewed in this chapter. The work done on foam breaking, which was of an experimental and empirical nature, is discussed elsewhere (Goldberg and Rubin, 1967; Rubin and Golt, 1970).

II. ADSORPTION TO GAS–LIQUID INTERFACES

A. General

Foam fractionation is based on the tendency of surface-active solutes to adsorb to gas–liquid interfaces. The generation and collection of controlled foams serves as an effective means for obtaining large solute-concentrated gas–liquid interfaces, and is the basis of foam fractionation.

Many fundamental studies of adsorption of surface active solutes to the gas–liquid interface, as well as foam fractionation studies, have been conducted, reported, and reviewed. These studies have been limited almost exclusively to solutions containing only one surface-active solute. For foam fractionation of solutions containing a single surface-active solute the equilibrium relationship between surface excess Γ and the experimental variables is given (Rubin and Gaden, 1962) by

$$\Gamma = (fD_{32}/6)(C_y - C_x). \tag{1}$$

However, in many cases actual solutions may contain several surface-active solutes. In such cases, selective adsorption to the gas–liquid interface may be expected. Only a few studies of selective adsorption have been reported in the literature. Most of these studies were concerned with static plane gas–liquid interfaces and, in addition, only the concentration of the selectively adsorbed solute was measured (Aniansson, 1951; Nilsson, 1957; Sobotka, 1954; Wilson *et al.*, 1957).

Foam fractionation can also be used to separate surface-inactive solutes, such as metal ions, by using ionic surfactants of opposite charge or surfactants capable of forming complexes with the surface-inactive solutes. The utilization of foam fractionation for the removal of metallic ions was reported in the literature, particularly in conjunction with removal of radioactive ions from low-level radioactive wastes (Rubin *et al.*, 1962).

Theoretical and experimental studies were conducted at the Technion using solutions containing two ionic surface active solutes. These studies were undertaken with the realization that the potential applicability of foam fractionation could be extended considerably if it would be applied to the separation or fractionation of two or more surface-active solutes. Additional experimental and theoretical studies conducted at the Technion included studies of the selectivity of surface adsorption of metallic counter ions, and prediction of surface hydrolysis effects and the inevitable pH change associated with foam fractionation of ionic surfactants.

B. Solutions Containing Two Surfactants

In a solution containing several surface active solutes, all the solutes tend to adsorb to the gas–liquid interface. Borrowing from other separation processes, a relative distribution coefficient for two surface-active solutes, A and B, may be defined as the ratio of their individual distribution coefficients (Rubin and Jorne, 1969):

$$\alpha_{A,B} = \frac{\Gamma_A/C_{x_A}}{\Gamma_B/C_{x_B}}. \tag{2}$$

The distribution coefficient for component A, Γ_A/C_{x_A}, is the ratio of its concentration at the gas–liquid interface and that in the bulk solution. It is analogous to the distribution coefficient in extraction or the volatility in distillation. For the case of foam fractionation with stable foams, under equilibrium conditions, $\alpha_{A,B}$ can be easily calculated from concentrations of the two components in the bulk liquid and collapsed foam liquid (foamate). Writing Eq. (1) for each of the two components yields

$$\alpha_{A,B} = \frac{C_{y_A}/C_{x_A} - 1}{C_{y_B}/C_{x_B} - 1}. \tag{3}$$

For foam fractionation with stable foams under equilibrium conditions, $\alpha_{A,B}$ depends only on the concentration of the solute and is independent of foam properties.

Theoretical and experimental work was conducted in order to determine the effect of concentration on $\alpha_{A,B}$ and to examine the possibility of its prediction from experimental data obtained with solutions containing the single solutes. Three possible theoretical approaches were evaluated: the Gibbs adsorption isotherm, the Langmuir adsorption isotherm, and the long-chain ions adsorption isotherm (Rubin and Jorne, 1969).

1. Gibbs Adsorption Isotherm

Consider a solution of two ionic surfactants NaA and NaB which dissociate completely in solution:

$$NaA \rightarrow Na^+ + A^-,$$

$$NaB \rightarrow Na^+ + B^-.$$

The Gibbs equation for all ions present in solution is

$$-d\gamma = \Gamma_{Na^+}\, d\mu_{Na^+} + \Gamma_{A^-}\, d\mu_{A^-} + \Gamma_{B^-}\, d\mu_{B^-}, \tag{4}$$

from which the following equation can be obtained (Rubin and Jorne, 1969):

$$-d\gamma/RT = \Gamma_{A^-}\, d\ln(C_{Na^+}C_{A^-}) + \Gamma_{B^-}\, d\ln(C_{Na^+}C_{B^-}). \tag{5}$$

Two extreme cases similar to those usually considered for solutions containing single solutes yield the following in the present case: For $C_{A^-} + C_{B^-} = C_{Na^+}$ a very complex expression is obtained for α_{A^-,B^-} indicating that α_{A^-,B^-} is not necessarily constant. For $C_{Na^+} \gg C_{A^-} + C_{B^-}$ (excess of C_{Na^+}),

$$\alpha_{A^-,B^-} = \frac{(\partial\gamma/\partial \ln C_{A^-})_{C_{B^-}}}{(\partial\gamma/\partial \ln C_{B^-})_{C_{A^-}}}, \tag{6}$$

which again indicates that α_{A^-,B^-} is not necessarily constant. In both cases the determination of α_{A^-,B^-} requires an exceptionally large number of surface tension–concentration determinations.

2. Langmuir Adsorption Isotherm

Assuming no interaction between the surfactants, the approach through the Langmuir adsorption isotherm yields

$$\alpha_{A^-,B^-} = k_B/k_A = const, \tag{7}$$

where k_A and k_B are the two desorption constants which can be obtained from experiments with solutions containing the single solutes.

3. Adsorption Isotherm for Long-Chain Ions

The adsorption isotherm for long-chain ions considers the effect of the potential of the electrical double layer and the cohesion forces at the gas–liquid interface not accounted for in the Langmuir isotherm. The adsorption isotherm for solutions containing a single long-chain solute has been derived and discussed (Davies and Rideal, 1961). Assuming no interaction at the surface in the case of solutions containing two ionic surface-active solutes, the theory for single solutes can be extended, resulting in the following equation:

$$\alpha_{A^-,B^-} = K \exp\left[\left(m_A - m_B\right)\left(\frac{521}{RT} + \frac{1200 \times 7750}{kT}\overline{\Gamma}^{1/2}\right)\right]. \tag{8}$$

According to Eq. (8), α_{A^-,B^-} is not constant, but a correlation of Log α vs. $\overline{\Gamma}^{1/2}$ should yield a straight line. It should be pointed out that the constants in Eq. (8) can be obtained from experiments with solutions containing the single solutes.

Experiments were conducted with solutions of sodium dodecylbenzene sulfonate (NaDBS) and sodium lauryl sulfate (NaLS) (Rubin and Jorne, 1969). The experimental results indicated that foam fractionation techniques may be successfully applied to separate two surface-active solutes, one from the other. For the system studied, the relative distribution coefficient was not constant. Its value increased with concentration below the critical micelle concentration up to $\alpha_{NaDBS,NaLS} \cong 4.5$, and decreased drastically at and above the critical micelle concentration. Thus, to obtain maximum separation between surface active solutes, it is advantageous to work just below the critical micelle concentration, where $\alpha_{A,B}$ has its maximum value.

The experimental results also indicated that, as might be expected, none of the theoretical derivations applied above the critical micelle concentration. Below the critical micelle concentration, calculated values of $\alpha_{A,B}$ based on equilibrium foam fractionation experiments with solutions containing the single solutes and the long-chain ions isotherm are in good agreement with experimental values obtained with solutions containing the two solutes. Calculations based on the Langmuir isotherm may be used for a conservative estimate of $\alpha_{A,B}$ at very low concentrations.

C. Metallic Ions

Consider an aqueous solution of an anionic surfactant. Unless the surfactant is in the H^+ form, there will be at least two cations present in the solution. Usually anionic surfactants are Na salts. Thus, in a solution of anionic surfactants at least Na^+ and H^+ ions are present. Other metallic ions from various origins may also be present in the solution.

If in an aqueous system a surfactant anion is locally concentrated, i.e., by adsorption, the concentration of all other counterions also will be increased locally. If more than one cation species is present, selective concentration of certain counterionic species at the interface may be expected.

1. pH Effects

In the specific case of H^+ it may be expected that the pH of the solution will rise during foaming and the pH of the foamate will be lower than that of the bulk liquid, even if hydrogen ions are not adsorbed preferentially. Changes in pH affect surface tension and foam properties, and hence also the extent of surfactant separation obtained by foaming. In addition, they also affect the separation of the metal ions and may cause the formation of the free acid form of the surfactant in the foamate when the pH decreases below the equivalence point.

Changes in pH associated with foaming may be enhanced by surface hydrolysis of surfactants at interfaces. For anionic surfactants, hydrolysis implies the preferential adsorption of hydrogen ions over the counterion (usually Na^+) of the surfactant salt, with or without the formation of un-ionized acid.

In a solution containing only H^+ and Na^+ as counterions, the preference of the negatively charged surface for H^+ relative to Na^+ can be expressed by an equation similar to Eq. (2) (Rubin and Jorne, 1970):

$$\alpha_{H^+, Na^+} = \frac{\Gamma_{H^+}/C_{x_{H^+}}}{\Gamma_{Na^+}/C_{x_{Na^+}}}. \tag{9}$$

Values of α_{H^+, Na^+} larger than unity indicate preferential adsorption of H^+.

Under steady state conditions, when a stable foam is in equilibrium with the bulk solution, equations similar to Eq. (1) can be written for each cation, and therefore

$$\alpha_{H^+, Na^+} = \frac{C_{y_{H^+}}/C_{x_{H^+}} - 1}{C_{y_{Na^+}}/C_{x_{Na^+}} - 1}. \tag{10}$$

The relation between bulk liquid and collapsed foam pH is given (Rubin and Jorne, 1970) by

$$pH_x - pH_y = \log_{10}\left[\frac{\alpha_{H^+, Na^+}\Gamma_{A^-}}{\delta(C_{x_{A^-}} + C_{x_{Na_2^+}} + C_{x_{H^+}})} + 1\right], \tag{11}$$

where

$$\delta = fD_{32}/6.$$

Note that in Eq. (11) $C_{x_{Na_2^+}}$ represents the excess Na^+ concentration, i.e., Na^+ not originating from the anionic surfactant but from another source, such as NaCl. Equation (11) is useful particularly if α_{H^+, Na^+} is constant. Experiments conducted with NaDBS solutions having pH values that were approximately neutral indicated that α_{H^+, Na^+} is essentially a constant equal to 3.7. Calculations using an equation similar to Eq. (11) and the data reported by James and Pethica (1960) based on surface tension measurements of NaLS solutions also indicated that α_{H^+, Na^+} is essentially constant.

Taking $pH_x - pH_y > 1.0$ as an indication of appreciable pH effects, order of magnitude calculations indicate that pH effects should be considered in most cases of foam fractionation of anionic surfactants when surface hydrolysis is appreciable. At low surface hydrolysis, pH changes can usually be ignored except under very extreme conditions, i.e., very dry foams and relatively low surfactant concentrations (Rubin and Jorne, 1970).

2. Selectivity among Different Cations

Ionic charge and size of the species present govern the selectivity of the surface adsorption of counterions. A theory was developed from which the selectivity of adsorption of the counterions can be predicted by using basic properties such as ionic charge and surface charge density (Jorne and Rubin, 1969). The theory is based on the Gouy–Chapman model of the diffuse double layer, with the restriction that the closest approach to the charged surface is determined not just by the size of the ions, but rather by the size of the hydrated ions.

The theory enables one to predict the distribution coefficient of each species between a solution of mixed electrolytes and a surface layer, and to calculate the selective adsorption coefficient between two ions of identical or of different valencies. The Theoretical equations, which are quite lengthy, agreed well with available experimental data (Jorne and Rubin, 1969).

III. PROPERTIES OF DYNAMIC FOAMS

A. General

For understanding the phenomena involved and for proper design of foam fractionation equipment, knowledge of foam flow is needed in addition to mass transfer and adsorption data. Until a few years ago, almost all the data available regarding the properties of foams pertained to static foams. Only fairly recently did the interest in foam fractionation stimulate theoretical and experimental studies of dynamic foams.

At low foam flow rates, plug flow prevails. The flow regime changes at higher rates. Most of the published experimental work and all the theoretical work reported in the literature pertain to the plug flow regime.

Experimental and theoretical work was therefore conducted at the Technion directed at gaining a better understanding of the foam flow regimes. Special studies of the plug flow regime were also conducted.

B. Flow Regimes in Vertical Columns

An investigation was conducted in order to study the flow regimes prevailing under different conditions, to find correlations between the relevant physical quantities in each flow regime, and to find out what causes a particular flow regime to prevail under a particular set of conditions (Hoffer and Rubin, 1969). The investigation was carried out experimentally in vertical columns of constant cross section. The experiments were restricted to those cases where there is slip between the foam and the solid column wall, i.e., glass columns. The experimental results showed the existence of three flow regimes for foams moving in vertical glass columns. In order of increasing velocity, they are plug flow, turbulent flow, and bubble column regime (where foam as such no longer exists).

Because of the multiple and complex factors involved, the work did not lead to a unique expression, equivalent to a critical Reynolds number, from which the transition from plug to turbulent flow can be accurately predicted in all cases. It did provide, however, important general correlations and insight into the mechanisms and behavior of dynamic foams.

It was found that the transition from one flow regime to the other, in particular from plug to turbulent, can be easily determined visually and is characterized by breaks in the curves of $\Delta P/\Delta Z$ (pressure drop per unit length of foam) vs. G, and f vs. G. It was also found that a general correlation for spinnerets (used as gas distributors), independent of D_o and H over wide ranges, is obtained by plotting $f D_{32}^2$ vs. G (Fig. 1). The curve breaks at $G = 160$ cm^3/cm^2 min, which is in good agreement with the values of G, where transition from plug to turbulent flow was visually observed. The equation of the solid curve in Fig. 1 is

$$f D_{32}^2 = K G^n, \tag{12}$$

where $K \cong 7.0 \times 10^{-6}$ and $n \cong 1.0$ for plug flow ($G < 160$ cm^3/cm^2 min) and $K = 5.5 \times 10^{-9}$ and $n \cong 2.5$ for turbulent flow.

Experiments with sintered porous spargers as gas distributors resulted in a similar equation but with a different value of K for plug flow and a less distinct break in the curve of $f D_{32}^2$ vs. G.

Experimental observations indicated that, in addition to G, the prevailing flow regime depends on a large number of other factors, of which the following have been investigated, at least qualitatively: bubble size and shape, bubble size distribution, foam ratio, foam height, and column geometry. A few qualitative experiments indicated that important parameters in foam flow are

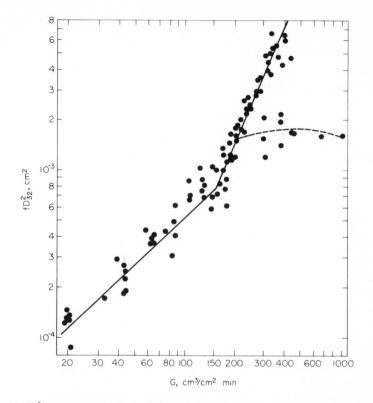

FIG. 1. fD_{32}^2 versus gas flow rate: Column height, 56–1500 cm; spinnerettes D_o, 0.004–0.01 cm; solid lines, plug and turbulent regime; dotted line, bubble column regime for $D_o = 0.004$ cm (Hoffer and Rubin, 1969).

uniformity of column cross section and the material of construction of the column. Details of these experimental observations have been reported (Hoffer and Rubin, 1969).

C. Plug Flow Regime

Two theoretical models have been reported in the literature for predicting the properties of foams moving in vertical columns under plug flow conditions.

Haas and Johnson (1967) developed a foam drainage model consisting of capillaries representing Plateau borders between foam bubbles. The flow in the assumed cylindrical capillaries is due to the gravitational head of the liquid, while the reduced pressure in the Plateau borders results in drainage of films between bubbles to the capillaries. The walls of the capillaries are elastic so that the internal pressure at any point equals the external pressure minus the capillary pressure difference at that point.

A different and more fundamental approach was used by Lemlich and co-workers (Leonard and Lemlich, 1965; Shih and Lemlich, 1967). For foams with low liquid contents and reasonably uniform bubble sizes, the foam structure can be approximated by bubbles in the shape of dodecahedra. Most of the liquid in such foams is concentrated in capillaries (or Plateau borders) comprising a network of channels. Geometrically, a Plateau border was assumed to be channel bounded by three cylinders of equal diameters (smaller than the bubble diameter) with axial liquid flow. Mathematically, this model involved solving the differential momentum conservation equation for flow through a representative Plateau border of curved triangular cross section with nonrigid boundaries characterized by a Newtonian surface viscosity.

The experimental data collected at the Technion as well as the analysis of additional experimental data (Hoffer and Rubin, 1969; Rubin et al., 1967) indicated that within the experimental range ($G = 10$–160 cm^3/cm^2 min, $D_o = 0.003$–0.010 cm, $H = 40$–150 cm, and $D_c = 1.2$–5.7 cm), the following empirical equation applies to plug flow in vertical columns of constant cross section:

$$fD_{32}^2 = KG/D_c^{0.2}, \qquad (13)$$

where K is a constant which seems to depend on the surfactant properties and has to be determined experimentally.

Comparing Eq. (13) with the two theoretical models mentioned above indicated that both are lacking in that they do not account for wall effects, i.e., $D_c^{0.2}$. Equation (13) seems to agree somewhat better with the Haas and Johnson model. However, this model is lacking in that it requires the experimental determination of a constant [as does Eq. (13)]. The model of Lemlich and co-workers requires somewhat more tedious calculations; however, it enables one to predict of foam properties without experiments if surface viscosity is known.

It is the opinion of the present author that in the absence of surface viscosity data one should use Eq. (13) which requires a few experimental points. When surface viscosity data is available one can use the more fundamental model of Lemlich and co-workers which requires no experimental points.

IV. EQUIPMENT DESIGN AND PERFORMANCE

A. General

Several published reviews of foam fractionation include some information on equipment design and performance (Rubin and Gaden, 1962; Lemlich,

1968a). In general, foam fractionation columns can be operated as single-stage (or simple) columns, stripping columns, reflux columns, or full columns consisting of stripping and reflux combined. In a simple column, feed enters the liquid pool under the foam, and exiting foam is in equilibrium with the relatively low concentration of the pool. In a stripping column, feed enters the foam directly. The exit foam can then be in equilibrium with a higher concentration than that of the bottoms liquid, and a lower bottoms liquid concentration (and hence a better separation) is obtained. In a reflux column some of the collapsed foam liquid (foamate) is returned to the foam, resulting in an increased concentration in the foamate. For the best performance with stripping and reflux columns, the foam should move in plug flow.

The operation and performance of a single stage column is well understood and equations for its performance agree well with experimental data. The work at the Technion was directed primarily at stripping and reflux columns using solutions containing a single surfactant and two surfactants.

B. Solutions Containing a Single Surfactant

A theoretical and experimental investigation was conducted primarily with stripping columns. For stripping operation, equations have been previously published describing mathematically infinite columns, where the distance between bottoms pool and liquid entry into the foam is large (Rubin *et al.*, 1962; Fanlo and Lemlich 1965; Lemlich, 1968b). For short stripping columns, equations have been written describing mass transfer between countercurrent streams within the foam using the transfer unit approach (Lemlich, 1968a; Haas and Johnson, 1965). In experimental data reported on stripping columns, either the effect of the column length has not been investigated fully or the inadequacy of the theoretical equations has been ascribed to detrimental flow effects. In any case, the published theory has not been successful in predicting the behavior of short stripping columns.

At the Technion, attempts were made to compare two models of foam fractionation columns (Goldberg, 1968; Goldberg and Rubin, 1970). One model is based on the hitherto assumed continuous solute transfer between countercurrent streams within the foam, and the other on "end effects" with hardly any solute transfer within the foam (Fig. 2). The latter proposed model, corresponding to new experimental results, predicts that separation should be independent of stripping column length (or height) to a large extent.

In this model it is assumed that most of the stripping is due to "end effects" originating from mixing at the feed entry point and at the bottom of the foam. This assumption is borne from the fact that practically all the liquid moving down flows in the Plateau borders and therefore only a small fraction of the

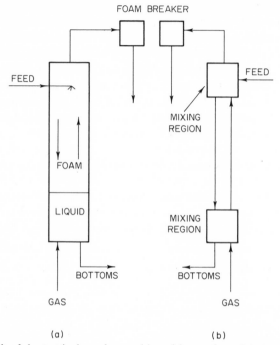

FIG. 2. Models of short stripping columns: (a) model assuming solute transfer in counter-current region; (b) model assuming no solute transfer in countercurrent region (Goldberg and Rubin, 1970).

gas–liquid interfacial area is close enough to the falling liquid to effect any appreciable concentration changes within the residence times involved in the stripping section of the foam column. Therefore, unless the falling liquid is mixed with the rising liquid, solute transfer between the streams is negligible. Thus, the stripping column length (i.e., the distance between feed entry point and the bottom of the foam) can be divided into three regions (Fig. 2). A mixing region around the feed entry point, a mixing region at the liquid pool (including primarily the upper part of the liquid pool and the very bottom of the foam) at the bottom of the foam column, and a countercurrent region in between. Solute transfer in the countercurrent region is assumed to be negligible. The term *mixing region* is used here quite liberally to indicate good contact between falling and rising liquid streams. The good contact around the feed entry point is due to a much wetter foam relative to the countercurrent region, and not necessarily to backmixing of foam.

Analysis of the experimental data obtained with solutions of monobutyl-diphenyl monosodium sulfonate (RWA 240, Roberts Chemical Co.) (Gold-berg, 1968; Goldberg and Rubin 1970) indicated that within the range of variables studied (feed concentration, 1.75×10^{-3}–6.0×10^{-3} M; height of stripping section, 5–152 cm; air flow rate, 1.7–3.3 liter/min; column diameter, 7.5 cm; and feed flow rate, 35–155 cm³/min) the height of a stripping foam fractionation column over 10 cm has practically no effect on separation. Generally, analysis of the experimental findings indicated that they are in agreement with the newly proposed model. It seems that the stripping length does not affect separation as long as the countercurrent region exists. Below a certain stripping column height, under given experimental conditions, the countercurrent region disappears, the two mixing regions merge, and stripping column height affects separation. Under the above-mentioned experimental conditions this happened at foam heights below 10 cm.

C. Solutions Containing Two Surfactants

A theoretical and experimental study was conducted with stripping and refluxing columns (Rubin and Melech, 1970). Due to the present state of the art and particularly due to the lack of information regarding liquid holdup and flows in the stripping and reflux sections of the foam, the only significant theoretical derivations obtained refer to the infinite stripping and refluxing columns. The following equations, obtained by material balances, apply to each solute:

For an infinite stripping column:

$$C_{y_i} = C_{f_i} + \frac{6(1 - f_t)}{D_{32} f_t} \Gamma_{f_i}. \tag{14}$$

For an infinite refluxing column:

$$C_{y_i} = C_{b_i} + \frac{6(1 - f_t)}{D_{32} f_t} \frac{1}{\overline{R}} \Gamma_{b_i}. \tag{15}$$

Figure 3 shows the ratio between $[\Gamma_A/(\Gamma_A + \Gamma_B)]_{exp}$ and $[\Gamma_A/(\Gamma_A + \Gamma_B)]_{theo}$ vs. $6V/f D_{32}$, where the subscripts A and B refer to NaDBS and NaLS, respectively. The term $[\Gamma_A/(\Gamma_A + \Gamma_B)]_{theo}$ was calculated assuming infinite stripping columns. The factor $6V/f D_{32}$ represents very closely the surface area generated per unit volume of feed. From the figure it can be seen that below $6V/f D_{32}$ of approximately 1000 the stripping column behaves very much as an infinite stripping column and Eq. (14) will apply.

For the system studied (solutions of NaDBS and NaLS) it was also found that the utilization of an average value of $\alpha_{NaDBS, NaLS}$ enables fairly accurate determination of the fractionation of one surfactant relative to the other in infinite stripping and refluxing columns.

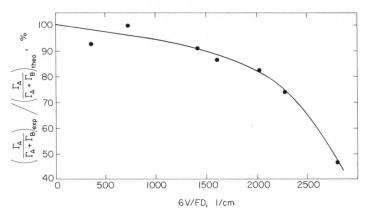

FIG. 3. Effect of gas and feed flow rates on separation in stripping column: surfactants, A = NaDBS; B = NaLS (Rubin and Melech, 1970).

D. The Total Reflux Column

Consider a foam fractionation column operating under total reflux and under such conditions that it behaves as an infinitely long reflux column. Equation (15) then indicates that at equilibrium C_{y_i} should approach infinity. In practice C_{y_i} cannot reach infinity. However, the column will never reach steady state and the concentration in the foamate will increase with time trying to reach infinite concentration.

Figure 4 (Rubin and Melech, 1970; Melech 1969) shows schematically the structure of a total reflux column. In particular, a U-shaped tube is located in the reflux line containing a relatively small volume of liquid. The experimental column was operated under total reflux for several hours. Indeed, the concentrations never reached steady state constant values. The concentration in the liquid pool decreased with time until eventually it was so low that no more foam could reach the top of the column. Practically all of the surfactant was concentrated in the small volume of liquid in the U tube located on the reflux line. As a matter of fact practically any surfactant concentration could be obtained in the U tube, depending on its volume.

Typical data from one run are as follows:

Original feed: 76×10^{-5} M NaLS + 45.6×10^{-5} M NaDBS

Final concentration in liquid pool: 9.2×10^{-5} M NaLS + 4.9×10^{-5} M NaDBS.

Final concentration in U tube: 9800×10^{-5} M NaLS + 6200×10^{-5} M NaDBS.

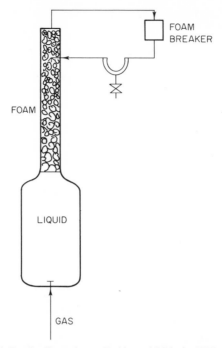

FIG. 4. Total reflux column (Rubin and Melech, 1970).

The total reflux foam fractionation column as described above seems to show particular promise as a means for total removal and for obtaining highly concentrated solutions of surface active solutes. Another possibility is the utilization of a surfactant carrier for removal and concentration of surface-active solutes present originally in very low concentrations.

SYMBOLS

C concentration
C_f feed concentration
C_x concentration in bulk liquid
C_y concentration in collapsed foam (foamate)
D_{32} average bubble diameter, $\Sigma n_i D_i^3 / \Sigma n_i D_i^2$
D_c foam column diameter
D_o diameter of holes in gas distributor
f foam ratio, cm³ liquid/cm³ foam
F Feed flow rate

G gas flow rate per unit cross section (empty column), cm³/cm² min
H height of foam column
k, K constants
$\bar{k}$ Boltzmann constant
m_A, m_B constants [Eq. (8)]
n constant
n_i number of bubbles of diameter D_i
R gas constant
$\bar{R}$ reflux ratio (leaving liquid/returning liquid)

T	absolute temperature	*Subscripts*
V	air flow rate to foam column, cm^3/min	
$\alpha_{A,B}$	relative distribution coefficient	A,B surfactant A or B
γ	surface tension, dyn/cm	t at the top of the column
Γ	surface excess, moles/cm^2	f at the feed point to the column
$\bar{\Gamma}$	average surface excess, moles/cm^2	b at the bottom of the column

REFERENCES

Aniansson, G. (1951). *J. Phys. Colloid Sci.* **55**, 1286.

Davies, J. T., and Rideal, E. K. (1961). "Interfacial Phenomena." Academic Press, New York.

Fanlo, S., and Lemlich, R. (1965). *Amer. Inst. Chem. Eng. Inst. Chem. Eng. Symp. Ser. No. 9*, **75**, 85.

Goldberg, M. (1968). Ph.D. thesis, Technion, Israel.

Goldberg, M., and Rubin, E. (1967). *Ind. Eng. Chem. Process Des. Develop.* **6**, 195.

Goldberg, M., and Rubin, E. (1970). In preparation.

Haas, P. A., and Johnson, H. F. (1965). *A. I. Ch. E. J.* **11**, 319.

Haas, P. A., and Johnson, H. F. (1967). *Ind. Eng. Chem. Fundam.* **6**, 225.

Hoffer, M. S., and Rubin, E. (1969). *Ind. Eng. Chem. Fundam.* **8**, 483.

James, J. W., and Pethica, B. A. (1960). *Proc. Int. Congr. Surface Activity, 3rd, Mainz,* p. A227.

Jorne, J., and Rubin, E. (1969). *Separ. Sci.* **4**, 313.

Lemlich, R. (1968a). *Ind. Eng. Chem.* **60** (10), 16.

Lemlich, R. (1968b). Principles of Foam Fractionation. *In* "Progress in Separation and Purification," (E. S. Perry, ed.), Vol. 1. Wiley (Interscience), New York.

Leonard, R. A., and Lemlich, R. (1965). *A. I. Ch. E. J.* **11**, 18.

Melech, D. (1969). M.Sc. thesis, Technion, Israel. Unpublished data.

Nilsson, G. (1957). *J. Phys. Chem.* **61**, 1135.

Rubin, E., and Gaden, E. L. (1962). Foam Separation, *In* "New Chemical Engineering Separation Techniques" (H. M. Schoen, ed.), Wiley (Interscience), New York.

Rubin, E., and Golt, M. (1970). *Ind. Eng. Chem. Process Des. Develop.* **9**, 341.

Rubin, E., and Jorne, J. (1969). *Ind. Eng. Chem. Fundam.* **8**, 474.

Rubin, E., and Jorne, J. (1970). *J. Colloid Interface Sci.* **33**, 208.

Rubin, E., and Melech, D. (1970). In preparation.

Rubin, E., Schonfeld, E., and Everett, R. (1962). Oak Ridge Nat. Lab. Rep. RAI-104.

Rubin, E., LaMantia, C. R., and Gaden, E. L. (1967). *Chem. Eng. Sci.* **22**, 1117.

Shih, F. S., and Lemlich, R. (1967). *A. I. Ch. E. J.* **13**, 751.

Sobotka, H. (1954). Monomolecular Layers, Amer. Assoc. Advan. Sci., Washington, D.C.

Wilson, A., Epstein, M. B., and Ross, J. (1957). *J. Colloid. Sci.* **12**, 345.

C. Jacobelli-Turi, F. Maracci, A. Margani, and M. Palmera

Laboratorio di Metodologie Avanzate Inorganiche del Consiglio
Nazionale delle Ricerche
Rome, Italy

I. ION ADSORPTION

A. Competing Phenomena

Since ion extraction by foam involves the adsorption of the surface-inactive ion at the liquid–air interface via the surface-active agent, it is necessary to consider the principal factors which govern this process of adsorption.

First, we will confine our attention to some competing phenomena which can hinder the adsorption of the surface inactive ion at the liquid–air interface (Jacobelli-Turi and Palmera, 1966). As a basis for discussion, reference can be made to the system cetyltrimethyl ammonium bromide–sodium phenate ($CTABr-C_6H_5ONa$) as a typical example of the adsorption of a surface-inactive ion on a charged liquid monolayer. Several phenomena to support the above assumption were found from the observations of experimental results obtained by changing the concentration of the surfactant or of C_6H_5OH, provided that the phenate concentration was smaller than the surfactant concentration. The experimental data are expressed in terms of Eq C_6H_5OH/Eq CTABr to indicate both phenol and CTABr equivalents extracted by foam, and in term of enrichment factor E for phenol, defined as the ratio of phenol concentration in the foam to that in the initial solution.

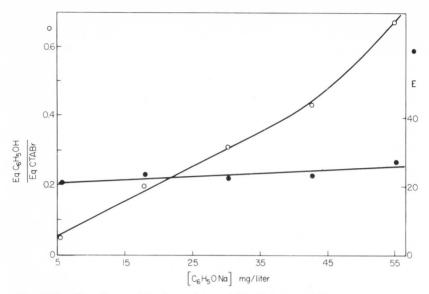

FIG. 1. The effect of competition for surface positions. E is the enrichment factor for phenol.

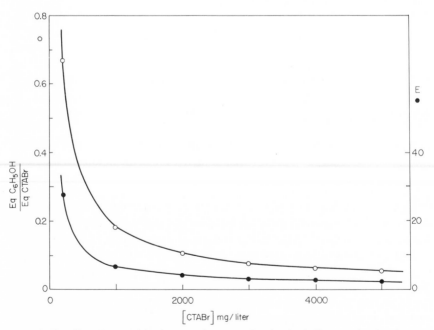

FIG. 2. The effect of competition between the unassociated and micellar-associated surfactant for the counterion.

Competitive adsorption of CTABr at the interface is evidenced by the marked decrease in Eq C_6H_5OH/Eq CTABr and the decrease of C_6H_5ONa concentration, both of which are shown in Fig. 1. In the experiments described, the CTABr concentration was assumed to be just below the critical micelle concentration. In the absence of competing effects, the curve would show a constant slope over the whole concentration range of C_6H_5OH, as the time of each run was short enough not to cause the exhaustion of the phenate ion in the bulk solution. Again, the constant behavior of E, indicating that phenate adsorption is independent of its initial concentration, further supports the above assumption.

The existence of competing effects also has to be assumed in relation to the occurrence of micelles in the bulk since the unassociated and micellar-associated surfactant compete with each other for the surface-inactive ion. This explains the reason why the marked decrease of Eq C_6H_5OH/Eq CTABr, shown in Fig. 2, takes place near the critical micelle concentration of CTABr.

Micelle formation, therefore, is antagonistic to ion enrichment. The situation illustrated above appears to be quite general in the case of ions adsorbed on the charged monolayer, since the surface activity of the surfactant is not markedly related to the nature of the counterion. The results obtained by Mazzella (1960) in strontium extraction by means of the chelating surfactant (DBTA) were found to be in agreement with the situation described above. Conversely, in the case of the Zn–dodecylimminodipropionic acid system, the complexed form of the surfactant is adsorbed preferentially at the interface (Jacobelli-Turi et al., 1966a). This means that the surface activity of the whole compound will depend on the nature of the surface-active complex.

B. Operating Variables

A crucial problem in the success of extraction by foaming is the correct choice of operating variables. The following discussion will be purposely kept short since references on this subject have already appeared in the literature (Gaden and Rubin, 1962).

Figure 3 illustrates the effect of temperature on the degree of extraction. Table I shows the effect of gas-flow rate. For convenience, experimental data were referred to the CTABr–C_6H_5ONa system.

Basically, the available evidence suggests that at low gas rates the foam produced rises slowly, draining well and yielding a high enrichment factor. At high gas-flow rates much liquid is entrapped in the rapidly rising foam, thus lowering the concentration of the solute in the foam. The increase of enrichment at higher temperatures can be ascribed to lower foam density resulting from lowered viscosity and hence high drainage rates. As might be expected, the pH is an important factor in foam processes and must be

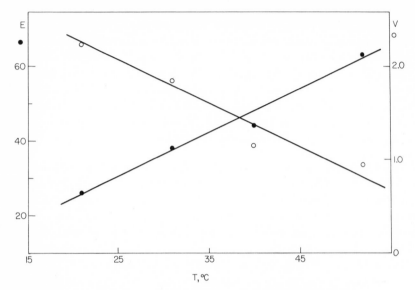

FIG. 3. Effect of temperature on enrichment factor and collapsed foam volume.

TABLE I

EFFECTS OF GAS-FLOW RATE ON EXTRACTION BY FOAMING

v	V	E	$\dfrac{\text{Eq } C_6H_5OH}{\text{Eq CTABr}}$
15	Unstable foam		
19.0	0.17	105.0	0.63
26.0	0.60	45.0	0.66
34.0	1.50	27.6	0.68
46.8	9.00	12.0	0.65

controlled as it affects both the charge of the ions in solution and the stability of the foam.

The effectiveness of the foam process at very low concentrations has been evidenced in the study of the influence of concentration on enrichment factor E. The plots of Fig. 4 and Fig. 5 refer to the CTABr–C_6H_5OH system where the molar ratio of CTABr to C_6H_5OH was equal to 1, and kept constant to avoid the presence of competitive phenomena (Jacobelli-Turi and Palmera, 1966). In Fig. 4 the results from such experiments are compared with surface tension data obtained at corresponding concentrations. We can see that the extraction behavior depends in a complicated manner on the slope of surface tension versus concentration.

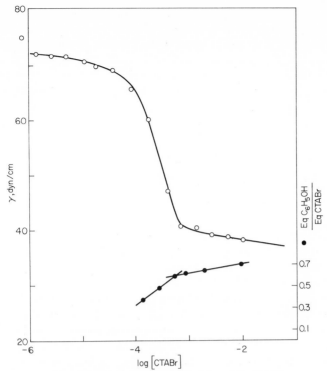

FIG. 4. Comparison of surface tension with extraction.

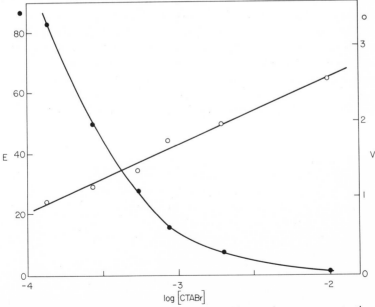

FIG. 5. Dependence of enrichment factor and collapsed foam volume on concentration in the absence of competing phenomena.

The variation of E shown in Fig. 5 is the result of the combined effects of two phenomena. If we assume a proportional relationship between E and Γ_1/c_1, then the variation of E should be closely related to the gradient of the surface tension curve. This is verified by the high enrichment values in the range of low concentrations. Noting the variation of volume of the collapsed foam, it is evident that the surfactant concentration affects the foam density, causing a decrease in E in the range of concentration where the gradient of the surface tension curve is constant.

II. REPRESENTATIVE SEPARATIONS

Because foam techniques will play an increasingly important role in the field of separation technology, a relatively large number of papers have been published during the past ten years.

The present section will be restricted to a brief survey of work carried out in our own laboratory; experimental results, apparatus and details of operation have been published and, for the sake of brevity, will not be described here.

Because separation by foaming is of the greatest value in the region where other methods run into economic limitations, it was found to be an attractive alternative to solvent extraction for the recovery of valuable metals. The practical convenience of using foam processes was demonstrated in the recovery of uranium from carbonate leach liquors (Barocas et al., 1963) and in the separation of uranium from thorium at high concentrations of hydrochloric acid (Jacobelli-Turi et al., 1967). To obtain higher uranium decontamination from thorium contained in the bulk liquid entrapped in the foam, a column equipped with a washing device was used. Since the recovery of the extracted ions from the foam requires further treatment, a process has been developed whereby the extraction of ions and their recovery from the foam can be accomplished in one operation. The chief feature of this method is the stripping of the extracted ions by bubbling foam through a series of single devices connected to the foam column. Each single device contains selective stripping solutions by means of which both selective scrubbing of the ions from the foam and economic recovery of the surfactant can be accomplished. An example of the last process is the uranium and vanadium extraction from carbonate solution, the uranium being stripped from the foam with $2N$ HCl after removal of vanadium with $0.2\ M$ versene (Barocas et al., 1967).

A further use of the foam process, which can be considered an economic possibility, is the purification of industrial effluents (Aronica and Grieves, 1966; Grieves, 1968). Attention has been given to the extraction of phenolic compound because of their deleterious effect on aquatic life (Jacobelli-Turi *et al.*, 1966b). Thus, foaming provides an effective process for the extraction of phenol and nitrophenols, employing CTABr as the foaming agent.

Complexing surfactants can play a very important role in the selective extraction of ions from very dilute solutions (Ghosh *et al.*, 1960). A selective extraction of a particular metal can be devised even when it is present with metals which also form extractable complexes with the complexing surfactant employed. From the known value of the stability constant of the system employed, the best conditions for the separation of many metals can easily be calculated. Thus, at pH 9 it has been found possible to extract both zinc and copper from aqueous solutions, even if very dilute, by the complexing surfactant dodecylimminodipropionic acid (Jacobelli-Turi *et al.*, 1966a). The zinc–copper separation can be achieved by changing the pH of the solutions, as illustrated in Fig. 6. At pH values above 10.5 the copper surfactant complex is stable while the zinc complex is less stable, as it preferentially forms non-extractable complexes of the type $Zn(OH)_4^{2-}$.

In some cases the overall degree of separation depends on the changes in surface activity determined by the formation of the chelated surface-active compound. Thus, for example, at a suitable pH value, zinc can be separated from the alkali earth metals by using the chelating surfactant dodecylimminodipropionic acid. The Zn^{2+} ion is extracted principally as a chelate, while the mechanism whereby the alkaline earth metals are extracted can, in the main, be ascribed to the electrostatic interaction of ion and surfactant at the liquid–air interface. The higher surface activity of the zinc

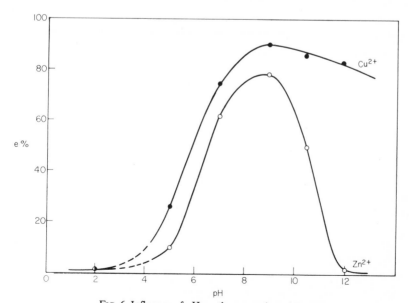

FIG. 6. Influence of pH on the extraction of Zn and Cu.

chelate leads to its separation from the alkali earth metals as it is almost entirely extracted in the first fraction of the foam (Di Lello et al., 1969).

Finally, mention should be made of an interesting class of chelating surfactants of the general formula

$$R-N(-COOH)-(CH_2)_2-N(-COOH)-(CH_2)_2-N(-COOH)_2$$

where R represents an alkyl radical having from 8 to 16 carbon atoms.

The synthesis of these compounds was carried out in our laboratory, and at the present time we are determining their physiochemical characteristics (Jacobelli-Turi et al., 1970).

SYMBOLS [1]

c_1 concentration of component in liquid
E enrichment factor
v gas flow rate

V volume of collapsed foam
γ surface tension
Γ_1 surface excess of component

REFERENCES

Aronica, R. C., and Grieves, R. B. (1966). *Nature* **210**, 901.

Barocas, A., Jacobelli-Turi, C., and Salvetti, F. (1963). *Gazz. Chim. Ital.* **93**, 1493.

Barocas, A., Jacobelli-Turi, C., and Terenzi, S. (1967). *Ind. Eng. Chem. Process Des. Develop.* **6**, 161.

Di Lello, M. C., Jacobelli-Turi, C., Margani, A., and Palmera, M. (1969). *Ric. Sci.* **39**, 264.

Gaden, L. G., and Rubin, E. (1962). *In* "New Chemical Engineering Separation Techniques" (H. M. Schoen, ed.), pp. 319–385. Wiley (Interscience), New York.

Ghosh, D., Mazzella, G., Hook, S., Sanford, R., and Schonfeld, E. (1960). AEC, NYO 9577, Radiation Applications, Inc.

Grieves, R. B. (1968). *Brit. Chem. Eng.* **13**, 77.

Jacobelli-Turi, C., and Palmera, M. (1966). *Gazz. Chim. Ital.* **96**, 1432.

Jacobelli-Turi, C., Margani, A., and Palmera, M. (1966a). *Ric. Sci.* **36**, 1198.

Jacobelli-Turi, C., Palmera, M., and Perinelli, S. (1966b). *Ric. Sci.* **36**, 1219.

Jacobelli-Turi, C., Palmera, M., and Terenzi, S. (1967). *Ind. Eng. Chem. Process Des. Develop.* **6**, 162.

Jacobelli-Turi, C., Margani, A., Maracci, F., and Palmera, M. (1970). *Ann. Chim.* (*Rome*) **60**, 674.

Mazzella, G. (1960). Radiation Applications, Inc., Long Island City, New York, N.Y. 2552.

[1]Symbols refer to those used in text.

CHAPTER **18**
SEPARATIONS OF PARTICLES, MOLECULES,
AND IONS BY FOAMING:
STUDIES IN JAPAN

T. Sasaki

Department of Chemistry, Faculty of Science
Tokyo Metropolitan University
Setagaya, Tokyo

I. GENERAL ASPECTS OF SEPARATION BY BUBBLING

The separation of solute or suspended matter in water has drawn increasing attention with regard to industrial wastewater (Nakamura *et al.*, 1967; Okura *et al.*, 1959 ; Noda, 1968a, b ; Ito and Shinoda, 1962), sewage treatments (Mimasaka *et al.*, 1964, Kobayashi and Ishii, 1964) and concentration of rare elements from seawater (Yamabe and Takai, 1969). Academic studies have also been made on the technique of separation (Kishimoto, 1963, Sasaki, 1970), the apparatus (Kishimoto, 1963), reagents (Sasaki and Yamasaki, 1965), substances to be separated, the types of reactions available for the separation (Sasaki, 1967c, 1970 ; Aoki and Sasaki, 1966), and nomenclature and classification of the methods to be adopted (Sasaki, 1970; Koizumi, 1963 ; Noda, 1962b; Noda, 1968c).

As for the nomenclature, the term *adbubble* or *adubble method* has been recommended for use (Sasaki, 1970), rather than "adsubble method" proposed by Lemlich (1966), since the bubble separation method involves both *ads*orption and *adh*esion processes. Adbubble method was eventually classified into foam chromatography, macro-, micro-, and precipitation flotations, adsorbing particle flotation, and solvent sublation.

Parallel to this classification, which is based mainly on the mechanism of flotation, another classification according to the nature of the substance to be separated is needed, that is, mineral, molecule, and ion flotations. These two kinds of classifications can be used in combination, in such a manner as "ion precipitation flotation" or "ion adsorption flotation," if necessary, to make clear the type or the mechanism of flotation and the state of the substances to be separated.

II. ION FLOTATION

A. Adsorption Flotation

The method consists of the flotation of ions in a solution by adsorbing them on ascending bubbles. Typical instances were seen in the flotation of dye ions by the addition of oppositely charged ionic surfactants, as in the case of the flotation of crystal violet cations (CV^+) by sodium dodecylsulfate anions (SDS^-) (Sasaki, 1967c, 1969). Simple shaking of this blue solution and breaking of the foam so produced left a thin, but deep, blue layer of concentrated CV on a colorless bulk solution. A uniform blue coloration of the whole solution, which was restored by a gentle stirring of this solution, proved simple adsorption of CV^+ to be the main factor causing the flotation in this case. Flotations of malachite green, methyl orange, rhodamin B, and phenolphthalein by SDS or octadecyltrimethylammonium chloride (OTAC) have been found to be of a similar type (Aoki and Sasaki, 1966; Sasaki, 1967c, 1969), where the Coulombic force was considered to play a dominant factor (Sasaki, 1967c, 1969; Wadachi et al., 1965).

In the case of inorganic ions, however, the effect of pH is marked in the process of flotation and the phenomena could not be distinguished from the precipitation flotation mentioned in Section II, B. To this group belonged the flotation of Na^+, Ca^{2+} (Shinoda and Ito, 1961), Sr^{2+}, Cs^+ (Matuura, 1964) by SDS; UO_2^{2+} by alkylphosphate (Yamabe and Takai, 1969); I^- by cetylpyridinium chloride (Koizumi, 1963); Zn^{2+} by alkylbenzenesulfonate (ABS) (Koizumi, 1967, 1969); the relative adsorbability of $CH_3CO_2^-$, Cl^-, Br^-, ClO_3^-, NO_3^-, SO_4^{2-} by amine salts (adsorbabilities increasing in this order as a whole) (Shinoda and Fujihira, 1968); selective flotation of Ca^{2+} and Na^+ by the aqueous SDS foam (Ca^{2+} being about 200 times greater than Na^+ at 6×10^{-3} mole/liter concentration) (Shinoda and Ito, 1961); foam column chromatography of NH_4^+, Mg^{2+} and Al^{3+} (Koizumi, 1962); and removal of phenol (anion) from industrial wastewater by a cationic surfactant (Yoshida and Isobe, 1966).

B. Precipitation Flotation

In this procedure, ions in the solution are rendered insoluble by the addition of proper substances or by the control of pH. The precipitates thus formed are then floated by adding a surface-active agent. Often the surface-active agent alone has proved sufficient to cause the precipitation flotation. For instance, Fe^{3+} was floated by anionic or cationic surfactants such as SDS (Sasaki, 1967c, 1969), sodium oleate (Koyanaka, 1965), or OTAC (Sasaki, 1967a, 1969). In these cases, the coexistence of Ca^{2+} enhanced the flotation (Koyanaka, 1965), and the addition of EDTA (Aoki and Sasaki, 1966) or Na_2SiO_3 (Sasaki, 1967c, 1969) has also proved effective. Cation Co^{2+} was floated by sodium oleate (Koyanaka, 1965) and alkali xanthate (Oyama et al., 1962, Yamasaki et al., 1963), and cobalt complex varying in valency by SDS (Matuura, 1966). Nuclear fission products like Ce, Zr–Nb, and Ru–Rh were first precipitated with Fe or Co ions and the precipitates were floated by sodium oleate (Koyanaka, 1965). Flotations of Ni^{2+}, Cu^{2+}, and Ag^+ have also been attempted (Aoki and Sasaki, 1966; Sasaki, 1969; Yamasaki et al., 1963). MnO_4^- ions were floated by the addition of OTAC, Na_2SiO_3 and sodium polyacrylate in this order, but no flotation occurred when OTAC was added after the addition of Na_2SiO_3 (Sasaki, 1967a, 1969).

C. Adsorbing Particle Flotation

This method consists of a simple adsorption or exchange adsorption of ions from a solution onto adsorbent particles like bentonite or synthetic ion exchanger which are subsequently floated by the addition of a suitable surfactant. Adsorbing particle flotation is perhaps the most effective method among other methods of ion flotation. Thus, about 98 % of Co^{2+} ions were floated from 1×10^{-5} mole/liter solution at pH 10 by adding 130 ppm of bentonite, 30 ppm of polyacrylamide, 5×10^{-5} mole/liter of hexadecyl-dimethylbenzylammonium chloride, and passing nitrogen gas for 2 min (Sasaki, 1968, 1970). Under similar conditions, Fe^{3+} and Cu^{2+} ions could be floated in the region of pH greater than 4 and 6, respectively, with an intermediate zone of pH for the separation of both ions (Sasaki, 1969). Dye ions such as rhodamin B could be floated by this method more efficiently than by the simple adsorption flotation described in Section II, A. Cu^{2+} ions have been floated by the addition of styrene sulfonate ion exchanger or sulfonated coal (Sasaki and Yamasaki, 1965; Yamasaki et al., 1963). Flotation of I^- ions has also been demonstrated by the successive addition to its solution of $CuSO_4$, bentonite, polyacrylamide, and OTAC. In this case, insoluble Cu_2I_2, as well as I_2, was observed to float (Sasaki, 1969).

III. MOLECULE AND PARTICLE FLOTATIONS

Molecule flotation is indistinguishable in its mechanism from ion flotation when it undergoes ionization. For instance, flotation of I_2 with the addition of cetylpyridinium chloride is believed to be due to the intermolecular compound formation between them (Koizumi, 1963). From 67 to 95% of I_2 has been collected at optimum conditions.

The selective adsorptivity α of solutes from mixed aqueous solutions of $R_{8\sim12}OH$ and $R_{10}SO_4Na$, $R_{12}SO_4Na$, or $R_{11}COOK$ (where R_n represents a hydrocarbon chain containing n carbon atoms) has been measured. The α value, for instance, of $R_{12}OH$ was found to be 1000 times as large as that of $R_{11}COOK$ (Shinoda and Nakanishi, 1963). The α values have been measured for several other systems (Shinoda and Mashio, 1960). These data will be useful as a measure of molecule flotation. Further studies were made for the adsorption of surface-active substances from their mixed solutions by using ratiotracer methods (Ito and Shinoda, 1962 ; Tajima et al., 1969 ; Seimiya, 1969).

An interesting case of molecule flotation has been reported in the selective concentration in the foam of polyvinyl alcohol of syndiotactic stereoregularity, compared with that showing no such regularity (Imai and Matsumoto, 1963). This seems to show some relation between the stability of a foam film and the molecular structure of the film-forming substance. Flotation of various types of surfactants has been studied from the viewpoints of their use as a flotation agent and the removal of alkyl benzene sulfonate from sewage (Mimasaka et al., 1964 ; Yoshida et al., 1966 ; Kobayashi and Ishii, 1964).

As for particle flotation, attempts have been made to float K and Na salts from their saturated solutions by the addition of surfactant of the alkylsulfate type. Potassium salt was selectively floated in the case of chlorides, non-selective flotation was observed for sulfates, and no flotation occurred in the case of nitrates (Shinohara, 1955).

The flotation of ion-exchanger particles has been studied with relation to the adsorbing particle flotation mentioned in Section II, C, namely, the flotation of polystyrene sulfonate (Yamasaki et al., 1963), sulfonated coal (Sasaki and Yamasaki, 1965), and bentonite (Sasaki, 1969). These substances are anionic in nature, but cationic surfactants of amine derivatives were also employed as coagulating, floating, and foaming agents. Here, dense crosslinking of exchanger molecules, and large size of amine molecules turned out to be favorable for effective flotation (Sasaki and Yamasaki, 1965). It was also confirmed that the amount of surfactant greatly affects the state of suspension of the particles, which changed from coagulated precipitation through agglomerated flotation to redispersion, with an increase in the amount of surfactant added (Sasaki, 1969).

In conclusion, it may be added that considerable interest has also been shown in the operation and arrangement of the apparatus for ion flotation with regard to its efficiency (Kishimoto, 1963; Yoshida and Isobe, 1966).

General reviews (Asada, 1964; Kishimoto, 1961, 1963; Noda, 1962a, b, c; Sasaki, 1961, 1967a; "Special issue on flotation," 1964; Yamasaki and Sasaki, 1966; Yoshida, 1970), and a monograph (Sasaki, 1967b) on ion flotation have been published and are available for further information on ion flotation problems.

REFERENCES

Asada, H. (1964). *Mizushori Gijutsu* **5,** No. 3, 1.
Aoki, N., and Sasaki, T. (1966). *Bull. Chem. Soc. Japan* **39,** 939.
Imai, K., and Matsumoto, M. (1963). *Bull. Chem. Soc. Japan* **36,** 455.
Ito, K., and Shinoda, K. (1962). *Kogyo Kagaku Zasshi* **65,** 1226.
Kishimoto, H. (1961). *Koroido to Kaimen Kasseizai* **2,** 494.
Kishimoto, H. (1963). *Kolloid-Z.* **192,** 66.
Kobayashi, Y., and Ishii, H. (1964). *Kogyo Yosui* **70,** 35.
Koizumi, I. (1962). *Kogyo Kagaku Zasshi* **65,** 1343.
Koizumi, I. (1963). *Kogyo Kagaku Zasshi* **66,** 912.
Koizumi, I. (1967). *Kogyo Kagaku Zasshi* **70,** 1641.
Koizumi, I. (1969). *Kenkyu Hokoku* (Technical Institute of Kanagawa Prefecture), No. 25, 13.
Koyanaka, Y. (1965). *Nippon Genshiryoku Gakkaishi* **7,** No. 11, 621.
Lemlich, R. (1966). *Chem. Eng.* **73** (21), 7.
Matuura, R. (1964). Studies on Removal of Radioactive Contamination. Collected Reports on Scientific Researches of Government of Education of Japan, 159.
Matuura, R. (1966). Studies on Removal of Radioactive Contamination. Collected Reports on Scientific Researches of Government of Education of Japan, 243.
Mimasaka, Y., Ogoshi, T., and Imafuku, M. (1964). *Kogyo Yosui* **70,** No. 7, 42.
Nakamura, M., Kachi, K., and Sasaki, T. (1967). Fundamental and Applied Studies on Interfacial Phenomena. Report on Co-operative Research, Grant in Aid of Scientific Research of Government of Education of Japan.
Noda, M. (1962a). *Kagaku Kogyo* No. 12, 52.
Noda, M. (1962b). *Mizushori Gijutsu* **3,** No. 1, 40.
Noda, M. (1962c). *Mizushori Gijutsu* **3,** No. 7, 51.
Noda, M. (1968a). *Kagaku to Kogyo* **42,** 515.
Noda, M. (1968b). *Yosui to Haisui* **10,** No. 5, 303.
Noda, M. (1968c). *Kagaku Kogaku* **32,** No. 6, 506.
Okura, T., Suzuki, H., and Sano, K. (1959). *Kogyo Kagaku Zasshi* **62,** 337.
Oyama, T., Shimoizaka, J., Yamasaki, T., Baba, I., and Ikeuchi, S. (1962). *Nihon Kogyo Kaishi* **78,** No. 887, 391.
Sasaki, H., and Yamasaki, T. (1965). *Nippon Kogyo Kaishi* **81,** No. 924, 391.
Sasaki, T. (1961). *Koroido to Kaimen Kasseizai* **2,** 59.
Sasaki, T. (1967a). *Kagaku to Kogyo* **20,** 862.
Sasaki, T. (1967b). *In* "Bunri to Seisei," Jikken Kagaku Koza, Ser. 2, Vol. 2, Section 15, pp. 565–603, Maruzen Co.
Sasaki, T. (1967c). Nikkoho (Nikko Chemicals Co. Ltd.) No. 43.
Sasaki, T. (1968). Studies on Removal of Radioactive Contamination. Collected Reports on Scientific Researches of Government of Education of Japan, 259.
Sasaki, T. (1969). *Nippon Kogyo Kaishi* **85,** No. 976, 610.

Sasaki, T. (1970). Fundamental and Applied Studies on Interfacial Phenomena. Report on Co-operative Research, Grant in Aid of Scientific Research of Government of Education of Japan, 58.

Seimiya, T. (1969). *Genshiryoku Kogyo* **15,** 68.

Shinohara, I. (1955). *Kogyo Kagaku Zasshi* **58,** 907.

Shinoda, K., and Fujihira, M. (1968). *Advan. Chem. Ser. No.* 79, 198.

Shinoda, K., and Ito, K. (1961). *J. Phys. Chem.* **65,** 1499.

Shinoda, K., and Mashio, K. (1960). *J. Phys. Chem.* **64,** 54.

Shinoda, K., and Nakanishi, J. (1963). *J. Phys. Chem.* **67,** 2547.

Special Issue on Flotation (1964). *Kogyo Yosui* **70,** No. 7.

Tajima, K., Maramatsu, M., and Sasaki, T. (1969). *Bull. Chem. Soc. Japan* **42,** 2471.

Wadachi, Y., Takada, K., Yamaoka, Y., and Noguchi, S. (1965). *J. Nucl. Sci. Technol.* **2,** 104.

Yamabe, T., and Takai, N. (1969). *Seisan Kenkyu* **21,** No. 9, 530.

Yamasaki, T., and Sasaki, H. (1966). *Kemikaru Enjiniyaringu* No. 7, 669.

Yamasaki, T., Hayamizu, H., and Misawa, S. (1963a). *Nippon Kogyo Kaishi* **79,** No. 896, 91.

Yamasaki, T., Shimoizaka, J., and Sasaki, H. (1963b). *Nippon Kogyo Kaishi* **79,** No. 899, 313.

Yoshida, T. (1970). *Hyomen* **8,** 65.

Yoshida, T., and Isobe, A. (1966). *Nenryo Oyobi Nensho* **33,** No. 6, 545.

CHAPTER **19**

SEPARATION OF SURFACTANTS AND METALLIC IONS BY FOAMING: STUDIES AT RADIATION APPLICATIONS, INC., AND OAK RIDGE NATIONAL LABORATORY, U.S.A.

W. Davis, Jr., and P. A. Haas

Chemical Technology Division
Oak Ridge National Laboratory
Oak Ridge, Tennessee

I. ORIGIN AND OBJECTIVES OF FOAM FRACTIONATION WORK AT RAI AND ORNL

The work described in this chapter was an outgrowth of the studies of Gaden and Kevorkian (1957) and Schnepf and Gaden (1959) at Columbia University. These studies were concerned with the foam fractionation of proteins and with a description of the relationships between pertinent fractionation column variables such as: surface tension γ, excess solute concentration at the surface of the foam Γ, and foam–solution enrichment ratio. The potential utility of foam fractionation in the nuclear energy industry arose (Schnepf *et al.*, 1959) from two factors: first, metallic ions in solution can be incorporated in surface-active chelates or complexing agents; second, large volumes of waste solutions contaminated with ^{137}Cs and ^{90}Sr are generated during the processing of irradiated nuclear fuels (Irish, 1957). These two radioactive fission products are emphasized because they have long half-lives and because they are retained within animal bodies after they are ingested.

The possibility of removing cesium and strontium, present in very low concentrations, from highly salted (1 M $NaNO_3$) process solutions, such as those generated at Hanford and Savannah River, was very attractive from the standpoint of reducing the potential biological hazards of these nuclear fuel processing wastes. Thus, toward the end of February 1958, the U.S. Atomic Energy Commission and Radiation Applications, Inc. (RAI), New York, entered into a contract to study the removal of strontium and cesium from nuclear waste solutions by foam separation. Somewhat later, the USAEC and RAI entered into a second contract for the purpose of investigating foam separation in isotope recovery. The responsibility for administering and expanding both aspects of the foam separation study was transferred by the Atomic Energy Commission to Oak Ridge National Laboratory (ORNL) on July 1, 1960. Shortly thereafter it was agreed that the specific goal of the work would be assessment of the technical and economic feasibility of developing a foam separation process for decontaminating the slightly radioactive process waste water that is generated in large quantities at all nuclear reactor and fuel processing sites. This particular type of solution was chosen because it was considered the most likely candidate for successful application of the foam fractionation concept. Assessment of the process included laboratory and small-scale engineering tests; it concluded with a pilot plant demonstration with a 2 by 2 by 8 ft foam column.

II. FRACTIONATION OF METAL IONS FROM NITRATE SOLUTIONS

The early work at RAI was concerned (Schonfeld et al., 1960; Gaden and Schnepf, 1962a, b) with the removal of radioactive nuclides, particularly the long-lived, biologically hazardous ^{137}Cs and ^{90}Sr, from nitrate solutions such as those obtained by neutralizing (with sodium hydroxide) the first-cycle raffinates generated by the processing of nuclear fuels by the Purex process (Irish, 1957). The high sodium content of these solutions, the very low cesium concentration (on the order of 10^{-6} M), and the similarity of cesium and sodium in their reactions with most anionic surfactants necessitated the search for surfactants that combine selectively with cesium. Because of its expected low concentration ($< 10^{-5}$ M), strontium, even though it is complexed much more strongly by most anionic surfactants than sodium, also required the use of highly specific surfactants to compensate for the mass action effects of sodium. For these reasons, many commercial surface-active agents were tested. The tests included the measurement of critical micelle concentration, foamability, foam stability, and, finally, the single-stage enrichment factor. Some of these variables were studied at three pH levels:

TABLE I

TYPICAL SINGLE-STAGE DATA FOR THE FRACTIONATION OF CESIUM AND STRONTIUM IN NITRATE SOLUTIONS

Surfactant	Surf. conc. (gm/liter)	NaNO₃	NaOH	NaAlO₂	Cs	Sr	pH	Gas, V	foam, Q	Foam cond., F	Foam density, ρ_f (gm/liter)	Average bubble diameter, $\bar{D}$ (cm)	Foam specific area, A (cm²/cm³)	$10^{-4}AQ/F$ (cm²/cm³)	ER	$10^5\Gamma/C_r$ (cm)
Aerosol AY	6.5	0	0.025		10^{-5}	0		154	176	0.197	1.12	0.06	100	8.94	1.54	0.61
					0	10^{-5}									1.10	21
Alipal CO-436	0.375	0	0.025		10^{-5}	0		154	186	0.950	5.10	0.05	120	2.35	1.94	4
					0	10^{-5}									2.60	54.6
Alipal LO-529	0.4	0	0.025		10^{-5}	0		154	174	0.415	2.40	0.06	100	4.2	1.35	0.84
Deriphat 170C	0.5	0	0.025		10^{-5}	0		154	60	4.92	74	0.025	240	0.29	1.02	0.60
					0	10^{-5}									7.10	104
Igepon CN-42	0.12	0	0.025		10^{-5}	0		154	72	1.6	24	0.038	158	0.71	1.03	0.38
Tergitol 7	2.0	0	0.025		10^{-5}	0		154	202	0.763	3.77	0.05	120	3.2	1.78	2.46
					0	10^{-5}									1.75	11.5
Ultrawet SK	0.08	0	0.025		10^{-5}	0		154	173	0.137	0.79	0.10	60	7.6	1.59	0.78
					0	10^{-5}									1.45	86.8
NaTφB+DBDTTA	1+0.02	1	0.025		10^{-6}		11									28.8
NaTφB+DBDTTA	1+0.04	1	0.02	0.024	10^{-6}	0		94							2.69	32
NaTφB+DBDTTA	1+0.04	1	0.02	0.024	10^{-6}	10^{-6}		91							3.06[1]	31[1]
															25.7[2]	374[2]

[a] Multiply by 0.214 to obtain superficial velocity in cm/min, by 0.43 to obtain ft/hr.

[b] Except for the last example, values for ER and $10^5\Gamma/C_r$ apply to Cs or Sr, as indicated by columns 6 and 7. In the last case, 1 refers to Cs, 2 to Sr.

1, 7, and 12. A few of the test results are presented in Table I to indicate the magnitudes of the various variables. The most effective separation of cesium was obtained by using sodium tetraphenylboron (NaTϕB) as the complexing agent and the sodium salt of dodecylbenzyldiethylenetriaminetetraacetic acid (DBDTTA) as the principal foaming agent. The latter agent is also one of the best surfactants for removing strontium and the rare earth element samarium.

It was recognized (Schonfeld et al., 1960) that a continuous system of feeding and foaming radioactive waste solutions would probably be very desirable for large-scale treatment. For this reason, studies of the many variables influencing the performance of continuous countercurrent foam columns accounted for a major portion of the early work at RAI.

Metal ions such as Cs^+ and Sr^{2+} can be removed by foam fractionation because they form stable complexes with anionic surfactants under certain conditions. For example, at a pH greater than 6, Sr^{2+} forms a complex with DBDTTA; however, when the pH is less than about 4, most of this surfactant is in the acid form. In general, the use of neutral or basic solutions of the anionic surfactants, which are superior to neutral surfactants, solves the problem of the competition between the acid form and the metal form.

The concentration of the surfactant is important to the extent suggested by the Gibbs equation, which gives the excess concentration of a dilute single solute i, at the surface of the foam as

$$\Gamma_i = -(1/RT) \, \partial\gamma/\partial \ln a_i. \tag{1}$$

Most anionic surfactants tested at RAI had critical micelle concentrations of about 10^{-3} M and showed an essentially linear variation of γ with log C_i at concentrations below 10^{-4} M ($a_i \approx C_i$). Metals were removed most efficiently from solution by foaming when the ratio of the contained equivalent of surfactant to the equivalents of metal was not grossly in excess of 1. This corresponds to a situation in which a significant portion of the surfactant is complexed with the metal of interest, rather than with H^+ or Na^+ or etc.

A continuous countercurrent column of the type that evolved from parametric studies (Schonfeld et al., 1960) is shown in Fig. 1. Although the feed solution contains a considerable portion of the surfactant, some surfactant is also fed into the liquid pool. A foam washing section (frequently not used for foam washing) is located above the feed point, and an expanded foam drainage section is located above the foam washing section. From the drainage section, the foam passes to a centrifugal foam breaker where the foam is condensed. When reflux is used, part of the foam condensate is returned to an appropriate place above the feed point.

Mass flow in a countercurrent foam column can be described in a manner analogous to a distillation column, except that an excess concentration of

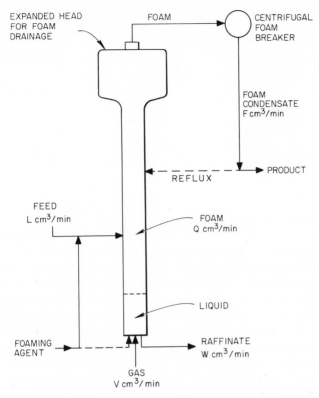

FIG. 1. Foam column used in most experiments.

solute(s) at the surface of the foam in foam separation replaces the enhanced vapor concentration in distillation. The surface excess is calculated as

$$\Gamma/C_r = \left(\frac{C_f - C_r}{C_r}\right)\Big/A = \lambda_1(\text{ER} - 1)\overline{D}\rho_f, \tag{2}$$

where

$$A = 6\rho_l/\overline{D}\rho_f \tag{3}$$

for a foam containing equal-sized bubbles; however, for any particular bubble size distribution in a column

$$A = (\lambda_2\overline{D}\rho_f)^{-1}. \tag{4}$$

The quantity $1/6\rho_l$ is replaced by λ_2 ($\simeq 0.15$).

The operation of the foam column can be described in terms of stages or transfer units, using a modified Kremser equation (Souders and Brown, 1932) or Colburn's (1937) concept of the transfer unit. These two methods are essentially equal from the standpoint of describing the foam separation because the product $(\beta\delta)$ assumes values in the range of 1–2. Here

$$\beta = (V_f A\rho_f/\rho_l + E/\delta_i)/W, \tag{5}$$

and $\delta_i = \Gamma_i/C_i$ is the slope of the operating line at stage i.

Single-column decontamination factors (DF's) in the range 500–1000 were obtained for Sr^{2+} under optimum conditions with DBDTTA as the surfactant. Other DF's, lower than these values, were used to determine the height of a transfer unit for Sr^{2+} separation to be about 1 cm of foam. Schonfeld et al. (1960) have described the effects of many parameters on solution decontamination, volume reduction, and the cascading of several foam columns for the purpose of obtaining DF's up to 10^6 and higher.

III. SCREENING OF SURFACE-ACTIVE AGENTS

The concentration region in which foam separation can be used for the removal of surface-active materials from salted or unsalted solutions is, generally, less than 10^{-3} M. This is a concentration region in which ion exchange also has many applications. Thus, foam fractionation is in competition with the more technologically advanced ion exchange as a method for removing low concentrations of radioactive nuclides from solution. In the case of removing strontium from highly radioactive, nitrate waste solutions, a solvent extraction process (Horner et al., 1963) also appeared to be very promising. Therefore, we felt that it would be desirable to use a solution other than a nitrate waste solution in the development and evaluation of foam separation. Process waste water, which, at many atomic energy installations, consists of slightly impure drinking water, was chosen for this purpose. By "slightly impure" water, we mean water which has a very low radioactivity level, and contains small amounts of laboratory chemicals, cleaning formulations (for example, those used to clean laboratory glassware), and organic materials. At ORNL, process waste water containing about 120 ppm of total hardness is discharged at the rate of about 500,000 gal/day.

Even in process waste water, ^{137}Cs and ^{90}Sr are the most significant problems from the standpoint of public health because of the need to discharge such water to the environment, particularly to rivers. Therefore, several commercial surfactants were evaluated using cesium and strontium as check variables. Each surfactant was tested (Weinstock et al., 1961, 1963; Rubin et al., 1962; Weinstock, 1965; ORNL, 1961, 1962) for solubility, foamability, cation enrichment at three different pH levels (1–2, 7, and 12), and critical micelle concentration. Of 107 surfactants tested, 73 provided a reasonably good foam at one or more of the pH values. Tests for cation enrichment showed that 20 were promising in removing cesium, 33 in removing strontium, and 10 in removing cerium. Generally, only the anionic surfactants were found to be effective, and each one of the three metal ions was found to be separated most effectively within a specific pH range. Judged on the basis of performance and economic considerations, dodecylbenzenesulfonate (DBS) was the best surfactant tested for decontaminating ORNL process waste water. For this reason, this surfactant was used in most of the succeeding experiments and the pilot plant test. Armosul 16 (α-sulfopalmitic acid) proved to be one of the best surfactants for removing cesium by foaming.

IV. DECONTAMINATION OF PROCESS WASTEWATER

Process wastewater is a dilute solution whose composition usually varies from plant to plant but remains fairly constant at a particular plant. Our studies of water decontamination were performed primarily with ORNL process wastewater having the composition shown in Table II (Davis and Schonfeld, 1962; Davis et al., 1963a, b, 1964, 1965a; ORNL, 1963, 1964; Schonfeld et al., 1962; Schonfeld and Davis, 1963, 1964). Of the radionuclides present in this water, only ^{90}Sr exceeds the maximum permissible concentration (MPC, U.S. Dep. of Commerce, 1959) for disposal to the environment. A DF of about 170 for this nuclide is needed to make discharge of the water to a river acceptable. Since the cesium concentration is about equal to the MPC, its removal is not mandatory; however, its removal might be necessary at other sites.

A ^{90}Sr DF in excess of 10 or so is difficult to achieve by a foam fractionation operation because of interference by calcium. Figure 2 shows that the calcium $(6.7 \times 10^{-4} \ M)$ in ORNL wastewater reduces the distribution coefficient (Γ/C_r) by a factor of 10 to 100.

Preliminary chemical treatment, necessitated by the presence of calcium which interferes with strontium removal during foam separation, consisted in (Schonfeld, 1964; Davis et al., 1965b; Schonfeld and Davis, 1965) adding sufficient sodium carbonate and sodium hydroxide to the water to give a concentration of 0.005 M each. Then a slowly stirred, suspended-bed sludge column was used to remove (by a filtering action) the resulting precipitates, calcium carbonate and magnesium hydroxide. Using this procedure and adding Fe^{3+} (a few parts per million) as a coagulant resulted in the reduction of the residual hardness of the water to 5 ppm (as $CaCO_3$) or less. At this concentration, calcium does not interfere with the removal of strontium by foam separation. In order to improve the removal of cesium, which is not extensively separated by foam fractionation, about 0.5 lb of baked grundite clay was added per 1000 gal of water.

TABLE II
COMPOSITION OF ORNL PROCESS WASTEWATER

Component	Concentration	Component	Concentration
HCO_3^-	100 ppm	^{137}Cs	370 cpm/ml
Ca^{2+}	27 ppm	^{89}Sr	21 cpm/ml
	$(6.7 \times 10^{-4}\ M)$	^{90}Sr	310 cpm/ml
Mg^{2+}	7 ppm	Rare earths	
Na^+	1.5 ppm	(primarily ^{90}Y)	80 cpm/ml
Cl^-	1.5 ppm	^{60}Co	40 cpm/ml
pH	7.7		

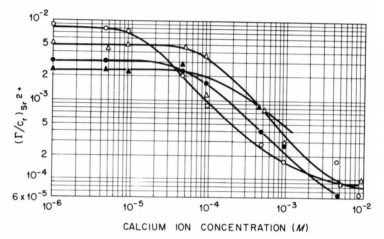

CALCIUM ION CONCENTRATION (M)

FIG. 2. Effect of calcium concentration on the distribution coefficient of strontium: (▲) Aerosol OS, 1.33 gm/liter; (△) Sipon LT-6, 1.24 gm/liter; (●) Tergitol 7, 2.0 gm/liter; (○) Veripon 4C, 0.85 gm/liter.

The overall flow sheet summarizing the two-step process for removing strontium from process waste water is shown in Fig. 3 (Davis *et al.*, 1965b). The sludge column and the foam column were operated at 12–13 gal/ft² hr and at 40–42 gal/ft² hr, respectively. Average bubble diameters were 0.06–0.08 cm; sodium dodecylbenzenesulfonate was added to the aqueous feed to the foam column and to the bottom pool at a volumetric ratio of between 5 and 5.5. Under these conditions, the Sr^{2+} DF across the foam column exceeded 200 when grundite clay was added as indicated in Fig. 3, but was only about 12 when grundite was not used. Foam removed about 75% of the remaining ruthenium and 80–85% of the ^{144}Ce.

Pilot plant studies (Blanco *et al.*, 1966; King *et al.*, 1968) were performed in equipment shown schematically in Fig. 4. Processing rates of 300 gal/hr and 120 gal/hr were achieved for scavenging precipitation and for foam separa-

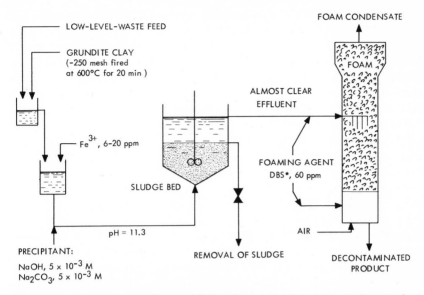

FIG. 3. Flow sheet for scavenging–precipitation foam separation process to decontaminate process waste water. DBS*, dodecylbenzenesulfonate.

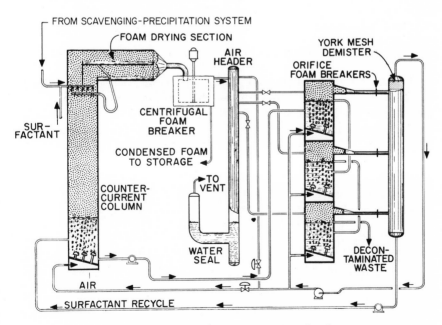

FIG. 4. Flow sheet for the foam separation pilot plant.

tion, respectively. The scavenging precipitation was performed essentially as indicated in the preceding section ; the foam column measured 2 by 2 by 8 ft. It was more difficult to generate a stable foam in this large column than in the smaller columns used in laboratory and engineering studies. Pall rings were installed in the foam section to prevent circulation of the foam, and a much higher concentration of surfactant was used in the liquid pool at the bottom of the column than was used previously in the smaller columns.

The use of the parameter $V/L\bar{D}$ as a basis for process control appeared to be valid for a large system even though the foam contained bubbles that varied widely in size. In a 93-hr test of the entire process at steady state, using ORNL low-level process waste as feed, the strontium DF was about 10^2 across the foam column and 10^3 overall. The concentration of strontium in the treated waste was 2.3 % of the MPC_w (168 hr). Table III summarizes typical results from the pilot plant equipment.

At the time that the pilot plant tests were completed, the scavenging precipitation–foam separation process did not appear to be economically competitive with other low-level waste treatment processes. A preliminary cost estimate (King et al., 1968) had indicated that : (1) a more efficient foam contactor must be developed to increase processing rates and decrease air flow requirements, and (2) increased volume reduction and more economical foam-breaking methods were needed. Increased volume reduction can be achieved, as described later in Section VII.

V. COMBINED FOAM SEPARATION–FROTH FLOTATION PROCESS

Phosphate–iron coagulation is known (Nesbitt et al., 1952 ; Frost, 1962a, b) to be an effective method for reducing the concentrations of calcium and magnesium dissolved in water. For this reason, an attempt was made to combine coagulation and foam fractionation into a one-step process for decontaminating process waste water. Experimentally, the process (Rubin et al., 1963) involved adding sodium phosphate, ferrous sulfate, sodium hy-droxide, a surfactant (such as sodium dodecylbenzene sulfonate), and a froth flotation agent (such as Armeen Z) to the wastewater and feeding the resulting mixture into a foam separation column. The column acts both as a froth flotation unit, removing the floc, and as a foam fractionation unit, removing dissolved metal ions.

This "one-step" process was optimized with respect to many variables. In particular, the Sr^{2+} DF's were in the range of 10^2–10^3 when the following conditions were used :

Na_3PO_4 :   60–70 mg/liter of ORNL wastewater
$FeSO_4$:   64 mg/liter—freshly prepared

TABLE III

Typical Decontamination Factors for Scavenging Precipitation–Foam Separation Process

Conditions:	Foam height	$V/L\bar{D}$ (cm^{-1})	Grundite clay (lb/1000 gal)	Time (hr)
Laboratory:	22 in	202	0.5	10
Pilot plant:	8 ft	140	0.7	93

	Scavenging precipitation		Foam column		Overall	
	Laboratory	Pilot Plant	Laboratory	Pilot Plant	Laboratory	Pilot Plant
Liquid flow (gal/ft^2 hr)	12	24	39	15		
			Decontamination factors			
Ca	50	60	~2	—	~100	—
Sr	16	10	220	36	>3500[a]	1080[b]
Cs	21	8	1	1	18	8.0
Co	2.3	—	1	—	2	1.3
Ru	1.8	—	4	—	>7[a]	4.2[a]
Ce	24	—	4	—	100	>20[a]
Zr–Nb	—	—	—	—	—	>50[a]

[a]Close to lower limit of analytical determination.
[b]Includes a decontamination factor of ~3.0 obtained in foam recovery columns.

WSR-35 coagulant: 5 mg/liter
UCAR C-149 coagulant: 3 mg/liter
DBS: 17.5 mg/liter fed to bottom of column
                             133 mg/liter fed to the mixing tank
Liquid residence time in mixing tank, 20 min

The one-step process was finally abandoned because of its complexity and the resulting buildup of radioactive solids on the foam column and foam breaker walls. Considerations of safety and operational feasibility would make such a buildup and the attendant possibly of plugging small-diameter process and instrument lines intolerable.

VI. SOME COMPARISONS BETWEEN STATIC AND DYNAMIC SURFACES

The separation of cationic fission products from an aqueous solution requires that the cation become part of a surface-active molecule. For example, when the sodium form of a surfactant (for example, an alkyl or an aryl sodium sulfonate) is added to a solution containing strontium, the strontium becomes part of the surfactant:

$$2RSO_3^- + Sr^{2+} \rightleftharpoons (RSO_3)_2Sr. \tag{6}$$

The model that best describes the excess concentration of a surfactant at a gas–liquid interface of a solution containing several surfactants is based on the Langmuir equation, which defines conditions for monolayer adsorption (Weinstock et al., 1961; ORNL, 1961):

$$\frac{\Gamma_i'}{\Gamma_{mi}} = \frac{\alpha_i a_i}{1 + \alpha_1 a_1 + \alpha_2 a_2 + \cdots + \alpha_n a_n} = \frac{\Gamma_i + C_i \tau}{\Gamma_{mi}} \simeq \frac{\Gamma_i}{\Gamma_{mi}}. \tag{7}$$

According to Eq. (7), Γ_i is the concentration of species i at the surface, due to its surface activity, in excess of the concentration $C_i \tau$ it would have if it were not surfaceactive. Since Γ_i is several orders of magnitude greater than $C_i \tau$ for surfactants of interest in foam separation, $\Gamma_i + C_i \tau \simeq \Gamma_i$.

Combining Eq. (7) with the Gibbs equation for a multicomponent system,

$$-d\gamma/RT = \Gamma_1 \, d \ln a_1 + \Gamma_2 \, d \ln a_2 + \cdots + \Gamma_n \, d \ln a_n, \tag{8}$$

and assuming that the values of Γ_{mi} are essentially constant, gives a Gibbs–Langmuir equation:

$$\gamma_0 - \gamma = \Gamma_m RT \ln(1 + \alpha_1 a_1 + \alpha_2 a_2 + \cdots + \alpha_n a_n). \tag{9}$$

Thus, if values of a_i for a solution are known, the degree to which surfactants can be separated from solution or from each other can be calculated from the surface tension measurements, γ and γ_0, of the solution and of pure water, respectively.

The Gibbs–Langmuir model was verified by comparing measured surface tensions of the solutions of several surfactants with values calculated from α_i's of the pure individual surfactants, and by comparing the separations calculated from surface tension and surface radioactivity with experimental values obtained in a total recycle foam column with the same chemical systems. Agreement of the values of Γ_m or Γ/C_r obtained from the different experimental methods was very good. Similarly, agreement was good between measured and calculated surface tensions, in the range of 35–70 dyn/cm, of solutions containing three or four of the following alcohols: n-pentanol, n-hexanol, n-heptanol, and n-octanol. Calculated surface tensions of the multicomponent solutions were based on values of α and Γ_m of the individual alcohols in water.

With aged solutions of RWA-100 (a butylphenyl phenol sodium sulfonate) in water or in 0.1 M NaNO$_3$, values of Γ_m calculated from surface tension measurements and values of Γ/C_r calculated from surface radioactivity measurements agreed (Weinstock et al., 1961) with values determined from foam column experiments. Surface tensions decreased as much as 10 % as the age of the surface prior to measurement increased from 0 to 24 hr; however, beyond this time period, they were essentially constant. These comparisons of Γ_m indicate that the foam in the column attained at least 90 % of the surface monolayer concentration, even though the bubbles of gas were in contact with the solution only a few seconds.

VII. DEVELOPMENT OF COUNTERCURRENT FOAM COLUMNS

Stable foams can flow upward, countercurrent to a liquid, to give counter-current differential separations with stripping, enrichment, or scrubbing, similar to those obtained in solvent extraction. Experimental engineering studies were made to optimize the design and operation of foam column systems. Six- and 24-in. i.d. foam columns were operated with DBS solutions since this surfactant was proposed for use in a pilot plant demonstration (Haas, 1962, 1963; Haas and Johnson, 1965; ORNL, 1963, 1964).

Two major objectives of this engineering study were to measure and to improve the efficiency of mass transfer between the countercurrent flows of foam surface and liquid. This efficiency is expressed in terms of the height of a transfer unit (HTU). The HTU values for stripping ^{89}Sr were determined (Haas, 1965) for more than 100 combinations of conditions. The smallest HTU's were found to be about 1 cm. These low values were obtained when there was plug flow of foam up the column, as indicated by visual observations and velocity measurements through the glass column wall, and uniform flow of liquid down through the foam. These results are in accordance with true differential flow and diffusion-controlled mass transfer in the liquid. Conditions that are more practical for a large-scale system give experimental HTU values of 5–20 cm. The bubble size and, in turn, the surface flow rate, or slope of the operating line, are difficult to measure and control. Therefore, a large number of transfer units is rarely of practical significance because the operating line and the equilibrium line will be selected to pinch at one end to favor the preferred separation.

Uniform distribution of the liquid feed across the column cross section and the formation of uniformly sized foam bubbles are important in the achievement of efficient operation and low HTU values. The best liquid feed distributors were found to be multiple capillaries sized to divide the flow and introduce the liquid at moderate velocities. Use of a single capillary in a 6-in. i.d. column, uneven flow through multiple capillaries, or a high liquid-jet velocity gives much larger HTU values. The formation of uniformly sized gas bubbles tended to stabilize the foam and to prevent channeling. Spinnerettes with 50–80-μ diameter holes, as used in the rayon industry, were distinctly superior to glass frits or sintered metal filters. However, the glass frits or sintered metal filters having average pore diameters of 65–100 μ and the best uniformity available are adequate; in fact, they are more practical for large-scale applications.

The experimentally observed effects of flow rates were a complex combination of effects resulting from channeling, liquid feed distribution differences, and bubble diameter changes. Channeling is excessive for flow rates that give volume fractions of liquid in the foam, ε, greater than 0.26 (see Section VIII).

For the decontamination of radioactive wastes (and for nearly all foam separation processes), a low volume of foam condensate (foamate) or a high volumetric ratio of feed to condensate is highly desirable. The foam leaving the countercurrent section of the column is passed through a drainage section without countercurrent flow to promote drainage of liquid from the foam. The effectiveness of these drainage sections can be predicted from the model and the experimental results described in the next section.

The volume reduction can also be increased by reflux of part of the condensed foam to provide an enrichment section. The maximum enrichment, and thus the maximum volume reduction, is limited by the maximum surface excess of the surface-active material [about 1.5×10^{-10} mole/cm^2 for Sr(DBS)$_2$, Ca(DBS)$_2$, or Mg(DBS)$_2$]. In the ORNL LLW pilot plant, the maximum volume reductions predicted for aqueous feed about 1×10^{-4} M in (Sr + Ca + Mg) would be 30–75 for foam densities of 0.005–0.002 gm/cm^3. This agrees with the experimentally observed values obtained during operation of the LLW pilot plant. For aqueous feed 10^{-6} M in (Sr + Ca + Mg), the maximum volume reductions predicted would be 3000 and 7500, respectively, for the two density limits. A value of 3700 was demonstrated experimentally using ^{85}Sr with 10^{-6} M Sr as carrier (Schonfeld and Kibbey, 1967). These experimental studies demonstrate more efficient foam separation as the concentrations decrease and demonstrate enrichment ratios, ER, for ^{85}Sr as high as 3×10^7.

After the foam leaves the drainage section, it must be condensed and separated into a liquid solution, or slurry, and a gas. Centrifuge and critical-orifice type foam breakers were used on both 6- and 24-in. foam columns; air-operated sonic whistles were used on the 6-in. columns, and cyclones were tested (Haas, 1965).

VIII. FOAM DRAINAGE MODEL AND EXPERIMENTAL RESULTS

A foam separation column should discharge a dry foam of low density in order to obtain a high volume reduction (in other words, to obtain the separated surface-active materials in a minimum volume of condensed foam). The foam leaving the countercurrent part of a foam column has a relatively high liquid content; therefore, a drainage section is necessary to promote drainage of liquid from the foam. As one part of the foam separation development program, foam drainage was measured using DBS solutions. The results were correlated by equations derived from a model of foam structure (Haas, 1965; Haas and Johnson, 1967). Drainage that results from the loss of surface area in an unstable foam is a different phenomenon and is not correlated by these equations.

The model for foam drainage postulates that most of the flow takes place in a network of capillaries (Plateau borders). The structure of dry foam may be visualized as polyhedral-shaped gas bubbles having rounded edges. Each Plateau border is enclosed by three bubbles. The flow of liquid in a Plateau border is laminar and is described by the Poiseuille equation. An equation for foam drainage in the network of Plateau borders was derived, using simplifying assumptions, and was applied with the boundary conditions for four foam drainage situations (Haas, 1965; Haas and Johnson, 1967).

For the region of steady state, countercurrent liquid foam flow in a foam separation column, all properties are constant and independent of both time and position. The equation derived to relate L, v, and ε is

$$L + [v/(1 - \varepsilon)]\varepsilon = (\rho_l g/32\mu)(4/\pi nk^2)\varepsilon^2. \tag{10}$$

The results for the NaDBS solution were correlated by

$$\varepsilon^2 \bar{D}^2 \left(L + \frac{v}{1 - \varepsilon}\varepsilon\right)^{-1} = 14 \times 10^{-4}. \tag{11}$$

Vertical drainage sections that may have an increased cross section to reduce the velocity of the foam (Fig. 5a) are simple and effective for foam columns of small diameter. In the case of a stable foam, drainage approaches equilibrium after several inches of vertical drainage since the downward liquid

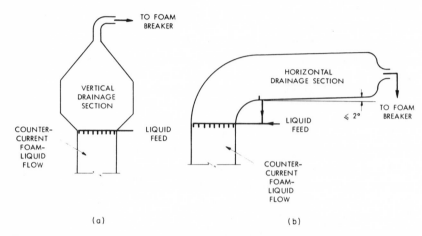

FIG. 5. Configurations of two foam column drainage sections: (a) vertical drainage; (b) horizontal drainage.

velocity, with respect to the foam, is opposed to the upward flow of foam. The equation derived for the equilibrium foam density is

$$\varepsilon_p = (k - 1)(32\mu/\rho_l g)(n\pi/4)[v/(1 - \varepsilon)]. \tag{12}$$

The results for the DBS solution were correlated by

$$\varepsilon_p = 3.2 \times 10^{-4} v/\bar{D}^2. \tag{13}$$

A horizontal drainage section (Fig. 5b), as proposed and demonstrated at ORNL (Haas, 1965; King et al., 1968), is more effective than a vertical drainage section for large-diameter columns. Drainage in a horizontal section is not limited by a velocity equilibrium between the liquid and the foam. Horizontal baffles may be used to decrease the effective height H through which drainage occurs, and channeling of foam through the drainage section is less likely. The equation derived for the volume fraction of liquid in the exit foam ε_p is

$$(\varepsilon_p)^{-1} - (\bar{\varepsilon}_0)^{-1} = (4/H)(\rho_l g/32\mu)(4/\pi n k^2)(t - t_0). \tag{14}$$

For the DBS solution and $\varepsilon_p \gg \bar{\varepsilon}_0$, the experimental results are correlated by

$$H/[\varepsilon_p \bar{D}^2(t - t_0)] = 2800. \tag{15}$$

In the drainage of solution between the foam bubbles, the liquid flows principally through the Plateau borders as shown by a comparison of theoretical equations with experimental data for four foam drainage situations. These include the time–steady state drainage relationships given above for continuous foam column systems and the non-steady-state drainage of a stationary foam (Haas, 1965 ; Haas and Johnson, 1967), which does not occur in a continuous foam column. The relationships among the experimental variables (namely, superficial liquid velocity, superficial gas velocity, foam bubble diameter, fractional volume of liquid in the foam, time, and configuration) are in agreement with the theoretical equations. The experimental coefficients differ by a factor of 2 or less than those calculated from the properties of the solution. This is considered to be good agreement in light of the approximations required for the derivations and the accuracy of the experimental measurements. Rubin *et al.* (1967) also found agreement between their independent measurements of vertical foam drainage and Eq. (12). By using these equations, the maximum flow rates and drainage conditions that are practical for countercurrent foam separation columns may be estimated.

SYMBOLS

A specific surface area of foam, cm^2/cm^3 of foam condensate

a_i thermodynamic activity of component i, moles/liter

C_0 concentration of species of interest in the feed solution, moles/liter

C_f concentration of species of interest in the foam condensate, moles/liter

C_r concentration of species of interest in the column aqueous raffinate, moles/liter (the r is deleted when referring to one of several species ; in this case, we write C_i)

$\bar{D}$ average bubble diameter, cm

DF   decontamination factor $(= C_0/C_r)$

E flow rate of solution mechanically entrained (as contrasted with being part of the surface layer) with the foam, cm^3/min

ER   enrichment ratio $(= C_f/C_r)$

F foam condensate flow rate, cm^3/min

g gravitational constant, cm/sec^2

H height of column of foam, cm

k constant defined by the foam drainage model (Haas, 1965) and (Haas and Johnson, 1967); $k \simeq 1.5$

L liquid flow rate, cm^3/min

l superficial liquid velocity, cm/sec

n a value such that $n(1 - \varepsilon)$ is the number of Plateau borders/cm^2 of foam area; $n \simeq 2.5/\bar{D}^2$

Q foam flow rate $(= V + F + E)$, cm^3/min

R gas constant, 8.314×10^7 ergs/(gm mole) °K

RF   recovery factor $(= C_f/C_0)$

T absolute temperature, °K

$t, (t_0)$ time (initial value), sec

V gas flow rate, cm^3/min

v superficial gas velocity, cm/sec; the foam bubble velocity is $v/(1 - \varepsilon)$

W raffinate flow rate, cm^3/min

α_i a constant in the Gibbs–Langmuir equation for surfactant i, liters/mole

β foam area/raffinate volume ratio, cm^2/cm^3 [see Eq. (5)]

Γ_i' concentration of species i in the surface phase, $moles/cm^2$

Γ_i excess surface concentration of component i, $moles/cm^2$ (the subscript i is omitted when omission would not cause confusion)

Γ_{mi} (or Γ_m if only one surfactant is present)   the concentration of surfactant i in a (saturated) monolayer, $moles/cm^2$

γ_0 surface tension of pure water, $ergs/cm^2$

γ surface tension, $ergs/cm^2$

δ_i Γ_i/C_i (= distribution coefficient at stage i), cm

ε volume fraction of liquid in the foam

ε_p average volume fraction of liquid in exit foam

λ_1, λ_2 constants

μ viscosity, poise

ρ_f density of foam, gm/cm^3

ρ_l density of solution, gm/cm^3

τ thickness of surface phase, cm

REFERENCES

Blanco, R. E., Davis, W., Jr., Godbee, H. W., King, L. J., Roberts, J. T., Yee, W. C., Alkire, G. J., Irish, E. R., and Mercer, B. W. (1966). Oak Ridge Nat. Lab. Rep. ORNL-TM-1289; *Proc. Symp. Practices Treatment of Low- and Intermediate-Level Radioactive Wastes* Int. At. Energy Agency, Vienna, December 6–10, 1965.

Colburn, A. P. (1937). *Trans. Amer. Inst. Chem. Eng.* **33**, 197.

Davis, W., Jr., and Schonfeld, E. (1962). Oak Ridge Nat. Lab. Rep. ORNL-TM-482.

Davis, W., Jr., Schonfeld, E., and Kibbey, A. H. (1963a). Oak Ridge Nat. Lab. Rep. ORNL-TM-516.

Davis, W., Jr., Rubin, E., Schonfeld, E., and Kibbey, A. H. (1963b). Oak Ridge Nat. Lab. Rep. ORNL-TM-603.

Davis, W., Jr., Schonfeld, E., Wace, P. F., and Kibbey, A. H. (1964). Oak Ridge Nat. Lab. Rep. ORNL-TM-830.

Davis, W., Jr., Kibbey, A. H., Schonfeld, E., and Wace, P. F. (1965a). Oak Ridge Nat. Lab. Rep. ORNL-TM-1081.

Davis, W., Jr., Kibbey, A. H., and Schonfeld, E. (1965b). Oak Ridge Nat. Lab. Rep. ORNL-3811.

Frost, C. R. (1962a). Aust. At. Energy Comm. Rep. AAEC-E-74.

Frost, C. R. (1962b). Aust. At. Energy Comm. Rep. AAEC-E-75.

Gaden, E. L., Jr., and Kevorkian, V. (1957). *A. I. Ch. E. J.* **3**, 180.

Gaden, E. L., Jr., and Schnepf, R. W. (1962a). U.S. Pat. 3,054,747.

Gaden, E. L., Jr., and Schnepf, R. W. (1962b). U.S. Pat. 3,054,746.

Haas, P. A. (1962). Oak Ridge Nat. Lab. Rep. ORNL-TM-482.

Haas, P. A. (1963). Oak Ridge Nat. Lab. Rep. ORNL-TM-516.

Haas, P. A. (1965). Ph.D. dissertation, Univ. of Tennessee; Oak Ridge Nat. Lab. Rep. ORNL-3527.

Haas, P. A., and Johnson, H. F. (1965). *A. I. Ch. E. J.* **11**, 319.

Haas, P. A., and Johnson, H. F. (1967). *Ind. Eng. Chem. Fundam.* **6**, 225.

Horner, D. E., Crouse, D. J., Brown, K. B., and Weaver, B. (1963). *Nucl. Sci. Eng.* **17**, 234.

Irish, E. R. (1957). In *Symp. Reprocessing Fuels Brussels, Belgium,* USAEC, TID-7534, pp. 83–106, May 20–25, 1957.

King, L. J., Shimozato, A., and Holmes, J. M. (1968). Oak Ridge Nat. Lab. Rep. ORNL-3803.

Nesbitt, J. B., Kaufman, W. J., McCauley, R. F., and Eliassen, R. (1952). U.S. At. Energy Comm. Rep. NYO-4435.

Oak Ridge Nat. Lab. (1961). Chem. Technol. Div., Rep. ORNL-3153.

Oak Ridge Nat. Lab. (1962). Chem. Technol. Div., Rep. ORNL-3314.

Oak Ridge Nat. Lab. (1963). Chem. Technol. Div., Rep. ORNL-3452.

Oak Ridge Nat. Lab. (1964). Chem. Technol. Div., Rep. ORNL-3627.

Rubin, E., Schonfeld, E., and Everett, R., Jr. (1962). Radiation Applications, Inc., Rep. RAI-104.

Rubin, E., Schonfeld, E., and Weinstock, J. J. (1963). Radiation Applications, Inc., Rep. RAI-116.

Rubin, E., LaMantia, C. R., and Gaden, E. L. Jr. (1967). *Chem. Eng. Sci.* **22,** 1117.

Schnepf, R. W., and Gaden, E. L., Jr. (1959). *J. Biochem. Microbiol. Technol.* **1,** 1.

Schnepf, R. W., Gaden, E. L., Jr., Mirocznik, E. Y., and Schonfeld, E. (1959). *Chem. Eng. Progr.* **55,** 42.

Schonfeld, E. (1964). *J. Amer. Water Works Assoc.* **56,** 767.

Schonfeld, E., and Davis, W., Jr. (1963). Oak Ridge Nat. Lab. Rep. ORNL-TM-260.

Schonfeld, E., and Davis, W., Jr. (1964). Oak Ridge Nat. Lab. Rep. ORNL-TM-835.

Schonfeld, E., and Davis, W., Jr. (1965). *Health Phys.* **12,** 407.

Schonfeld, E., and Kibbey, A. H. (1967). *Nucl. Appl.* **3,** 353.

Schonfeld, E., Sanford, R., Mazzella, G., Ghosh, D., and Mook, S. (1960). U.S. At. Energy Comm. Rep. NYO-9577.

Schonfeld, E., Davis, W., Jr., and Haas, P. A. (1962). Oak Ridge Nat. Lab. Rep. ORNL-TM-376.

Schonfeld, E., Kibbey, A. H., and Davis, W., Jr. (1963). Oak Ridge Nat. Lab. Rep. ORNL-TM-757.

Souders, M., and Brown, G. G. (1932). *Ind. Eng. Chem.* **24,** 519.

U.S. Dept. of Commerce (1959). "National Bureau of Standards Handbook" Vol. 69. Nat. Bur. Std., Washington D.C.

Weinstock, J. J. (1965). Radiation Applications, Inc., Rep. RAI-350.

Weinstock, J. J., Rubin, E., Schonfeld, E., and Mook, S. (1961). Radiation Applications, Inc., Rep. RAI-100.

Weinstock, J. J., Mook, S., Schonfeld, E., Rubin, E., and Sanford, R. (1963). Radiation Applications, Inc., Rep. NYO-10038.

CHAPTER **20**

SEPARATION OF SURFACTANTS AND IONS FROM SOLUTIONS BY FOAMING: STUDIES IN THE U.S.S.R.

V. V. Pushkarev

The Urals Kirov Polytechnical Institute
Sverdlovsk, U.S.S.R.

I. INTRODUCTION

The production of synthetic surface-active agents (surfactants) and detergents has developed very rapidly in recent years. Nowadays one can hardly find a branch of industry that does not use surfactants. Such an extensive application of surfactants led to problems with the separation of surfactants and their associated components from aqueous solutions.

In principle, it is possible to separate surfactants from solutions. This is based on the ability of surfactants to lower the surface tension of water, adsorb at the gas–liquid interface, and produce stable foam. These particular properties of surfactants were used as the basis for the procedure of separation of dyes and protein substances from solutions.

The interaction between certain surfactants and mineral particles and the transition of the aggregates being formed at the interface is the basic principle of ore flotation. Colloid flotation may become an effective means of removing colloidally dissolved substances and solution components sorbed by sol particles from solutions. Recently, ion flotation, which is based on the interaction between the ionic component of the solution and surfactant molecules or ions, has been greatly developed.

Some results of investigations of the foam separation of surfactants and ions are presented here.

II. SURFACTANT SEPARATION

A. Kinetics of Surfactant Separation from Aqueous Solutions by Blowing with Dispersed Air

Surfactant separation by foaming is based on the phenomenon of surfactant adsorption by the solution–gas interface and continuous removal of the surface layer. At least two conditions should be satisfied here: (a) surfactants being separated should possess adequate foaming properties; (b) the foam should be stable (Rebinder, 1966). Foam separation of surfactant may be carried out by blowing small air bubbles through the solution. This process is accompanied by the formation of foam, which is separated from the solution when necessary.

In the process of surfactant separation by foaming quite a number of factors are involved, such as: (a) the change of the solution surface tension, (b) the reduction of volume of the solution, (c) the presence of diffusion effects. Therefore, the kinetics of the process is exceedingly complicated.

All this precludes the application of the Gibbs equation for the calculation of the amount of surfactant on the surface of ascending bubbles. However, in solving some particular problems, when there are grounds to suppose that only one of the above-mentioned factors prevails, the Gibbs equation may be justly applied. For example, with a long column of surfactant solution, the time of air bubble ascension is quite sufficient for establishing equilibrium distributions of the surfactant between the phase interface and the solution (Shinoda et al., 1966). In this case diffusion processes may be neglected (Ward and Torda, 1944, 1946).

It can be assumed that the change in the amount of surfactant $d\Gamma$ in the surface layer at a very small period of time dt will be proportional to the total amount of surfactant Γ in the surface layer at a given time t, i.e.,

$$-d\Gamma = \alpha\Gamma \, dt, \tag{1}$$

where α is the proportionality factor.

From the Gibbs equation

$$\Gamma = -(C_p/RT)(-\partial\sigma/\partial C_p), \tag{2}$$

it follows that

$$\Gamma = [(1/RT)(-\partial\sigma/\partial C_p) + (C_p/RT)(-\partial^2\sigma/\partial C_p)]\,dC_p, \tag{3}$$

where C_p is the equilibrium concentration of the surfactant, σ is the surface tension, R is the gas constant, and T is the temperature in degrees Kelvin.

Substituting Eq. (3) into Eq. (1) and integrating, one obtains

$$\ln\frac{C_p}{C_0} = \alpha t - \ln\left[\frac{(\partial\sigma/\partial C_p)}{(\partial\sigma/\partial C_0)}\right], \tag{4}$$

where C_0 is the initial surfactant concentration.

Expression (4) is the equation of the kinetics of surfactant separation from the solution by foaming, which also takes into account the change of the surface tension of the solution with changing surfactant concentration. This equation shows that the formal order of the surfactant separation process is governed by the manner in which the surface tension of the solution depends on the concentration of dissolved surfactant. If the dependence is linear, Eq. (4) becomes a kinetic equation of the first order:

$$\ln C_p/C_0 = \alpha t. \tag{5}$$

In some cases an equation similar to Eq. (5) described the process of colloid foam chromatography with good approximation (Skrilev and Mokrushin, 1964).

Assuming as a basis the conditions which made it possible to obtain Eq. (5), one can derive the kinetic equation for the process of foam separation of surfactant, taking into account the volume change of the system. For systems in which some processes are connected with the change of volume, the kinetic equation can be written in the form (Veylas, 1964)

$$-(1/V_t)(dm/dt) = km_t^n, \tag{6}$$

where V_t is the volume of the system at the time t, m_t is the amount of surfactant at the same time, k is the process constant, and n is the order of the process. One can use Eq. (6) only when the dependence of volume on time and the change in the amount of surfactant are known.

Suppose that at any given instant the volume of the solution is proportional to the amount of surfactant

$$V_t = V_0 - A(m_0 - m_t), \tag{7}$$

where V_0 is the initial volume, A is the amount of water bound with the surfactant in the foam product, and m_0 is the initial amount of the surfactant in the solution.

Substituting Eq. (7) into Eq. (6) one obtains

$$-(V + Am_t)^{-1}\,dm_t/dt = km_t^n, \tag{8}$$

where $V = V_0 - Am_0$ is the volume of the solution remaining upon complete surfactant separation, i.e., at $m_t = 0$.

The previous assumption allows one to consider a simpler case when $n = 1$. As has already been shown, it will be justified if diffusion processes are neglected and the surface tension is allowed to be linearly dependent on surfactant concentration. In real systems the latter may be valid for solutions with a very low surfactant concentration or for processes accompanied by limited changes in surfactant concentration.

As a result, one has

$$-(V + Am_t)^{-1} \, dm_t/dt = km_t. \tag{9}$$

Upon integration of Eq. (9) and finding the integration constant, one obtains

$$\ln m_t/m_0 - \ln (V + Am_0)/(V + Am_t) = -kVt. \tag{10}$$

At the very beginning of the separation process $m_t \simeq m_0$. In this case the second term in Eq. (10) is approximately unity and the equation becomes

$$\ln m_t/m_0 = -kVt. \tag{11}$$

The equation obtained (11) differs from the kinetic equation for the first order process by only the constant factor V. On a plot of $\ln m_t/m_0$ vs. t the initial shape of the kinetic curve is that of a straight line passing through the coordinate origin, with a slope of kV.

In the final stage of the process m_t is very small and Eq. (10) becomes

$$\ln m_t/m_0 = -kVt - \ln V/(V + Am_0). \tag{12}$$

This limiting equation in the coordinates $\ln m_t/m_0$ vs. t also has the form of a straight line but it is displaced with respect to the origin along the ordinate axis by a segment equal to $\ln V/(V + Am_0)$.

Thus, the real shape of the kinetic curve for the foam separation of surfactant is between two asymptotes.

In principle, the curves obtained experimentally make it possible to determine the process constant k, the parameter A characterizing the amount of water carried with the surfactant, and the parameter V.

One of the author's papers (Khrustalev et al., 1968) gives instances of the application of Eq. (10) for the description of different surfactant separation processes (nonionogenic and anionactive).

B. Parameters Describing Processes in Real Systems

Unfortunately, in the final stage of surfactant separation in real systems a small amount of surfactant still remains (the foam formed becomes unstable). In this instance the curve in most cases fails to reach the asymptote. Hence, a graphical solution of Eq. (10) entails serious errors in calculations.

On the other hand, even if the experimental data allow the determination of the parameters k, A, and V, the solution of Eq. (10) will be too cumbersome and hardly acceptable for finding optimal conditions in solving applied problems.

For industrial purposes it is more convenient to find optimal means of conducting the process with the help of parameters which are easily established experimentally. One such parameter is the percentage of surfactant separation,

$$S = [(m_0 - m_t)/m_t] \times 100. \tag{13}$$

The distribution of surfactant between the foam product and the solution is of essential significance in the technique of surfactant separation. The process should be conducted in such a way that a product of the least possible volume and the highest possible surfactant concentration would be obtained finally. The estimation of separation efficiency in this case can be made by means of the distribution coefficient

$$E = C_n/C_p, \tag{14}$$

where C_n is surfactant concentration in the foam product. The distribution coefficient greatly depends on the volume of the foam product and the rate of surfactant separation from the solution. The condition $E > 1$ is practically always fulfilled. The condition $E = 1$ is unlikely, being equivalent to the escape of the whole body of solution with the foam product.

In the course of a separation process, the surfactant concentrations in the foam product and in the solution, as well as the volumes, continually change. E appears to change too. Therefore, in the case under discussion, E is a kind of averaged distribution coefficient describing the resulting surfactant concentration in the foam at the end of the separation process.

Since the time for the air bubble to come to the surface in the experiments described below was several times that for establishing adsorption equilibrium, E can be regarded as the equilibrium distribution coefficient.

For the description of the separation process one more parameter is useful:

$$F = E/\tau, \tag{15}$$

where τ is the time necessary for the most complete surfactant separation. In this equation, F represents the rate of surfactant enrichment of the foam product per unit time. This is essentially the intensity factor of the process. Here one also obtains a time-averaged value for F.

The dimension of F $(\min^{-1})$ coincides with that of the flotation rate constant (Glembotsky et al., 1961). In the present case, F also characterizes the process rate. Indeed, with $E = $ const, the less the time spent on surfactant separation, the higher is the value of F. If, however, under changed conditions E also changes, this relation becomes more complicated.

In short, such integral concepts as S, E, and F characterize the process completely enough. They are valuable criteria for finding optimum conditions of conducting the process in practice.

C. Dependence of the Parameters S, E, and F on the Rate of Blowing with Dispersed Air

An increased rate of dispersed air feed W results in a higher frequency f of bubble formation. When the rate of air feed provides laminar bubble ascension, the following equation (Mahoney and Wensel, 1963)

$$f = W/(a + bW) \tag{16}$$

proves to be valid. Here a is a constant identical to the volume of a gas bubble at small values of W; b is a constant equal to the time necessary for the formation of one gas bubble.

As f increases, the total liquid–gas interface increases correspondingly and so does the amount of surfactant sorbed by it. Over a small range of change in f, S may be assumed in this case to be directly proportional to f:

$$S = lf, \tag{17}$$

where l is the proportionality coefficient. Then

$$S = lW/(a + bW). \tag{18}$$

From the above equation a plot of $1/S$ vs. $1/W$ is a straight line with an intercept of b/l on the ordinate axis, and a slope of a/l. As may be seen in Fig. 1, this relationship holds well over the range of air feed rate investigated for the surfactants studied.

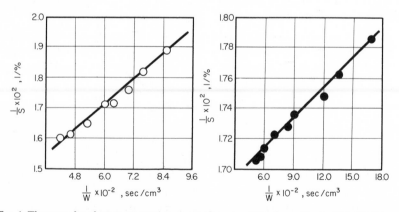

FIG. 1. The rate of surfactant separation versus the rate of blowing with air : (○) OP-7 (nonionogenic surfactant based on isooctylphenol with seven added molecules of ethylene oxide), pH = 2.3, V_0 = 0.5 liter, C_0 = 80 mg/liter ; (●) potassium palmitate, pH = 10.7, C_0 = 100 mg/liter.

As W increases, the volume of foam product sharply increases and surfactant concentration in the foam falls. As a result, with the increase of W the values of E and F decrease.

The OP-7 molecule has on the average seven polar hydrated groups, binding more water than the soap molecule having only one hydrated polar group. For this reason, with the same W for molecular solutions, the volume of foam product in the separation of OP-7 proves to be greater than in the separation of potassium palmitate. Both in the course of the potassium palmitate separation and the OP-7 separation the same kind of dependence is observed in the changes of E and F.

The minimum value of air feed rate for the beginning of potassium palmitate separation is lower than for the case of OP-7 separation. This can be accounted for by the different air penetrability of OP-7 and potassium palmitate films. Smaller volumes of foam product and greater E values for potassium palmitate result in a potassium palmitate concentration in the foam product twice as great as the corresponding concentration for OP-7. The surface layer develops a closer packing of potassium palmitate molecules and the foam film is less penetrable to air. A closer packing of potassium palmitate molecules in the surface layer is caused by physiochemical peculiarities of the molecules of this substance. Potassium palmitate has a long hydrophobic radical and a very small hydrophilic group. In OP-7 molecules the weight ratio of a comparatively small hydrophobic group (isooctylphenol) to a long hydrophilic group (polyoxyethylene chain) is rather low, which causes a disordered distribution of molecules of this substance in the surface layer.

D. Effect of Surfactant Concentration

The effect of the initial surfactant concentration on the rate of its separation from the solution is different for different surfactants. The separation rate of OP-7 increases from 50 to 70% as its concentration increases from 30 to 200 mg/liter, while the rate of potassium palmitate separation is maximum ($\sim 65\%$) in the concentration range below the critical micelle concentration ($\sim 180–200$ mg/liter). A greater volume of the foam product causes lower values of the distribution coefficient and the intensity factor. Thus, surfactant separation is most efficient at low surfactant concentrations in the solution.

E. Effect of Solution pH

The effect of the solution pH is governed entirely by acid–base properties of the surfactants separated. Potassium palmitate, being the salt of a weak acid and a strong base, is hydrolyzed in aqueous solution. The palmic acid formed by hydrolysis is not soluble in water and does not produce a stable foam.

Hence, potassium palmitate separation begins only at pH > 9.5 and reaches its maximum ($\sim 63\%$) at pH 12. With a further increase in pH, the separation rate decreases, added alkali acting as an electrolyte.

Greater electrolyte concentrations have an adverse effect on the process of foam separation of detergent. As hydrolysis equilibrium shifts towards salt formation, increasingly greater quantities of hydrated potassium counterions pass over into the foam product. This results in an increase of foam product volume. The values of E and F for potassium palmitate are maximum in the pH range 11–12. There is no definite opinion concerning acidic–basic properties of OP-7 and other oxyethylized compounds. On the strength of the total results obtained in the present study, such substances should be regarded as partial ampholytes.

In acid medium, oxonium compounds are formed and OP-7 molecules become positively charged. In alkali medium, OP-7 molecules are partially hydrolyzed to form negatively charged hydrolysis products. In accordance with this, the separation rate of OP-7 is minimum ($\sim 50\%$) in the neutral zone and increases up to 65–70% with a decrease in solution pH, and up to 80–85% with an increase in pH. The values of E and F change accordingly. The neutral zone accounts for the maximum volume of foam product.

F. Effect of Electrolytes

The electrolyte effect was investigated in the process of separation of the nonionogenic substance OP-7. Low electrolyte concentrations, as a rule, have a favorable effect on the separation process. However, the volume of foam product increases sharply as electrolyte concentration increases up to 10^{-3} M and over. On the assumption of salting-out action by the electrolyte, an increase of its concentration in the solution is equivalent to an increase in the amount of combined water. Hence, a more efficient concentration of surfactant results in an increased volume of foam product. The latter is corroborated by the comparative effect of different monovalent cations on the separation process. It turned out that according to their effect on the process of OP-7 separation these cations can be arranged in the following order: $Na^+ >$ $NH_4^+ > K^+$. The hydrating power of the cations also diminishes in this order. Under the same conditions the nature of the anion affects the process of OP-7 separation considerably less than the nature of the cations. Nevertheless, the anions investigated also can be arranged in order according to their decreasing effect on the surfactant separation process: $P_2O_7^{4-} > SO_4^{2-} >$ $Cl > NO_3^-$. It should be noted that calcium and aluminum cations have a specific effect. Small additions of these cations adversely affect the separation process. This is accounted for by the interaction between the above cations and OP-7 molecules.

G. Effect of Nonelectrolytes

The effect of ethanol and butyl alcohol as well as glycerine on the process of OP-7 separation was studied. With regard to their effect on the separation process they too can be arranged in series: butyl alcohol > glycerine > ethanol.

Small additions of organic diluents sharply increase the efficiency of the process, yet their great concentration in the solution causes a decrease in the OP-7 separation rate. This is due to the fact that the above-mentioned substances behave like surfactants and, their concentration in the solution being high, they oust OP-7 molecules from the surface of the ascending bubbles, which causes a drop in the OP-7 separation rate.

III. ION SEPARATION

The fact that ions and various surfactants can interact and together pass over to the phase interface was made use of by some investigators for the purpose of extraction, separation, and concentration of ions from solutions. The process of foam separation of these components is referred to as *ion flotation* (Sebba, 1962).

A specific feature of ion flotation is that it can be used to concentrate different components from dilute solutions. However, it cannot be used for the separation of ions in large quantities because it demands considerable amounts of surfactant. Due to intensive foaming, the diluent is carried away with the foam product, and the concentration coefficient decreases sharply.

The term ion flotation does not describe the physiochemical nature of all the processes involved. That transition of ions to the foam product occurs not only is due to chemical interaction between ions and surfactants, but also is due to sorption bonds. For example, uncharged particles can also pass into the foam product. Apart from ion interaction, transition can be caused by the phenomenon of complexing between solution components and surfactants. In some instances there is a purely Coulombic interaction.

The scope of the present chapter precludes a detailed analysis of all the physiochemical peculiarities of ion flotation. Information about this process is just beginning to appear in the scientific literature. Some problems of theory for ion flotation with three kinds of surfactants (cationic, anionic, and ampholitic) are discussed in the following sections.

A. Nature of the Ion-Surfactant Bond

When the surfactant is an acid (such as an alkyl–aryl hydrogen sulfate) the phase interface is quite naturally thought of as an ion exchanger of the hydrogen type. In this case thermodynamic equilibrium between the foam product and solution cations will be expressed by the equation (Nickolsky and Paramonova, 1939)

$$a_{M_n}^{1/z}/a_{M_p}^{1/z} = K(a_{H_n}/a_{H_p}). \qquad (19)$$

According to Kurbatov (1951) and Yegorov et al. (1961), one can obtain the following dependence of E on the value of pH for cations:

$$\log(C_{M_n}/C_{M_p}) = \log E = B + z(\text{pH}). \qquad (20)$$

In the above equation a_{M_n} and a_{M_p} represent the activity of ions in the foam product and solution, respectively, a_{H_n} and a_{H_p} represent the same for hydrogen ions, K is the equilibrium constant, C_{M_n} and C_{M_p} are the ion concentrations in the foam product and solution, respectively, z is the ion charge, and B is a constant value depending on the hydrogen ion concentration, the cation charge, the exchange constant, and the activity coefficients of ions both in the foam product and in the solution.

If the assumption concerning the ion-exchange nature of interaction between ions and surfactants proves to be right, the plot of log E of the cations versus pH of the solution will give a straight line.

When considering the equilibrium between a cationic surfactant and solution anions Eq. (20) changes to

$$\log E = B - z(\text{pH}). \qquad (21)$$

This equation shows the inverse dependence of log E for anions on the pH of the solution. Equation (21) differs from the previously stated similar dependence of cation-exchange processes in the sign of the pH term only.

A direct application of Eqs. (20) and (21) for the description of experimental data becomes complicated due to certain circumstances. Very often surfactants are not chemically pure. They consist of molecules differing in the weight and structure of their hydrocarbon parts. This means that the foam product possesses polyfunctional ion exchanger properties.

Owing to electrolyte effect, on reaching some critical concentration every group of surfactant molecules forms micelles which have a considerably lower surface activity compared to dissolved molecules. The micelles, as well as the foam product, can be regarded as a pseudosolid ion exchanger of the hydrogen type. In the process of foaming a certain amount of surfactant molecules will remain in the solution. Then C_{M_n} determined experimentally will be lower, and C_{M_p} higher.

With experimental conditions unchanged, the final surfactant concentration in the solution can have the same value in all cases. Then Eq. (20) can be written as follows:

$$\log(Ew) = B + z(\text{pH}), \qquad (22)$$

where $w \simeq \text{const} < 1$ is the coefficient incorporating the final surfactant concentration in the solution.

Joining $\log w$ and B, one obtains

$$\log E = B_1 + z(\text{pH}). \tag{23}$$

Thus, Eq. (23) preserves the proportional dependence of $\log E$ on pH.

The values of C_{M_n} are affected by the volumes of foam product which change from experiment to experiment. To determine E one should always take into account the volume of foam product and solution upon completing the process of foaming.

Not only ion-bound, but also nonmodified surfactants pass into foam. Their surface-active properties being different, the extent of their concentration at the phase interface is also different. This fact causes some slight errors in calculations when using equations such as (23).

Special experiments were made to check the validity of Eqs. (20) and (21). The scope of this chapter being limited, only the experiments involving the extraction of SO_4^{2-} and PO_4^{3-} ions by $C_{14}H_{29}NH_2 \cdot HCl$ will be mentioned. The effect of ion concentration in foam can be tentatively accounted for by exchange of surfactant hydroxyl groups for an isotope anion.

Preliminary investigations revealed that the cationic surfactant studied has a low critical micelle concentration. Above this value, micelles are formed which noticeably reduce the ability of surfactants to pass over to the solution–air phase interface. To avoid this the experiments were carried out with aqueous ethanol instead of aqueous solutions, thus minimizing the possibility of micelle formation. First, concentrated alcoholic $C_{14}H_{29}NH_2 \cdot HCl$ solutions were made. Then aqueous alcoholic solutions of the desired surfactant and alcoholic concentrations were obtained from them.

Equation (21) was derived proceeding from the assumption of anion exchange with hydroxyl surfactant ions. That is why it can be reasonably applied only to the experimental data which show increased anion extraction rate with decreased solution pH. In our case for PO_4^{3-} this is for pH values $\geqslant 5.0$–6.0; and for SO_4^{2-}, pH $\geqslant 3.0$ (Fig. 2). Parts of the curves corresponding to a more acidic medium cannot be described by Eq. (21). This is evidently the result of the competitive effect of anions of the acid (HNO_3) which was used to increase the solution acidity. The way in which the addition of foreign anions affects the rate of foam separation of anions will be shown. As was to be expected, the treatment of data obtained from experiments with surfactant solutions containing micelles does not give satisfactory results.

In separating PO_4^{3-} from aqueous alcoholic solutions in weakly acidic and neutral media, the plot of $\log E$ vs. pH of the solutions gives a straight line (Fig. 3). Thus, the application of the mass action law to describe surfactant anion separation proved to be valid. Using solutions of surfactant concentra-

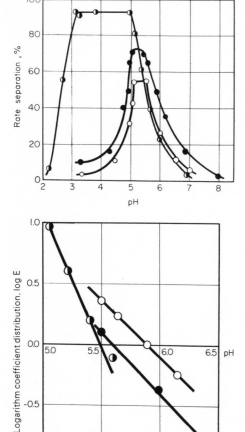

FIG. 2. Rate of foam separation of SO_4^{2-} (5 mg/liter) and PO_4^{3-} (5 mg/liter) with tetradecylamine hydrochloride from aqueous alcoholic solutions versus pH: (◑) SO_4^{2-}, surfactant concentration of 40 mg/liter; (●) PO_4^{3-}, surfactant concentration of 40 mg/liter; (○) PO_4^{3-}, surfactant concentration of 20 mg/liter.

FIG. 3. Coefficient of foam-solution distribution of SO_4^{2-} and PO_4^{3-} vs. pH using tetradecylamine hydrochloride–aqueous alcoholic solution: (◑) SO_4^{2-}, surfactant concentration of 40 mg/liter, tan $\beta = -2.04 \pm 0.03$; (○) PO_4^{3-}, surfactant concentration of 40 mg/liter, tan $\beta = -0.97 \pm 0.03$; (●) PO_4^{3-}, surfactant concentration of 20 mg/liter, tan $\beta = -1.01 \pm 0.05$.

tion of 40 mg/liter at pH $\geqslant$ 6.5 causes a deviation from linear dependence. The transition to the alkali zone again gives rise to micelle formation. A lower surfactant concentration (20 mg/liter) in the process of foaming helps to avoid micelle formation (micelles are formed on prolonged standing). The maximum separation rate in this case, however, is somewhat lower (56%) than that at 40 mg/liter surfactant concentration.

For the phosphorus anion, z in Eq. (21) (which shows the amount of exchanging anion charge) in the pH range 5.5–7.0 is about unity. The finding of phosphorus in that state under the above conditions is most likely. The plot of $\log[SO_4^{2-}]$ distribution between the foam product and the solution versus pH in the pH range 5.0–5.6 also gives a straight line. The slope of the line is 2.04, i.e., near the value of the charge on the anion SO_4^{2-}. At pH $\sim$ 5.6 some formation of micelles occurs and the plot will not give a straight line.

Similar results were obtained in the separation of cations with anionic surfactants, and both cations and anions with ampholitic surfactants.

B. Competitive Effect of Foreign Ions

It was interesting to investigate the competitive effect of the addition of different salts on the distribution of separated ions between the solution and the foam product.

In case the solution contains salts, thermodynamic equilibrium between separated ions and ions of salts formed during the dissociation will be expressed by a general equation of the form

$$a_{M_n}^{1/z}/a_{M_p}^{1/z} = K\, a_{M_{1n}}^{1/z_1}/a_{M_{1p}}^{1/z_1}, \qquad (24)$$

where $a_{M_{1n}}$ and $a_{M_{1p}}$ are the activities of the competitive ions in the solution and in the foam product, respectively; z_1 is the charge of the competitive ions.

After simple mathematical transformations and substituting the constant values from Eq. (25), one can obtain an equation to express the dependence of ion distribution between the foam product and the solution on the concentration of added electrolytes, as follows:

$$\log E = B - (z/z_1) \log a_{M_{1p}}. \qquad (25)$$

Direct application of the equation obtained for the quantitative description of the competitive effect of ions of added salts is possible if the above mentioned experimental conditions are fulfilled.

It is evident from Eq. (25) that the plot $\log E$ vs. $\log a_{M_{1p}}$ must give a straight line, the slope of which, characterized by the coefficient z/z_1, indicates the equivalence of exchange. Of course, all this may take place only when preconditions for the derivation of Eq. (25) prove to be valid or, at any rate, approximate the behavior of real systems.

In the experiments described the amounts of electrolytes added were comparatively small and the activity of the ions in the solution was set equal to their concentration.

Thus, if added ions of foreign salts compete with separated ions in an ion-exchange mechanism, the plot of $\log E$ vs. $\log a_{M_{1p}}$ (in our case $C_{M_{1p}}$) must give a straight line. The evidence in Fig. 4 does not contradict the idea of the interaction between SO_4^{2-} and $C_{14}H_{29}NH_2 \cdot HCl$ being of an ion-exchange nature. The cited graphical dependence gives a satisfactory straight line. Moreover, when almost all the surfactant ions available at the phase interface are replaced by anions of added salts, the coefficient z/z_1 (Fig. 4 and Table I) approaches the value of the ratio of SO_4^{2-} charge to the charge of added salt anions.

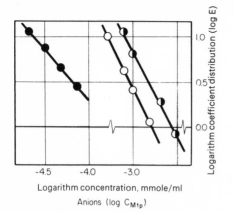

Logarithm coefficient distribution ($\log E$)

Logarithm concentration, mmole/ml

Anions ($\log C_{M1p}$)

FIG. 4. Effect of added salt anions on SO_4^{2-} (5 mg/liter) foam separation with tetradecylamine hydrochloride from aqueous alcoholic solutions: (◑) NaCl; (○) NaNO₃; (●) Na₂C₂O₄. pH = 3.50–3.67, surfactant concentrations of −40 mg/liter.

According to the degree of depression of SO_4^{2-} foam separation, the anions studied are arranged in series: $Cl^- < NO_3^- < C_2O_4^{2-}$. Other things being equal, it follows that the competitive effect of added anions is stronger, the greater their charge and radius.

Unfortunately, upon reaching certain concentrations of foreign anions, a further increase of their content in the solution causes the formation of surfactant micelles, and the application of Eq. (25) becomes difficult. For example, upon addition of foreign anions, a drop in the rate of SO_4^{2-} separation of up to ∼50 % takes place according to Eq. (25). However, the further addition of anions gives rise to surfactant micelles. The solution becomes visibly turbid and the SO_4^{2-} separation rate falls sharply to as low as zero.

The treatment of experimental data showing a decreased rate of anion foam separation with increased solution acidity according to Eq. (25) corroborated the idea that a decreased rate of anion transition to foam with decreased pH is accounted for by the competitive effect of the anions of the acid used for increasing the solution acidity rather than by C_{H_p} growth in the solution.

Previously it was established that PO_4^{3-} ions exchange with surfactant OH^- anions only with one valence bond as $H_2PO_4^-$. One might expect a similar situation also when there is a competitive effect of the acid anion at increased acid concentration in the solution. However, phosphoric acid

TABLE I

VALUES OF z/z_1 IN EQ. (25) FOR THE PROCESS OF ION
EXCHANGE OF SO_4^{2-} FOR ADDED SALT ANIONS
AT pH 3.50–3.67

Salt	NaCl	NaNO₃	Na₂C₂O₄
z/z_1	1.94 ± 0.05	2.02 ± 0.07	1.07 ± 0.04

anions exhibit maximum valence in the process of competitive exchange. Similar results were obtained in experiments with cations.

Thus, the equation resulting from the law of mass action proves to be suitable for the quantitative description of foam separation of ions under conditions of their competition with foreign ions.

REFERENCES

Glembotsky, V. A., Klassen, V. I., and Plaksin, I. N. (1961). In "Flotatsiya." Gosgortekhizdat, Moscow.

Khrustalev, B. N., Pushkarev, V. V., and Mikheev, E. A. (1968). Zh. Prikl. Khim.

Kurbatov, M. N., Wood, G. V., and Kurbatov, I. P. (1951). J. Chem. Phys. 2, 258.

Mahoney, J. F., and Wensel, L. A. (1963). A. I. Ch. E. J. 5, 641.

Nickolsky, B. P., and Paramonova, V. I. (1939). Usp. Khim. 10, 1935.

Rebinder, P. A. (1966). Zh. Vses. Khim. Obshchest. 11, 362.

Sebba, F. (1962). "Ion Flotation," American Elsevier, New York.

Shinoda, K., Nakagava, G., Tamamusi, B., and Isemura, T. (1966). In "Kolloidnie poverkhnost-no-aktivnie veschestva," p. 207. Mir, Moscow.

Skrylev, L. D., and Mokrushin, S. G. (1964). Zh. Prik. Khim. 1, 211.

Veylas, K. (1964). In "Kinetika i rascheti promishlennikh reaktorov," p. 12. Mir, Moscow.

Ward, F. A. H., and Torda, L. (1944). Nature 154, 146.

Ward, F. A. H., and Torda, L. (1946). J. Chem. Phys. 14, 453.

Yegorov, Yu. V., Krilov, E. I., and Tkachenko, E. V. (1961). Radiokhimiya 6, 654.

AUTHOR INDEX

Numbers in italics refer to the pages on which the complete references are listed.

SUBJECT INDEX

A

ABS, *see* Alkylbenzene sulfonate
Accra (Ghana), 226
Acetate, 95, 274
Acid, 62, 125, *see also* pH
Activated sludge, 227
Activator, 93, 120
Activity, 68
Adsoplet techniques, *see* Adsorptive droplet
    separation techniques
Adsorbing colloid flotation, 3
Adsorption, 34-36, 106-111, 122-125,
    135, 153, 162, 250-255, 290,
    308-311
  chemical, 108, 111, 122-125
  ionic competition, 311-313
  physical, 108
Adsorptive buble fractionation, *see*
    Bubble fractionation
Adsorptive bubble separation techniques,
    1-5
  classification of techniques, 2-3
  nomenclature, verbal, 1-3
Adsorptive droplet separation techniques,
    3-4
Adsubble techniques, *see* Adsorptive bubble
    separation techniques
Aging of protein, 160
Albumin, 45, 158
Alkylammonium acetate, 114
Alkylammonium-quartz, 108
Alkylbenzenesulfonate, 176, 228, 233, 274,
    276
Alkylmorpholines, 116
Alkyl phosphate, 274
Alkylphosphonic acids, 117
Alkylsulfonate-alumina, 108
Alkylsulfonates, 93
Alum, 202, 214, 239
Alumina, 76, 107, 118, 119
Aluminosilicates, 101, 113
Aluminum, 63, 71, 76, 122, 207, 274, 306

Aluminum hydroxide, 76
Amine, 62, 83, 93, 105, 117, 119, 125,
    210, 274, 276
Ammonium, 274, 306
Amylase, 164
Analytical applications, 66-68, 71, *see also*
    Ring test
Anisole, 147
Antifoaming agents, 48
Apatite, 124, 126
Applications to wastewaters, *see* Wastewater
Area of interface, 9
Aresket 300, 46, 138, *see also* Monobutyl-
    diphenyl sodium sulfonate
Aromatic alcohol, 93
Autoactivation, 122, 124

B

Bacteria, 193-195, 202, 214
Barbiturates, 71
Barite, 100, 113, 124
Barium, 107
Barium sulfate, 76
Bartstow (California), 234
Beer, 1, 157
Bentonite, 101, 195, 275
Benzene, 87
Benzoinoxime, 86
Benzoylacetone, 86
Beryl, 93
Beryllium, 71, 76
Black film, 46
BOD, 223
Boehmite, 101
Booster bubble fractionation, 3, 142
Bromide, 274
Bromophenol blue, 67
Bubble cap, 42
Bubble fractionation, 3, 34, 36, 133-143
  batchwise operation, 134-141
  combined with foam fractionation, 134